Elements of
Optimal Control

Elements of Optimal Control

STEPHEN J. CITRON

Purdue University

HOLT, RINEHART AND WINSTON, INC.
New York   Chicago   San Francisco   Atlanta
Dallas   Montreal   Toronto   London   Sydney

Library of Congress Catalog Card Number: 69–13795
SBN: 03–073370–7

Printed in the United States of America

9   8   7   6   5   4   3   2   1

To Barbara,
Todd, Devan, and Wendy Sue

Preface

It is the purpose of this text to provide an *introduction* to the development and utilization of techniques applicable to the solution of optimal control problems. Such problems are within the domain of system optimization theory.

System optimization theory has been of great interest in the professional journals devoted to applied mathematics and theoretical engineering for some years. It appears, however, that the usual filtration process by which this knowledge commonly reaches both the engineering student and the practicing engineer has not been as effective as might have been hoped. This may result from the impression sometimes given that an inordinately high level of mathematics is required to begin work in the field.

It is my belief that a level of mathematical sophistication equivalent to that possessed by a beginning graduate student or a senior in a good engineering program is sufficient to begin meaningful study of the area. Further, there is no need to embed the theory within a framework built up in previous courses in automatic control. Thus the material in the text is presented so that it may be learned or taught independent of the background ordinarily acquired during a classical control course.

It is felt that the text is a suitable *beginning* point for the engineering reader interested in the fields of optimal control and system optimization. No prerequisites in control theory are required and use of the text is not limited to any one special field of engineering. Several methods of formulating and solving deterministic optimal control problems are presented. Starting from the classical calculus of variations formulation the maximum principle of Pontryagin is obtained, providing a link to the vast literature existing on the calculus of variations. (It is personally more satisfying to tie modern control theory to the classical approach than to present it through the maximum principle in an independent manner.) Problems with control variable and state variable inequality constraints are treated, and techniques are developed and used to solve the two-point boundary value problems that arise. The direct methods of steepest descent and dynamic programming are developed, and the relationship between dynamic programming and the calculus of variations is shown. Finally, the solution of optimal feedback control problems by the Riccati transformation technique (sweep method) is treated. Several specific engineering examples containing numerical results are used to illustrate the theory presented.

The theory is developed in the text in a heuristic manner so as to make each necessary step seem reasonable, if not in some cases mathematically rigorous. Thus the point of the derivations undertaken is not great rigor but to show from where and in what manner the significant results arise. The functions used are assumed to have the required derivatives; the extremals found are assumed to be normal in the sense of Bliss with all that this implies. It is felt that in an introductory engineering text more is gained by not dwelling on these points than is lost.

In Chapter 1 a brief overview is given of the optimal control problem. With this general background, in Chapter 2 work is initiated on the development of the mathematical techniques required to solve optimal control problems. The subject is approached by considering problems in which the system may be optimized through use of the theory of the minima of functions. By treating this simple problem some intuitive feelings as to the requirements of the theory are developed. The problem is next broadened by consideration of problems requiring the minimization of functionals. In this manner the Euler equation and the Weierstrass-Erdmann corner conditions of the calculus of variations are found. By introduction of the Lagrange multiplier, the theory is then taken through the point where the system variables are related by differential constraints.

Chapter 3 continues the development of the necessary conditions of the theory. The case in which the end-points of the system are variable is studied leading to formulation of the general transversality condition for the problem. In addition, the necessary conditions of Weierstrass and Legendre are obtained. At this point the results obtained from the classical calculus of variations formulation are reexpressed using the state variable formulation of the control problem. The Weierstrass condition of the calculus of variations is shown to be equivalent to Pontryagin's maximum principle of modern optimal control theory.

Chapters 4 and 5 extend the theory to treat problems involving control variable and state variable inequality constraints. In addition, the maximum principle is used to solve several engineering problems to illustrate application of the theory. Among these are the optimal control of a chemical reactor under control constraints and the design of a minimum weight beam. The latter problem involves both a control variable and a state variable inequality constraint. One other example illustrates singular control.

The direct method of steepest descent is developed in Chapter 6. Following Bryson and Denham, a computational expression for the path of steepest descent to the minimum of a function or functional is obtained. Application is made to the satellite ascent problem previously solved analytically through use of the maximum principle.

In Chapter 7 dynamic programming is presented. The computational utility of the technique is shown by example. Also developed is the relation between dynamic programming and the maximum principle (calculus of variations). In so doing a physical interpretation is given to the multipliers of the maximum principle. The text concludes with the development in Chapter 8 of the solution to optimal feedback control problems through use of the Riccati transformation technique.

I have used the material presented in the text as the nucleus of a semester long course at Purdue University entitled System Optimization Techniques. An advanced group with fair knowledge of the material in Chapters 1 and 2 could cover much of the remaining six chapters in this time. A beginning group could generally cover the material in Chapter 1 through Chapter 6 during this period.

However, there is no reason why the chapters need be studied strictly in order. It is suggested for those interested in the heuristic development of the theory from the calculus of variations that Chapter 1 through Chapter 4 be covered in sequence to start the course. Beyond this the remaining chapters may be treated independently and covered in such a manner as to achieve the instructor's course goal. Those not interested in the development of the theory from the calculus of variations may skip after Chapter 1 to Section 3-8 where the results of the theory are expressed for control applications. Again, following Chapter 4 the remaining chapters may be treated in any order deemed desirable.

The author is indebted to Dean J. E. Gibson for use of material on which much of Chapter 1 is based, as well as for material relating to the example of Section 4-8. Dr. B. E. Coy, Mr. A. L. Crosbie, Dr. E. W. Hahn, and Professor W. E. Schmitendorf all made significant contributions to the examples presented in the text. I am most appreciative of the efforts of each one. The support of Purdue University over the years and specifically in providing sabbatical leave to complete the text is gratefully acknowledged as is the aid of the Georgia Institute of Technology where the leave was spent.

STEPHEN J. CITRON

Contents

1 INTRODUCTION   1

1-1   The Control Problem   1
1-2   Optimal Control versus Conventional Control   3
1-3   State-Variable Formulation of the System
      Equations   5
1-4   The Index of Performance   11
1-5   The Control Function versus The Control Law   12
1-6   Suboptimal Control   13
1-7   Controllability and Observability   14
1-8   Summary   15

2 FORMATIVE CONCEPTS   17

2-1   Preliminary Remarks   17
2-2   Minima of Functions   17
2-3   One Independent Variable   19
2-4   Two Independent Variables   22
2-5 N Independent Variables   27
2-6   Constraint Equations   30
2-7   Minima of Definite Integrals—The Calculus of
      Variations   33
2-8   The Euler Equation   35
2-9   The Case $f = f(x, \dot{x})$ 40

2-10 The Weierstrass–Erdmann Corner Conditions 40

2-11 The Second Variation 43

2-12 The Extension to N System Variables 50

2-13 Differential Constraints Among the System Variables 55

3 DEVELOPMENT THROUGH THE CONTROL FORMULATION   61

3-1 A More Realistic Problem 61

3-2 The First Variation 62

3-3 The Euler (Characteristic) Equations 67

3-4 The Transversality Condition 68

3-5 The Results for N System Variables 71

3-6 The Weierstrass and Legendre–Clebsch Conditions 80

3-7 Problems Not Involving Rates 86

3-8 The Results Expressed for Control Applications 88

3-9 The Lunar Ascent Problem 92

4 THE EXTENSION TO CONTROL-VARIABLE INEQUALITY CONSTRAINTS   102

4-1 A Simple Linear Plant 102

4-2 Control of a Chemical Reactor 106

4-3 A Two-Point Boundary-Value Problem 109

4-4 Control-Variable Inequality Constraints 112

4-5 An Alternative Formulation 119

4-6 The Weierstrass Condition with Control-Variable Inequality Constraints 121

4-7 The Chemical Reactor—Bounded Control 123

4-8 The Pure Inertia Plant 131

5 STATE-VARIABLE INEQUALITY CONSTRAINTS   139

5-1 Problems with Intermediate Conditions 139

5-2 State-Variable Inequality Constraints 143

5-3 Conditions for Problems with Discontinuous Controls 146

5-4 An Exercise in Road Building 147

5-5 Toward Smoother Roads 152

5-6 A Minimum-Weight Beam 154

5-7 A Minimum-Weight Beam—Solution Continued 159

**6 DIRECT METHODS—THE TECHNIQUE
OF STEEPEST DESCENT**   167

6-1   Introduction   167

6-2   A Simple Example from the Minima of
Functions   168

6-3   The Treatment of Constraints   170

6-4   Control Formulation   175

6-5   The Case of n State Variables and m Control
Variables—Formulation   181

6-6   The Case of n State Variables and m Control
Variables—Solution   184

6-7   The Adjoint Equations   188

6-8   Application to the Minimization of a Functional—
Formulation   193

6-9   Solution of the Problem   199

6-10 Application of the Technique to the Lunar Ascent
Problem   202

6-11 The Penalty-Function Approach   208

6-12 Concluding Remarks   209

**7 DYNAMIC PROGRAMMING AND
OPTIMAL CONTROL**   210

7-1   The Optimal Allocation Problem   210

7-2   Application to the Control Problem—Theory   216

7-3   A Perfect Integrator   219

7-4   The Computational Scheme   223

7-5   Relation to the Calculus of Variations   229

7-6   The Hamilton–Jacobi Equation   232

7-7   Conclusions   233

8 OPTIMAL FEEDBACK CONTROL   235

8-1   Introduction   235

8-2   Linear Systems with a Quadratic Performance
Index—Formulation   236

8-3   Linear Systems with a Quadratic Performance
Index—Solution   238

8-4   The Inertia Plant   242

8-5   Implementation of the Solution   245

8-6   A Tutorial Example   248

8-7   The Nonlinear Problem—A Guidance
Solution   250

8-8   A Transfer Problem   255

INDEX   263

Elements of Optimal Control

1

INTRODUCTION

1-1 THE CONTROL PROBLEM

To a number of readers this text will be a first introduction to the field of automatic control. To others with a background in conventional control theory the philosophy of optimal control will be unfamiliar. To provide a common basis, let us start by discussing the basic concepts of the control problem.

Consider a plant (system), assembled to meet specific objectives. These objectives are to be achieved while operating the system within limitations imposed by physical, legal, or moral constraints. For example, a boiler that supplies steam for electricity and other uses in an industrial plant must be operated within a limited temperature and pressure range. The objective or the product of the system, which is the steam produced, must be in the quantity required and must have the required number of degrees of superheat at the required pressure.

A commercial airliner must fly at the altitude and speed set by air-traffic controllers. Moving the passengers to their destination by following these specifications is the objective of the airliner. At the same time the airliner must be operated within limits that the aircraft and passengers can safely endure.

The control problem in these two examples is that of designing the system so as to achieve the objectives while satisfying the operational constraints.

Some systems are so constructed that once started they operate without requiring further knowledge of the process development. Such situations are examples of *open-loop control*. In the absence of meaningful disturbances and errors or when the natural, stable operational mode of the system, because of good design, coincides with the desired operating mode, open-loop control can be very successful.

Open-loop control is the simplest and cheapest form of control. It can generally be applied in situations in which demands on the system and its product are relatively relaxed. It is only when performance requirements are raised that open-loop control fails. However, more elaborate control schemes are costly and the engineer should carefully examine the possibility of using open-loop control before proceeding to consider the feedback concept and closed-loop control.

In *closed-loop control* the desired operating point† (input) of the system is compared to the actual operating point (output) and a knowledge of the difference (error) fed back to the input to drive the actual operating point toward the desired one. Thus closed-loop control requires determination (measurement) of the existing operating point of the system at various times resulting from the control previously applied. Since the process development is sampled at various times the terminology sampled-data control or discrete control might be used. As the sampling interval becomes large the control approaches open-loop control, while as the interval approaches zero we speak of continuous feedback control.

A system designed for open-loop operation should be stable in its natural operational mode. Small instantaneous disturbances should be smoothed out by the plant. The reaction of the plant to even large disturbances should be gradual and moderate. That is, the plant should have a long time constant. Let us examine a specific application to lend concreteness.

If a chemical process plant is to be designed for open-loop control the concept of a long time constant would imply the following:

(a) The vessels in which raw materials react together to form the products should be relatively large. "Relatively large" can be given a perfectly definite meaning here. The volume of the vessels should be such that the rates of inflow and outflow are small with respect to the rate at which the reaction progresses.

(b) The pressures and temperatures at which the reaction takes place should be set low enough to cause the reaction to proceed at a rate slow enough for a human operator to sense a difficulty and correct it before the reaction is ruined. (To the extent that the operator does intervene occasionally, the control is not open loop, but is closed loop with a long sampling time.)

We could add other design implications based on the long-time-constant concept, but these two should suffice to illustrate the point. Following this concept thus results in a physically large plant; it is not unusual for a chemical plant to occupy several square miles in area. The "holdup," that is the amount of material being processed by the plant at any one time, becomes a large proportion of a full day's production (in fact, it may be several days' production in

† The desired operating point may change with time.

extreme cases). This means that to start up a process or to effect a change in the products being produced may take several days. Not only is the time required a difficulty, but also the off-grade material produced during this time may be useless. Sometimes this waste may be reprocessed at additional expense but often the reaction is irreversible and the off-grade product is a total loss.

Alternatively, suppose that closed-loop control of the process is to be used. The design philosophy for the plant can now be changed. In a closed-loop system a short time constant is desirable to enable the system to respond quickly to nullify any error signal being fed back. Thus for closed-loop control:

(a) Reactor vessels should be made small, reducing the plant time constant. In fact, if possible, the system should be designed so that the reaction takes place in an in-line flow vessel which is physically simply a pipe, rather than in a tank into which the material is dumped, as below.

(b) Temperature, pressure, and other process variables should be chosen so as to cause the reaction to take place most efficiently. Often this means fast reactions at elevated temperatures and pressures.

The choice between open-loop and closed-loop control must be made early in the design process for maximum effectiveness. Reduction in plant size and material holdup as well as gain in reaction efficiency are all advantages one loses if closed-loop control is applied to a process initially designed for open-loop operation. One still obtains certain advantages such as closer control of quality and better reliability from closed-loop operation, but the other possibilities are sacrificed.

1-2 OPTIMAL CONTROL VERSUS CONVENTIONAL CONTROL

The purpose of designing a closed-loop control system is to decrease the sensitivity of the *plant* to external disturbances. The word "plant" in this text is a technical one and may be taken to mean the portion of the system fixed by considerations other than the control problem. Conventional frequency-domain techniques, with which the reader may have some acquaintance, can be used to design a *controller* provided the plant is linear. Thirty years of work by a wide variety of engineers such as Bode (*1*), Nyquist, and Black have established links between the frequency response of a control system and its closed-loop transient performance in the time domain. These links are tenuous, it is true, and often for convenience one argues by analogy to the response of a simplified model rather than the actual system at hand. But generally speaking the design of linear, stationary, single-loop control systems to frequency-domain specifications is a known art.

The philosophy of design is to adjust the value of various parameters of a controller (compensator) of given form by intelligent trial and error until it appears that the overall system will meet specifications. If it appears that a controller of the form given cannot be made to do the job, a more complex form is tried. The process is one of trial and error and requires a certain sophistication of attitude. Sophistication is needed because the design is carried out in the frequency domain by means of shifting poles and zeros or by observing asymptotic frequency-response plots of the elements while seeking to achieve a given complex and perhaps subjective performance characteristic in the time domain.

However, as this process is successful in a wide variety of practical control problems, there is an understandable reluctance to surrender it for the doubtful virtues of another technique. The reader of this text should retain this skepticism until the benefits of applying modern optimal control theory are made quite clear.

The essential features of the conventional methods, then, are first the establishment of somewhat ad hoc specifications in the frequency domain which are purported to be related to the desired time-domain performance, and, second, the trial and error adjustment of certain system parameters in an attempt to bring the system performance within the required bounds.

Among the most important reasons usually given control engineers for working in the frequency domain with Laplace transforms is the following: One is usually told that transform methods promote the solution of differential equations of the kind commonly met in control-system analysis. However, another reason of equal or even greater importance for use of frequency-domain methods is the fact that reliable experimental frequency-response data may be obtained in the laboratory on real hardware, thus connecting analysis with the real world. The experimental data obtained may be used in either the frequency domain or the time domain, depending on the particular problem and the prejudices of the engineer, to model the actual system mathematically.

Modern optimal control takes a different view of the problem. First, the problem is formulated in the time domain. However, the use of the time domain rather than the frequency domain as in classical control is not the important factor. The important feature of modern optimal control is the establishment of an analytic *index of performance* (IP) for the system and design of the system so as to optimize the index selected.

For example, consider a controller with parameter values to be chosen. The value of the IP will be influenced by the parameter values of the controller. One thus seeks the values of these parameters which will optimize the IP. This point of view is apparent in a very important germinal text by Newton, Gould, and Kaiser (2). Of course, one may formulate much more general forms of the optimization problem.

Within the last decade the classical mathematical techniques of the calculus of variations and steepest descent have been restated and extended to apply to the optimal control problem. In addition, other techniques, for example, "dynamic programming" and the "maximum principle," have been conceived to solve optimal control problems. In each of these the system is described in terms of a state-variable formulation of the system equations, that is, the laws of motion of the process. This formulation is considered next.

1-3 STATE-VARIABLE FORMULATION OF THE SYSTEM EQUATIONS

The systems to be considered in the text are continuous; that is, their operation may be modeled using differential equations. A discrete representation of these equations will be used when it is advantageous for computational purposes.

We shall assume that the physical systems to be studied may be described mathematically as lumped-parameter systems. This means that an ordinary differential equation will suffice for their description rather than the partial differential equation that would be required if the system were of the distributed-parameter type. A further assumption made is that the system is deterministic rather than stochastic. The problem of incorporating knowledge of the statistics of random disturbances into the optimum solution is treated in the field under the heading "stochastic optimal control." The subject will not be covered in this text.

No additional general restrictions, such as stationary as opposed to non-stationary, linear as opposed to nonlinear, first order as opposed to nth order, single input–output as opposed to multivariable, are needed, although when appropriate such limitations will be made as required to treat special cases.

There are a variety of ways in which one may write the differential equations that describe the "motion" (operation) of the system. One common method consists of writing the equations for each portion of the system. Then, by substitution, all but one of the variables is eliminated, yielding a single nth-order differential equation governing the behavior of the variable in the system. An alternative approach, and the one we shall use, consists of organizing the system equations into a set of n first-order equations of the form

$$\frac{dx_j}{dt} = f_j(t, x_1, x_2, \cdots, x_n, u_1, u_2, \cdots, u_m) \qquad j = 1, 2, \cdots, n \qquad (1)$$

The n differentiated variables appearing, the x_j, are termed the state variables, and the other m variables, the u_j, are termed the control variables.

The set of equations (1) along with a set of initial conditions at some time t_i,

$$x_j(t_i) = x_{j_0} \text{ specified} \qquad j = 1, 2, \cdots, n \qquad (2)$$

completely describe the operation of the system.

Let us examine some examples illustrating the preceding discussion.

Example 1–3 (1)   A simple *RLC* circuit is shown in Fig. 1-1. By Kirchhoff's law the following voltage equation may be written for the circuit:

$$e_a = L\frac{di}{dt} + Ri + \frac{1}{C}\int_{t_i}^{t} i\,dt \tag{3}$$

where R, L, and C are the resistance, inductance, and capacitance, respectively, and i is the circuit current. Introducing the new variable q, which may be interpreted as the electric charge density, where

$$\frac{dq}{dt} \equiv i \tag{4a}$$

$$q(t_i) = 0 \tag{4b}$$

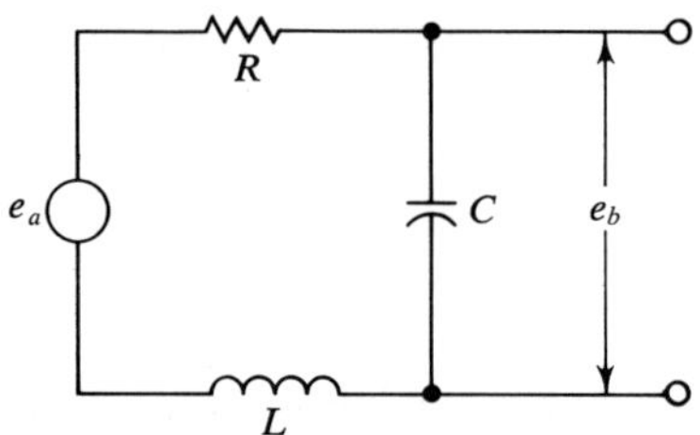

FIGURE 1-1   Simple *RLC* circuit

Equation (3) may be written

$$e_a = L\frac{d^2q}{dt^2} + R\frac{dq}{dt} + \frac{1}{C}q \tag{5}$$

Equation (5) is a second-order equation governing the behavior of the charge density.

To put Eq. (5) in the standard first-order form of Eq. (1) use is made of the definition of Eq. (4a). Thus Eq. (5) is equivalent to the two coupled first-order equations.

$$\frac{dq}{dt} = i$$

$$L\frac{di}{dt} = -Ri - \frac{1}{C}q + e_a \tag{6}$$

Defining

$$x_1 \equiv q \qquad x_2 \equiv i \qquad u_1 \equiv e_a \tag{7}$$

the system equations of (6) can be written

$$\frac{dx_1}{dt} = x_2$$

$$\frac{dx_2}{dt} = -\frac{1}{LC} x_1 - \frac{R}{L} x_2 + \frac{1}{L} u_1 \tag{8}$$

which is of the desired form.

It is sometimes convenient to use vector-matrix notation in our work as a shorthand. In terms of the state and control vectors

$$\mathbf{x}(t) = \begin{bmatrix} x_1 \\ x_2 \end{bmatrix} \qquad \mathbf{u}(t) = [u_1] \tag{9}$$

Eq. (8) may be written $[d/dt(\ ) \equiv (\ \dot{}\ )]$

$$\dot{\mathbf{x}} = \begin{bmatrix} 0 & 1 \\ -\dfrac{1}{LC} & -\dfrac{R}{L} \end{bmatrix} \mathbf{x} + \begin{bmatrix} 0 \\ \dfrac{1}{L} \end{bmatrix} \mathbf{u} \tag{10}$$

or, more neatly,

$$\dot{\mathbf{x}} = \mathbf{A}\mathbf{x} + \mathbf{b}\mathbf{u} \tag{11}$$

where $\mathbf{A}$ and $\mathbf{b}$ are the matrices

$$\mathbf{A} = \begin{bmatrix} 0 & 1 \\ -\dfrac{1}{LC} & -\dfrac{R}{L} \end{bmatrix} \qquad \mathbf{b} = \begin{bmatrix} 0 \\ \dfrac{1}{L} \end{bmatrix} \tag{12}$$

Example 1-3 (2)   Suppose that the output voltage $e_b(t)$ (see Fig. 1-1), is of particular interest. To keep track of $e_b(t)$ it is most convenient to define it to be a state variable. Thus in this case let us use as our state variables

$$x_1 = e_b = \frac{1}{C} \int_{t_i}^{t} i\, dt$$

$$x_2 = i \tag{13}$$

while retaining the same control variable,

$$u_1 = e_a \tag{14}$$

In terms of the variables of Eqs. (13) and (14), the governing equation of (3) written in standard form becomes

$$\frac{dx_1}{dt} = \frac{1}{C}x_2$$

$$\frac{dx_2}{dt} = -\frac{1}{L}x_1 - \frac{R}{L}x_2 + u_1$$

(15)

Using Eq. (9) one then obtains

$$\dot{\mathbf{x}} = \begin{bmatrix} 0 & \dfrac{1}{C} \\ -\dfrac{1}{L} & -\dfrac{R}{L} \end{bmatrix} \mathbf{x} + \begin{bmatrix} 0 \\ 1 \end{bmatrix} \mathbf{u}$$

(16)

Equations (10) and (16) both describe the same system, that given by Fig. 1-1. It is obvious on comparing the two equations that there is no unique state-variable description of a system. This is a powerful practical advantage of the state-variable point of view. *Any set of system variables (provided they completely describe the response of the system) may be chosen as the state variables.* Thus the engineer is free to choose as state variables those quantities which are easy to measure without having this choice dictated and restricted by the mathematics.

Example 1-3(3)   Now suppose the circuit of Fig. 1-1 is nonlinear. For example, imagine that the inductance is a function of the current, as it will be in many cases of physical importance. Then Eq. (6) becomes

$$\frac{dq}{dt} = i$$

$$L(i)\frac{di}{dt} = -Ri - \frac{1}{C}q + e_a$$

(17)

Using the state and control variables defined in Eq. (7) allows Eq. (17) to be written

$$\dot{x}_1 = x_2 = f_1(x_2)$$

$$\dot{x}_2 = -\frac{1}{CL(x_2)}x_1 - \frac{R}{L(x_2)}x_2 + \frac{1}{L(x_2)}u_1 = f_2(x_1, x_2, u_1)$$

(18)

It is impossible to put Eq. (18) in the simple linear form of Eq. (11), of course, but it does fit within the general form of Eq. (1).

Example 1-3 (4)   We consider next the state-variable representation of the dynamics of a vertical sounding rocket. A free-body diagram of the rocket is shown in Fig. 1-2. By *D'Alembert's principle*, the vector sum of all the forces acting on the vehicle, including the inertia force, must be zero.

$$F_{\text{Drag}} + F_{\text{Gravity}} + F_{\text{Thrust}} + F_{\text{Inertia}} = 0 \tag{19}$$

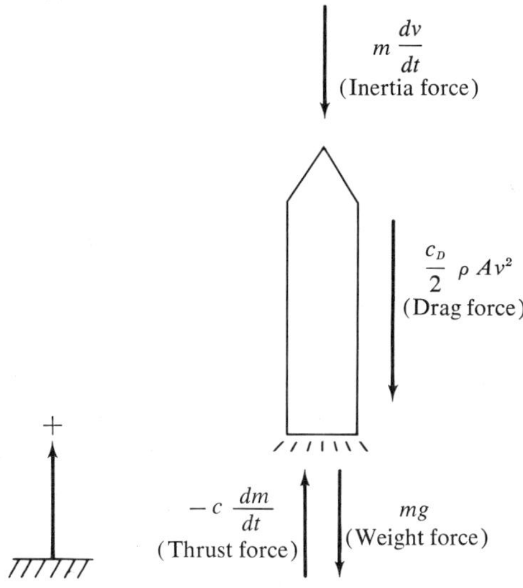

FIGURE 1-2   Vertical-sounding-rocket free-body diagram

As indicated in Fig. 1-2, the upward direction will be taken as positive.

The drag force is usually written in the form

$$F_{\text{Drag}} = -\frac{C_D}{2}\,\rho A v^2$$

where $C_D = C_D(\rho, v)$, drag coefficient
$\rho = \rho(h)$, air density
$h(t) = $ vertical height of the rocket
$v(t) = dh/dt$, absolute vertical velocity of the rocket
$A = $ cross-sectional area of the rocket

The gravitational force is given by

$$F_{\text{Gravity}} = -mg$$

where m is the mass of the vehicle and $g = g(h)$ is the gravitational acceleration.

The thrust of the vehicle can be written in terms of the mass ejected and its exhaust velocity (c) with respect to the vehicle as

$$F_{\text{Thrust}} = -c\,\frac{dm}{dt}$$

where the exhaust velocity will be assumed constant. Finally, the inertia force is given by

$$F_{\text{Inertia}} = -m\,\frac{dv}{dt}$$

Using these quantities, Eq. (19) becomes

$$\frac{C_D(\rho, v)}{2}\,\rho(h)Av^2 + mg + c\,\frac{dm}{dt} + m\,\frac{dv}{dt} = 0 \tag{20}$$

An obvious choice of state variables for this problem are the vehicle's height above the ground, $h(t)$; its vertical velocity, $v(t)$; and its mass, $m(t)$. Available to control the vehicle's velocity is the mass flow rate of material being ejected by the engine dm/dt, $(dm/dt \leqslant 0)$, which sets the thrust magnitude. Thus we define

$$\frac{dm}{dt} \equiv -u \tag{21}$$

To summarize, the system equations are the kinematic relation between v and h, the force balance of Eq. (20), and the control definition of Eq. (21). They are repeated below in the order named.

$$\frac{dh}{dt} = v$$

$$\frac{dv}{dt} = -\frac{C_D(\rho(h), v)}{2m}\,\rho(h)Av^2 - g + \frac{c}{m}\,u \tag{22}$$

$$\frac{dm}{dt} = -u$$

Equations (22) provide the desired representation of the system. This is easily seen by letting

$$\begin{aligned} x_1 &= h(t) \\ x_2 &= v(t) \\ x_3 &= m(t) \\ u_1 &= u \end{aligned} \tag{23}$$

In terms of these new variables the system equations become

$$\frac{dx_1}{dt} = x_2 = f_1(x_2)$$

$$\frac{dx_2}{dt} = -\frac{C_D(\rho(x_1), x_2)}{2x_3} Ax_2^2 - g + \frac{c}{x_3} u_1 = f_2(x_1, x_2, x_3, u_1) \qquad (24)$$

$$\frac{dx_3}{dt} = -u_1 = f_3(u_1)$$

Equations (24) are in the standard form of Eq. (1).

The choice of state variables as successive derivatives of an interesting quantity is an important special case. When this is done the variables are termed *phase variables*. Thus x_1 and x_2, (h, v) are phase variables in this example.

1-4   THE INDEX OF PERFORMANCE

The choice to be made for the system index of performance, IP, is of prime importance in finding a meaningful solution of the optimal control problem. The same care must be exercised in selecting the performance index as is used in determining the system model.

The IP is to be selected by the system designer. The choice is limited only by his knowledge of the mathematical techniques available to solve the problem posed by the selection made. In recent years the number and power of these techniques, as well as our understanding of them, has greatly increased, vastly enlarging the range of problems that can be handled.

Many classical system-control results involving, for example, overshoot or response time can be related to selection of a particular performance index in the formulation of the corresponding optimal control problem (3). Thus direct links exist between conventional control techniques and the optimal control problem. In this light, optimal control can be considered an extension of classical control to allow solutions to be obtained for a wider choice of performance indices than was previously permitted, while lifting restrictions formerly required on the plant model.

As an illustration of possible selections that might be made for the system index of performance, consider the problem of the vertical sounding rocket formulated in Example 1-3(4). The thrust of the rocket is to be controlled so as to optimize the index of performance. A number of likely candidates for the system index of performance exist. The choice to be made will, of course, depend on the goal sought. For example:

(a) Maximize burnout velocity (that is, vertical velocity of the rocket at the point where fuel is exhausted), v_{bo}.

(b) Maximize height at burnout, h_{bo}.

(c) Maximize final height (that is, height where vertical velocity is zero), h_f.

(d) Minimize fuel expended while achieving a specified burnout height or final height or burnout velocity.

(e) Maximize some functional relation between the variables involved in (a) through (d) plus any desired integral over the process development between the times of thrust initiation and process completion.

To standardize the formulation of the optimal control problem, optimization of the system will be taken to mean *minimization* of the system index of performance, IP. For those systems where optimization requires maximization of some quantity, the IP to be minimized will be chosen as the negative of the quantity to be maximized. For example, in (a) the index of performance would be given by $IP = -v_{bo}$.

1-5 THE CONTROL FUNCTION VERSUS THE CONTROL LAW

Once having formulated the desired problem, its solution will be found by the mathematical techniques to be developed. In so doing the required functional relationship which the control variable vector must obey to minimize the system index of performance is obtained.

The optimal control vector is conventionally designated $\mathbf{u}^*$. If the optimal control vector is found as a function of time, $\mathbf{u}^*(t)$, it is called the *optimal control function* vector. The trajectory resulting from using $\mathbf{u}^*(t)$ is designated $\mathbf{x}^*(t)$. If, on the other hand, the control vector is found as a function of the state and time, $\mathbf{u}^*(\mathbf{x}, t)$, it is called the *optimal control law* vector. In the absence of control errors and extraneous disturbances, application of either $\mathbf{u}^*(t)$ or $\mathbf{u}^*(\mathbf{x}, t)$ leads to the same nominal trajectory, $\mathbf{x}^*(t)$.

The optimal control function for the vertical-sounding-rocket example would consist of a predetermined program for controlling the mass flow rate (and through it the thrust) as a function of time. Hence use of the control function provides open-loop control. If no disturbances or errors occur, application of this program will result in optimal performance.

If, however, an unexpected gust of wind or momentary imperfection in the smooth operation of the rocket motor were to cause the vehicle to deviate from its planned flight program, the preprogrammed "optimal control function" would continue on, heedless of this fact. The result would not be optimum and possibly might not even be acceptable.

The optimal control law, on the other hand, is not a precomputed thrust program. Rather, implementation of the optimal control law $\mathbf{u}^*(\mathbf{x}, t)$ requires that the state variables be continuously measured by sensors during the flight with the control (thrust) to be applied computed from $\mathbf{u}^*(\mathbf{x}, t)$ using this knowledge. This design, when executed in hardware, will cause the vehicle to control its thrust so as to achieve an optimal trajectory from its current state. In principle at least, such a design can overcome unexpected disturbances during flight.

As will be seen, however, the solution of the optimal control problem is most easily obtained in the form of the control function and not the control law. If the control function $\mathbf{u}^*(t)$ is used, in general, because of errors or disturbances, the system will not stay on the nominal open-loop trajectory $\mathbf{x}^*(t)$ but will be in some neighborhood $\delta\mathbf{x}$ about it. Thus it is necessary to determine the control perturbation $\delta\mathbf{u}^*(t)$ to be added to $\mathbf{u}^*(t)$ so as to reach the desired terminal state in an optimal manner starting from the state $\mathbf{x}^* + \delta\mathbf{x}$ existing at time t. This area is termed optimal feedback control and will be discussed later.

1-6   SUBOPTIMAL CONTROL

As has been indicated, solutions to optimal control problems are not always easily obtained. In addition, even when it is possible to obtain the desired solution it may be impractical to implement it, that is, to construct a controller capable of duplicating the exact mathematical result. For example, determination of the optimal control law for a particular problem may require that a complex set of computations be carried out in real time or that a function with a complicated time-varying behavior be generated.

One way out of the dilemma is to solve a restricted problem in which the form of the controller that will be allowed is postulated. The problem then remaining is to choose the values of the controller parameters so as to achieve optimality within these constraints. This is the so-called *specific optimal control problem.*

One may ask how the specific form for the suboptimal controller is to be chosen? Here knowledge of the solution of the unrestricted original optimal control problem becomes valuable. It would seem that the suboptimal controller should be built to permit it to approximate the response that one would obtain if the optimal controller were built.

Knowledge of the exact solution to the optimal control problem has one further benefit. A comparison can be made between the optimum value of the system *index of performance* IP* and the value of the performance index obtained using the suboptimal controller. In this manner a judgment can be made on the acceptability of the suboptimal controller designed. Should the deviation be too large, further design refinements can be attempted to enable the suboptimal solution to more closely approximate the optimal one.

As previously indicated, it is many times not possible to find an analytic representation of the optimal control law for a problem. If an analytic representation of the control law is required, a satisfactory suboptimal one may sometimes be found in the following manner (4). It is assumed that the physical process can be approximated by a simple one, for which an analytic representation of the control law can be found exactly. The suboptimal control law for the actual process is then obtained as a modification of the optimal control law for the simplified process. The technique is such that the required correction can be computed without prior knowledge of the exact control law for the actual process.

It is important to emphasize that the value of the performance index obtained using a suboptimal controller should always be tested against that obtained using the optimal control for the actual process to check the degree of optimality of the approximate solution.

1-7   CONTROLLABILITY AND OBSERVABILITY

The concepts of controllability and observability were introduced into optimal control theory by Kalman (5), (6). They provide necessary and in some cases sufficient conditions for a given problem to possess a solution. In addition to the work by Kalman just cited, a useful independent development of the subjects of controllability and observability is presented by Kreindler and Sarachik (7).

The concepts have been referred to as "duals" because controllability involves the relation between the state variables of the system and the system inputs (the control vector), while observability involves the relation between the state variables and the system outputs. Controllability and observability may be defined for a general system; however, at the present time it is only for linear systems that conditions have been found to ensure that a system be controllable and observable. These concepts will be discussed briefly at this time.

Intuitively, state controllability is a property of a system which guarantees that any initial state can be transferred to any desired terminal state in finite time. A system that is output-controllable is one where any desired terminal output can be attained in finite time starting from an arbitrary initial state.

Observability is an expression of our ability to determine at t_i the state variables $\mathbf{x}(t_i)$ based upon measurement of the unforced system output over some interval $[t_i, t_1]$ where $t_1 \geq t_i$ and finite. It is obvious that if each state variable is available for measurement, the plant is observable. That this sufficient condition is not also necessary is what makes the concept interesting.

As a simple illustration, consider a plant defined by the two state variables x_1, x_2 and governed by the system equations

$$\dot{x}_1 = x_2 + u$$
$$\dot{x}_2 = x_2 \tag{1}$$

The measurable output of the plant is assumed to be

$$y(t) = x_1(t) \tag{2}$$

First note from Eq. (1) that starting from a specified initial state $[x_1(t_i), x_2(t_i)]$ it is not possible to reach any desired terminal state $x_2(t_f)$. This follows, because the control does not influence the x_2 equation in Eqs. (1). Thus the system is not state-controllable.

If the terminal conditions for the problems are such that only $x_1(t_f)$ is specified, then by Eqs. (1) it is apparent that the control can be made to drive the system to the desired terminal value. In general, when the system-control vector is such that the terminal-state manifold can be reached, even if this does not include all states, the system is said to be controllable in the reduced sense *(8)*.

Note, also, that measuring the unforced system output $y(t) = x_1(t)$ over some interval $[t_i, t_1]$ is sufficient to also determine $x_2(t_i)$, since by Eq. (1) $x_2(t_i) = \dot{x}_1(t_i)$ for the unforced system. Thus the system is observable. If, on the other hand, the measurable system output is not given by Eq. (2) but rather by

$$y(t) = x_2(t) \tag{3}$$

the system is not observable. This follows, because measurement of x_2 over the interval $[t_i, t_1]$ is not sufficient to determine $x_1(t_i)$ but only the change from it over the interval. Thus with Eq. (3) as the output, the system is not observable.

The conditions to be satisfied for a linear system to be controllable and observable can be precisely given, but this will not be done here.†

In the main no further reference to controllability and observability will be made. The systems studied will be assumed to possess these properties to the degree required by the solution.

1-8 SUMMARY

In this chapter a brief introduction and overview has been given to the optimal control problem. In Chapter 2 we begin to develop the mathematical techniques required to obtain the solutions so glibly spoken of here.

† See references (5), (6), and (7).

REFERENCES

1. H. W. Bode, *Network Analysis and Feedback Amplifier Design*, Van Nostrand, Princeton, N.J., 1945.
2. G. C. Newton, Jr., L. A. Gould, and J. F. Kaiser, *Analytic Design of Linear Feedback Controls*, Wiley, New York, 1957.
3. W. C. Schultz and V. C. Rideout, "Control System Performance Measures: Past, Present and Future," *IRE Trans.*, Feb. 1961, pp. 22–35.
4. B. Friedland, "A Technique of Quasi-Optimum Control," *ASME J. Basic Eng.*, June 1966, pp. 437–443.
5. R. E. Kalman, "Contributions to the Theory of Optimal Control," *Sociedad Matematica Mexicana. Boletin.*, Segunda Serie, Vol. 5, No. 1, April, 1960, pp. 102–119
6. R. E. Kalman, "On the General Theory of Control Systems," in *Proceedings of the First International Congress of the International Federation of Automatic Control*, Vol. 1, Butterworths, London, 1960, pp. 481–492.
7. E. Kreindler and P. E. Sarachik, "On the Concepts of Controllability and Observability of Linear Systems," *IEEE Trans. Automatic Control*, April 1964, pp. 129–136.
8. W. E. Schmitendorf and S. J. Citron, "On the Applicability of the Riccati Transformation Technique for Solution of Optimal Control Problems," *AA and ES Rept. 67–10*, Purdue Univ., Lafayette, Ind., 1967.

2

FORMATIVE CONCEPTS

2-1 PRELIMINARY REMARKS

It is the purpose of this chapter to develop some of the basic analytical tools and concepts required for the solution of optimal control problems. To help provide the physical insight, so useful in the formulation and solution of these problems, the subject is introduced by a discussion of the theory of the minima of functions. In this way a background is formed from which the theory of the minima of functions (of functions), or functionals, follows naturally. The particular function or functional to be considered is, of course, the index of performance (IP) of the system under design.

The determination from the theory of what conditions must be satisfied to obtain the minimum index of performance thus provides the criteria that must be met in the design of our system.

2-2 MINIMA OF FUNCTIONS

As stated above, we start our discussion by considering the index of performance to be a function of several system variables, $x_1, x_2, \cdots, x_N$. Thus $IP = IP(x_1, x_2, \cdots, x_N)$. It is the problem of the designer to minimize the IP by proper choice, within any system constraints, of the N values $x_1, x_2, \cdots, x_N$. Let us say that the designer believes that $x_1^*, x_2^*, \cdots, x_N^*$ are such values.

Relying on physical feeling, it seems apparent that if $IP(x_1^*, x_2^*, \cdots, x_N^*)$ is to be an absolute minimum, then changing *any* of the x_i by *any* amount consistent with system constraints should yield a larger IP than the IP corresponding to use of those values x_i^*, believed to be the minimum values. These intuitive comments lead us to form

Definition 1   Given that the index of performance is a function of the N variables $x_1, x_2, \cdots, x_N$, $\mathrm{IP} = \mathrm{IP}(x_1, x_2, \cdots, x_N)$, then the absolute minimum of the IP occurs at the point $x_1^*, x_2^*, \cdots, x_N^*$ if for *all changes* $\Delta x_1, \Delta x_2, \cdots, \Delta x_N$ *of any magnitude.* consistent with system constraints one obtains

$$\Delta \mathrm{IP}(x_1^*, x_2^*, \cdots, x_N^*) = \mathrm{IP}(x_1^* + \Delta x_1, x_2^* + \Delta x_2, \cdots, x_N^* + \Delta x_N)$$
$$- \mathrm{IP}(x_1^*, x_2^*, \cdots, x_N^*) > 0$$

Although this definition is conceptually useful, it is difficult to transform it into a procedure for finding the values $x_i = x_i^*$ which make the IP an absolute minimum. Therefore, lowering our sights for the moment, we introduce the concept of local minima of the IP. By a point of local minimum of the IP we mean a set of values x_i^* for which Definition 1 holds for Δx_i not of arbitrary magnitude but such that x_i is within some small neighborhood of x_i^*, or, more formally:

Definition 2   Given that the index of performance is a function of the N variables $x_1, x_2, \cdots, x_N$, $\mathrm{IP} = \mathrm{IP}(x_1, x_2, \cdots, x_N)$, then a local minimum of the IP exists at the point $x_1^*, x_2^*, \cdots, x_N^*$ if for *all infinitesimal changes* $\Delta x_1, \Delta x_2, \cdots, \Delta x_N$ consistent with system constraints, one obtains

$$\Delta \mathrm{IP}(x_1^*, x_2^*, \cdots, x_N^*) = \mathrm{IP}(x_1^* + \Delta x_1, x_2^* + \Delta x_2, \cdots, x_N^* + \Delta x_N)$$
$$- \mathrm{IP}(x_1^*, x_2^*, \cdots, x_N^*) > 0$$

As an illustration of these ideas consider Fig. 2-1 where an IP which is a function of the single variable x is shown in the region within the system constraint $x_l \leqslant x \leqslant x_u$. Local minima occur at $x^{(1)}$, $x^{(2)}$, and x_u with the absolute minimum at $x = x^{(2)}$.

The reason for the introduction of the concept of a local minimum can now be made clear. Imagine that a workable test for local minima could be found. Then to find the absolute minimum of the IP, rather than comparing the value of the IP at each point with its value at every other point, as would be required if Definition 1 were applied directly, one need only compare the value of the IP corresponding to each of the comparatively few local minima and pick the smallest.

To put the above scheme into effect we proceed to construct a test for local minima. The key is the fact that only infinitesimal changes need be considered.

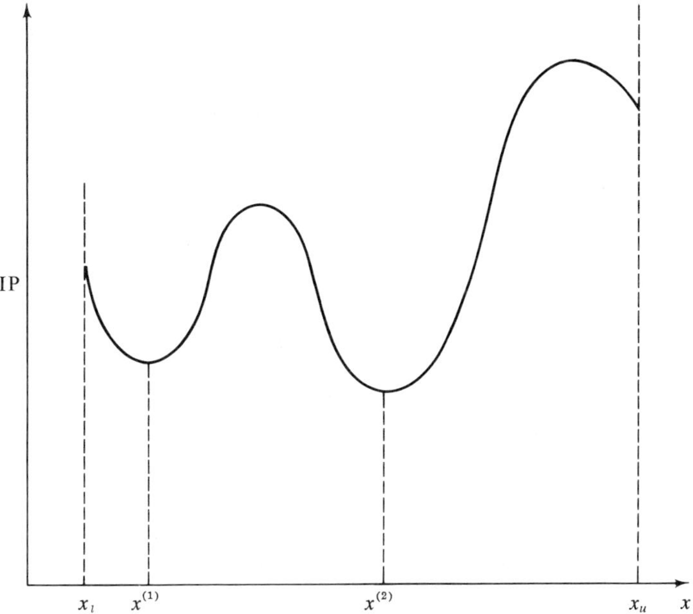

FIGURE 2-1   Location of local minima of an IP at $(x^{(1)}, x^{(2)}, x_u)$

2-3 ONE INDEPENDENT VARIABLE

Consider that the IP is a function of the single variable x_1, subject to the system constraint $x_{1_l} \leqslant x_1 \leqslant x_{1_u}$. Imagine further that a minimum is to exist at $x_1 = x_1^*$. Then by Definition 2, for a local minimum at x_1^*,

$$\Delta IP(x_1^*) = IP(x_1^* + \Delta x_1) - IP(x_1^*) > 0 \tag{1}$$

As Δx_1 is an infinitesimal, the quantity $\Delta IP(x_1^*)$ is easily formed by taking a series expansion of $IP(x_1^* + \Delta x_1)$ about $x_1 = x_1^*$. Thus

$$IP(x_1^* + \Delta x_1) = IP(x_1^*) + \frac{dIP}{dx_1}\bigg|_{x_1^*} \Delta x_1 + \frac{d^2 IP}{dx_1^2}\bigg|_{x_1^*} \frac{\Delta x_1^2}{2} + O(\Delta x_1^3)$$

To simplify the notation in the following work, quantities of the form $d/dx_1\big|_{x_1^*}$ will be represented by d/dx_1^*. Therefore,

$$\Delta IP(x_1^*) = \frac{dIP}{dx_1^*} \Delta x_1 + \frac{d^2 IP}{dx_1^{*2}} \frac{\Delta x_1^2}{2} + O(\Delta x_1^3) \tag{2}$$

Further, since Δx_1 can be made as small as we like, it is apparent that the sign of the right side in Eq. (2) can be made to depend on the first term provided

$dIP/dx_1^* \neq 0$. But if x_1^* is not on the boundary and if $dIP/dx_1^* \neq 0$, then the sign of $\Delta IP(x_1^*)$ will change with the sign of Δx_1, which is arbitrary, contrary to the condition given in Eq. (1). Thus if x_1^* is not on the boundary, a necessary condition for a local minimum of $IP(x_1^*)$ to exist is that

$$\frac{dIP}{dx_1^*} = 0 \qquad \text{for } x_{1_l} < x_1^* < x_{1_u} \tag{3}$$

If, on the other hand, $x_1^* = x_{1_u}$, the upper boundary, Δx_1, can be of only one sign, $\Delta x_1 < 0$. We then see from Eq. (2) that Eq. (1) requires

$$\frac{dIP}{dx_1}\bigg|_{x_{1_u}} \leqslant 0 \tag{4}$$

Correspondingly, a local minimum at the lower boundary requires that

$$\frac{dIP}{dx_1}\bigg|_{x_{1_l}} \geqslant 0 \tag{5}$$

If the equality sign is removed from Eqs. (4) and (5), then these conditions are sufficient to satisfy Eq. (1). If the equality sign is included, then on treating it the same situation exists as at an interior point, and at a minimum point Eq. (2) becomes

$$\Delta IP(x_1^*) = \frac{(\Delta x_1)^2}{2}\left[\frac{d^2IP}{dx_1^{*2}} + O(\Delta x_1)\right]$$

Assume† that $d^2IP/dx_1^{*2} \neq 0$. Then with $dIP/dx_1^* = 0$, since Δx_1 can be made as small as we like, the second-order terms dominate Eq. (2). As $(\Delta x_1)^2$ is always positive, the sign of the right side is determined by the sign of d^2IP/dx_1^{*2}. A sufficient condition to satisfy Eq. (1) when $dIP/dx_1^* = 0$ is thus $d^2IP/dx_1^{*2} > 0$.

To summarize: $IP = IP(x_1)$; $x_{1_l} \leqslant x_1 \leqslant x_{1_u}$.

INTERIOR POINT: $x_{1_l} < x_1 < x_{1_u}$, a necessary condition for a local minimum is

$$\frac{dIP}{dx_1^*} = 0 \tag{6}$$

If Eq. (6) is satisfied, then a sufficient condition for a local minimum at x_1^* is

$$\frac{d^2IP}{dx_1^{*2}} > 0 \tag{7}$$

† If $d^2IP/dx_1^{*2} = 0$, the argument would continue in the same way to higher-order terms.

BOUNDARY POINT: $x_1 = x_{1_l}$ or x_{1_u}, a sufficient condition for a local minimum is

$$\frac{d\text{IP}}{dx_1}\bigg|_{x_{1_l}} > 0 \quad \text{or} \quad \frac{d\text{IP}}{dx_1}\bigg|_{x_{1_u}} < 0 \tag{8}$$

If Eq. (6) is satisfied at a boundary point, then a sufficient condition for a local minimum at the point is that Eq. (7) also be satisfied there.

Example 2-3(1)   Find the absolute minimum of $\text{IP} = x_1 + \cos x_1$, within the system constraint $0 \leqslant x_1 \leqslant 2\pi$.

(a) Interior Point: Checking Eq. (6) one has

$$\frac{d\text{IP}}{dx_1^*} = 0 = 1 - \sin x_1^*$$

Thus in the interior of the admissible region the only possible solution is $x_1^* = \pi/2$. Forming.

$$\frac{d^2\text{IP}}{dx_1^2} = -\cos x_1$$

the solution is tested on the condition given in Eq. (7). Inserting $x_1^* = \pi/2$ into $d^2\text{IP}/dx_1^2$, one finds $d^2\text{IP}/dx_1^2 = 0$.

Equation (7) is not strictly satisfied. Therefore, on the basis of second-order terms† it cannot be concluded whether $x_1 = \pi/2$ is a point at which a local minimum occurs in the IP.

(b) Boundary Point: Checking Eq. (8) one has at the boundary points

$$\frac{d\text{IP}}{dx_1}\bigg|_{x_1=0} = 1 \qquad \frac{d\text{IP}}{dx_1}\bigg|_{x_1=2\pi} = 1$$

Then by virtue of Eqs. (4) and (5), a local minimum exists at the lower boundary $x_1 = 0$. To find the absolute minimum all that is required is to compare the value of the IP at the local minimum $x_1 = 0$ and at the only other possible location $x_1 = \pi/2$.

$$\text{IP}(0) = 0 + \cos 0 = 1$$

$$\text{IP}\left(\frac{\pi}{2}\right) = \frac{\pi}{2} + \cos\frac{\pi}{2} = \frac{\pi}{2}$$

One sees that the absolute minimum occurs at the boundary $x_1 = x_{1_l} = 0$ and has the value $\text{IP}^*(0) = 1$.

---

† Examination of third-order terms would show that $x_1 = \pi/2$ is not a point at which a local minimum occurs.

2-4 TWO INDEPENDENT VARIABLES

Consider now that the IP is a function of the two independent variables x_1, x_2 within the region defined by the system constraint, $\psi(x_1, x_2) \leqslant 0$. That is, $\text{IP} = \text{IP}(x_1, x_2)$. Then for a local minimum to exist at x_1^*, x_2^*, by Definition 2, we require

$$\Delta \text{IP}(x_1^*, x_2^*) = \text{IP}(x_1^* + \Delta x_1, x_2^* + \Delta x_2) - \text{IP}(x_1^*, x_2^*) > 0 \qquad (1)$$

As before, our procedure is to express the value of the IP at any neighboring point to (x_1^*, x_2^*) in terms of quantities evaluated at (x_1^*, x_2^*). By series expansion of $\text{IP}(x_1^* + \Delta x_1, x_2^* + \Delta x_2)$, one finds

$$\Delta \text{IP}(x_1^*, x_2^*) = \frac{\partial \text{IP}}{\partial x_1^*} \Delta x_1 + \frac{\partial \text{IP}}{\partial x_2^*} \Delta x_2$$

$$+ \frac{1}{2} \left[\frac{\partial^2 \text{IP}}{\partial x_1^{*2}} \Delta x_1^2 + 2 \frac{\partial^2 \text{IP}}{\partial x_1^* \partial x_2^*} \Delta x_1 \Delta x_2 + \frac{\partial^2 \text{IP}}{\partial x_2^{*2}} \Delta x_2^2 \right] + O(\Delta x^3) \quad (2)$$

INTERIOR POINT: $\psi(x_1^*, x_2^*) < 0$. Since Δx_1 and Δx_2 can be made as small as we like, it is apparent that the first-order terms will determine the sign of $\Delta \text{IP}(x_1^*, x_2^*)$ if either $\partial \text{IP}/\partial x_1^* \neq 0$ or $\partial \text{IP}/\partial x_2^* \neq 0$. However, at a point interior to the constraint boundary the changes Δx_1 and Δx_2 may have either positive or negative signs. Therefore, at an interior point, the sign of the right side will change with the sign of Δx_1, Δx_2 in violation of Eq. (1) if the first-order terms do not vanish. It is thus clear that to have a local minimum at an interior point a necessary condition is

$$\frac{\partial \text{IP}}{\partial x_1^*} = 0 \qquad \frac{\partial \text{IP}}{\partial x_2^*} = 0 \qquad \psi(x_1^*, x_2^*) < 0 \qquad (3)$$

The second-order terms in Eq. (2) remain to be considered to determine what additional conditions are sufficient to ensure that the minimum exists. With Eq. (3),

$$\Delta \text{IP}(x_1^*, x_2^*) = \frac{1}{2} \left[\frac{\partial^2 \text{IP}}{\partial x_1^{*2}} \Delta x_1^2 + 2 \frac{\partial^2 \text{IP}}{\partial x_1^* \partial x_2^*} \Delta x_1 \Delta x_2 + \frac{\partial^2 \text{IP}}{\partial x_1^{*2}} \Delta x_2^2 \right] + O(\Delta x^3) \quad (4)$$

Provided all the second-order derivatives are not zero, the second-order terms will predominate, since Δx_1 and Δx_2 can be made as small as we like. To determine the conditions required for $\Delta \text{IP} > 0$, assuming $\partial^2 \text{IP}/\partial x_1^{*2} \neq 0$, we write Eq. (4) in the form

$$\Delta \text{IP} = \frac{1}{2(\partial^2 \text{IP}/\partial x_1^{*2})}$$

$$\times \left[\left(\frac{\partial^2 \text{IP}}{\partial x_1^{*2}} \right)^2 \Delta x_1^2 + 2 \frac{\partial^2 \text{IP}}{\partial x_1^{*2}} \frac{\partial^2 \text{IP}}{\partial x_1^* \partial x_2^*} \Delta x_1 \Delta x_2 + \frac{\partial^2 \text{IP}}{\partial x_1^{*2}} \frac{\partial^2 \text{IP}}{\partial x_2^{*2}} \Delta x_2^2 \right] \quad (5)$$

Adding and subtracting the term $(\partial^2 \mathrm{IP}/\partial x_1^* \, \partial x_2^*)^2 \, \Delta x_2^2$ within the brackets in Eq. (5), one obtains

$$\Delta \mathrm{IP} = \frac{1}{2(\partial^2 \mathrm{IP}/\partial x_1^{*2})} \left[\left(\frac{\partial^2 \mathrm{IP}}{\partial x_1^{*2}} \Delta x_1 + \frac{\partial^2 \mathrm{IP}}{\partial x_1^* \, \partial x_2^*} \Delta x_2 \right)^2 \right.$$
$$\left. + \left(\frac{\partial^2 \mathrm{IP}}{\partial x_1^{*2}} \frac{\partial^2 \mathrm{IP}}{\partial x_2^{*2}} - \left(\frac{\partial^2 \mathrm{IP}}{\partial x_1^* \, \partial x_2^*} \right)^2 \right) \Delta x_2^2 \right] \quad (6)$$

Let us examine the sign of the bracketed quantity on the right. The first term within the brackets is obviously nonnegative. We proceed by examining the possibilities for the second term.

(a) Assume that

$$\frac{\partial^2 \mathrm{IP}}{\partial x_1^{*2}} \frac{\partial^2 \mathrm{IP}}{\partial x_2^{*2}} - \left(\frac{\partial^2 \mathrm{IP}}{\partial x_1^* \, \partial x_2^*} \right)^2 > 0$$

Then, if $\partial^2 \mathrm{IP}/\partial x_1^{*2} > 0$, one has that $\Delta \mathrm{IP}(x_1^*, x_2^*) > 0$ and the point (x_1^*, x_2^*) corresponds to a minimum of the IP.

(b) Assume that

$$\frac{\partial^2 \mathrm{IP}}{\partial x_1^{*2}} \frac{\partial^2 \mathrm{IP}}{\partial x_2^{*2}} - \left(\frac{\partial^2 \mathrm{IP}}{\partial x_1^* \, \partial x_2^*} \right)^2 = 0$$

Since by choosing the direction given by

$$\frac{\partial^2 \mathrm{IP}}{\partial x_1^{*2}} \Delta x_1 + \frac{\partial^2 \mathrm{IP}}{\partial x_1^* \, \partial x_2^*} \Delta x_2 = 0$$

one can make $\Delta \mathrm{IP} = 0$, we see that on the basis of second-order terms it is not possible to conclude that $\Delta \mathrm{IP} > 0$ and that a local minimum exists.

(c) Assume that

$$\frac{\partial^2 \mathrm{IP}}{\partial x_1^{*2}} \frac{\partial^2 \mathrm{IP}}{\partial x_2^{*2}} - \left(\frac{\partial^2 \mathrm{IP}}{\partial x_1^* \, \partial x_2^*} \right)^2 < 0$$

Since $\Delta \mathrm{IP}$ can be either positive or negative depending on whether the direction chosen is $\Delta x_2 = 0$ or

$$\frac{\partial^2 \mathrm{IP}}{\partial x_1^{*2}} \Delta x_1 + \frac{\partial^2 \mathrm{IP}}{\partial x_1^* \, \partial x_2^*} \Delta x_2 = 0$$

we conclude that a local minimum does not exist at this point.

BOUNDARY POINT: $\psi(x_1^*, x_2^*) = 0$. Consider next the possibility of a local minimum on the boundary of the region. The Δx_i are now not completely arbitrary, since crossing the boundary is not permitted, but must satisfy

$$\psi(x_1^* + \Delta x_1, x_2^* + \Delta x_2) \leqslant 0 \tag{7a}$$

Therefore, as all Δx_i are not allowed, one *cannot* conclude from Eq. (2) that the first-order terms must vanish. Expanding Eq. (7a) about (x_1^*, x_2^*), one finds that Δx_1 and Δx_2 must satisfy the relation

$$\frac{\partial \psi}{\partial x_1^*} \Delta x_1 + \frac{\partial \psi}{\partial x_2^*} \Delta x_2 + O(\Delta x^2) \leqslant 0 \tag{7b}$$

Again, since the magnitude of the Δx_i can be made as small as we like, it is the first-order terms in Δx_i which predominate. In this light, for Eq. (1) to be satisfied for neighboring points to (x_1^*, x_2^*) within $\psi(x_1, x_2) < 0$, the condition to be met is

$$\frac{\partial \mathrm{IP}}{\partial x_1^*} \Delta x_1 + \frac{\partial \mathrm{IP}}{\partial x_2^*} \Delta x_2 > 0 \tag{8}$$

where $\Delta x_1, \Delta x_2$ must be chosen so that

$$\frac{\partial \psi}{\partial x_1^*} \Delta x_1 + \frac{\partial \psi}{\partial x_2^*} \Delta x_2 < 0 \tag{9}$$

Multiplying Eq. (9) by -1 one obtains

$$-\frac{\partial \psi}{\partial x_1^*} \Delta x_1 - \frac{\partial \psi}{\partial x_2^*} \Delta x_2 > 0 \tag{10}$$

Compare Eq. (8) with Eq. (10). For a local minimum to exist on the boundary, Eq. (8) must hold for *all* $\Delta x_1, \Delta x_2$ for which Eq. (10) is valid. It is apparent that this requires

$$\begin{aligned} \frac{\partial \mathrm{IP}}{\partial x_1^*} &= -k \frac{\partial \psi}{\partial x_1^*} \\ \frac{\partial \mathrm{IP}}{\partial x_2^*} &= -k \frac{\partial \psi}{\partial x_2^*} \end{aligned} \qquad \text{where } k > 0 \tag{11}$$

Geometrically Eq. (11) can be interpreted as requiring the rate of increase of the IP to be largest perpendicular to $\psi(x_1^*, x_1^*) = 0$ and positively directed toward the interior of the admissible region.

Before concluding that a local minimum does exist at (x_1^*, x_2^*) on the boundary it is necessary to check if Eq. (1) is satisfied for changes $\Delta x_1, \Delta x_2$ about x_1^*, x_2^* such that one stays *on* the boundary; that is,

$$\psi(x_1^* + \Delta x_1, x_2^* + \Delta x_2) = 0$$

In this case

$$\frac{\partial \psi}{\partial x_1^*} \Delta x_1 + \frac{\partial \psi}{\partial x_2^*} \Delta x_2 + O(\Delta x^2) = 0 \tag{12}$$

and one sees that Δx_1 and Δx_2 are not both independent. Rather, providing $\partial \psi / \partial x_1^* \neq 0$,

$$\Delta x_1 = -\frac{\partial \psi / \partial x_2^*}{\partial \psi / \partial x_1^*} \Delta x_2 + O(\Delta x^2) \tag{13}$$

Since Δx_1 and Δx_2 cannot be varied independently, let us eliminate Δx_1 from Eq. (2) in favor of Δx_2.

$$\Delta \mathrm{IP}(x_1^*, x_2^*) = \frac{\partial \mathrm{IP}}{\partial x_1^*} \left[\frac{\partial \mathrm{IP} / \partial x_2^*}{\partial \mathrm{IP} / \partial x_1^*} - \frac{\partial \psi / \partial x_2^*}{\partial \psi / \partial x_1^*} \right] \Delta x_2 + O(\Delta x_2^2) \tag{14}$$

As Δx_2 can now be considered arbitrary, if its coefficient is not zero, a change in its sign will change the sign of the right side of Eq. (14). Thus to satisfy Eq. (1) a necessary condition is

$$\frac{\partial \mathrm{IP} / \partial x_2^*}{\partial \mathrm{IP} / \partial x_1^*} = \frac{\partial \psi / \partial x_2^*}{\partial \psi / \partial x_1^*} \tag{15}$$

which, one sees, is equivalent to Eq. (11). Therefore, having now considered all possible points in the neighborhood of a point on the boundary, $\psi(x_1^*, x_2^*) = 0$, one can conclude that Eq. (11) is only a necessary condition for a local minimum to exist at a boundary point. If Eq. (11) is not satisfied, the only other way a minimum might occur at x_1^*, x_2^* would be if $\partial \mathrm{IP} / \partial x_1^* = \partial \mathrm{IP} / \partial x_2^* = 0$.

To determine whether a local minimum actually exists on the boundary, the condition for the terms of $O(\Delta x^2)$ in Eq. (14) to be positive must be found. This will not be done here.

To summarize: $\mathrm{IP} = \mathrm{IP}(x_1, x_2)$, $\psi(x_1, x_2) \leqslant 0$.

INTERIOR POINT: $\psi(x_1^*, x_2^*) < 0$. A necessary condition for a local minimum is

$$\frac{\partial \mathrm{IP}}{\partial x_1^*} = 0 \qquad \frac{\partial \mathrm{IP}}{\partial x_2^*} = 0 \tag{16}$$

If Eq. (16) is satisfied, then a sufficient condition for a local minimum at (x_1^*, x_2^*) is that

$$\frac{\partial^2 \mathrm{IP}}{\partial x_1^{*2}} > 0 \quad \text{and} \quad \frac{\partial^2 \mathrm{IP}}{\partial x_1^{*2}} \frac{\partial^2 \mathrm{IP}}{\partial x_2^{*2}} - \left(\frac{\partial^2 \mathrm{IP}}{\partial x_1^* \, \partial x_2^*} \right)^2 > 0 \tag{17}$$

BOUNDARY POINT: $\psi(x_1^*, x_2^*) = 0$. A necessary condition for a local minimum to exist on the boundary is that

$$\frac{d\text{IP}}{dn}\bigg|_{x_1^*, x_2^*} \geq 0 \tag{18}$$

where dn is directed normal to $\psi(x_1^*, x_2^*) = 0$ toward the interior of the admissible region. A less elegant, but perhaps easier to use, statement equivalent to Eq. (18) is that on $\psi(x_1^*, x_2^*) = 0$ it is necessary that either

$$\frac{\partial \text{IP}}{\partial x_1^*} = -k \frac{\partial \psi}{\partial x_1^*} \qquad \frac{\partial \text{IP}}{\partial x_2^*} = -k \frac{\partial \psi}{\partial x_2^*} \qquad \boxed{k > 0} \tag{19a}$$

or

$$\frac{\partial \text{IP}}{\partial x_1^*} = 0 \qquad \frac{\partial \text{IP}}{\partial x_2^*} = 0 \tag{19b}$$

for a local minimum to exist.

Example 2-4(1)   Find the absolute minimum of $\text{IP} = x_1^2 - x_2^2/4$ within the system constraint $x_1^2 + x_2^2 \leq 1$; that is, $\psi = x_1^2 + x_2^2 - 1 \leq 0$.

(a) Interior Point: $\psi = x_1^{*2} + x_2^{*2} - 1 < 0$.

$$\frac{\partial \text{IP}}{\partial x_1^*} = 0 = 2x_1^*$$

$$\frac{\partial \text{IP}}{\partial x_2^*} = 0 = -\frac{x_2^*}{2}$$

The only possible local minimum interior point is thus $x_1^* = x_2^* = 0$. Testing to see if Eq. (17) is satisfied,

$$\frac{\partial^2 \text{IP}}{\partial x_1^{*2}} = 2 \qquad \frac{\partial^2 \text{IP}}{\partial x_2^{*2}} = -\frac{1}{2} \qquad \frac{\partial^2 \text{IP}}{\partial x_1^* \, \partial x_2^*} = 0$$

and

$$\frac{\partial^2 \text{IP}}{\partial x_1^{*2}} \frac{\partial^2 \text{IP}}{\partial x_1^{*2}} - \left(\frac{\partial^2 \text{IP}}{\partial x_1^* \, \partial x_2^*}\right)^2 = -1 < 0$$

The point $x_1^* = x_2^* = 0$ is not a local minimum.

(b) Boundary Point: $\psi = x_1^{*2} + x_2^{*2} - 1 = 0$. Since Eq. (19b) is not satisfied, we must satisfy the necessary conditions given by Eq. (19a), from which one obtains

$$2x_1^* = -k2x_1^*$$

$$-2x_2^* = -k\frac{x_2^*}{2}$$

$$k > 0$$

The only possible solution is seen to be $x_1^* = 0, \ k = 4$.

Substituting into $\psi(x_1^*, x_2^*) = 0$, one finds corresponding to $x_1^* = 0$ that $x_2^* = \pm 1$. Thus, using our theory we have shown that the points $(0, 1)$ and $(0, -1)$ are the only possible locations of local minima on the boundary. Since we have not developed sufficient conditions for a point to be a local minimum if it lies on the boundary, one might be tempted to stop and wait for the theory to be produced. In actuality the wait is unnecessary. Certainly there must be a minimum value for the IP within the region.

It is the role of the necessary conditions to pick from among all points only those that have the possibility of being the absolute minimum (points of local minima, maxima, and inflection). It is the role of the sufficient conditions to further restrict this list by picking out only points of local minima. In lieu of the sufficiency conditions we need only test each possible solution (those which satisfy the necessary conditions) in the IP and pick the one that makes the IP smallest.

For the points $(0, 1)$ and $(0, -1)$,

$$\mathrm{IP} = 0 - \tfrac{1}{4} = -\tfrac{1}{4}$$

and we see that the absolute minimum of the IP occurs at both points. $\mathrm{IP}^* = -\tfrac{1}{4}$.

2-5   *N* INDEPENDENT VARIABLES

In this section we shall attempt to generalize the work of Sections 2-3 and 2-4 to treat the case of an index of performance which is a function of N independent variables within a given region of N space. Thus $\mathrm{IP} = \mathrm{IP}(x_1, x_2, \cdots, x_N)$ within the region defined by $\psi(x_1, x_2, \cdots, x_N) \leqslant 0$.

By series expansion about any local minimum point, $x_1^*, x_2^*, \cdots, x_N^*$ we find

$$\Delta\mathrm{IP}(x_1^*, \cdots, x_N^*) = \frac{\partial\mathrm{IP}}{\partial x_1^*}\,\Delta x_1 + \frac{\partial\mathrm{IP}}{\partial x_2^*}\,\Delta x_2 + \cdots + \frac{\partial\mathrm{IP}}{\partial x_N^*}\,\Delta x_N + O(\Delta x^2) > 0 \quad (1)$$

where the first-order terms predominate for small enough Δx_i.

INTERIOR POINT: $\psi(x_1^*, x_2^*, \cdots, x_N^*) < 0$. Since the x_i are independent let us first set $\Delta x_2 = \Delta x_3 = \cdots = \Delta x_N = 0$. Then Eq. (1) becomes

$$\Delta\mathrm{IP}(x_1^*, \cdots, x_N^*) = \frac{\partial\mathrm{IP}}{\partial x_1^*}\,\Delta x_1 + O(\Delta x_1^2) \quad (2)$$

As the sign of Δx_1 is arbitrary, we see that for $\Delta\mathrm{IP} > 0$ a necessary condition is that $\partial\mathrm{IP}/\partial x_1^* = 0$. In a similar manner we conclude that all the other first-order terms must vanish. Thus

$$\frac{\partial\mathrm{IP}}{\partial x_1^*} = \frac{\partial\mathrm{IP}}{\partial x_2^*} = \cdots = \frac{\partial\mathrm{IP}}{\partial x_N^*} = 0 \quad (3)$$

Referring back to basic calculus, remember that the first-order terms represent the principal part of ΔIP and are equivalent, by definition, to

$$d\mathrm{IP}(x_1^*, x_2^*, \cdots, x_N^*) = \frac{\partial \mathrm{IP}}{\partial x_1^*} dx_1 + \frac{\partial \mathrm{IP}}{\partial x_2^*} dx_2 + \cdots + \frac{\partial \mathrm{IP}}{\partial x_N^*} dx_N \qquad (4)$$

In this notation, then, Eq. (3) is seen to be equivalent to the condition

$$d\mathrm{IP}(x_1^*, x_2^*, \cdots, x_N^*) = 0 \qquad (5)$$

That is, the first variation about the minimum point must vanish in the interior of the admissible region.

Sufficiency conditions would follow from investigation of the second-order terms in Eq. (1).

BOUNDARY POINT: $\psi(x_1^*, x_2^*, \cdots, x_N^*) = 0$. In this case, in treating Eq. (1) the allowable Δx_i must be restricted so that one stays within the admissible region. Namely, $\psi(x_1^* + \Delta x_1, x_2^* + \Delta x_2, \cdots, x_N^* + \Delta x_N) \leqslant 0$ or on expansion

$$\frac{\partial \psi}{\partial x_1^*} \Delta x_1 + \frac{\partial \psi}{\partial x_2^*} \Delta x_2 + \cdots + \frac{\partial \psi}{\partial x_N^*} \Delta x_N + O(\Delta x^2) \leqslant 0 \qquad (6)$$

Since first-order terms will always dominate if they are not identically zero, our requirements for the local minimum on the boundary may be put in the form

$$d\mathrm{IP}(x_1^*, x_2^*, \cdots, x_N^*) > 0$$
$$d\psi(x_1^*, x_2^*, \cdots, x_N^*) \leqslant 0 \qquad (7)$$

Expanding Eq. (7), we require

$$\frac{\partial \mathrm{IP}}{\partial x_1^*} dx_1 + \frac{\partial \mathrm{IP}}{\partial x_2^*} dx_2 + \cdots + \frac{\partial \mathrm{IP}}{\partial x_N^*} dx_N > 0 \qquad (8)$$

for all dx_i such that

$$\frac{\partial \psi}{\partial x_1^*} dx_1 + \frac{\partial \psi}{\partial x_2^*} dx_2 + \cdots + \frac{\partial \psi}{\partial x_N^*} dx_N \leqslant 0 \qquad (9)$$

Equation (8) can be ensured only if

$$\frac{\partial \mathrm{IP}}{\partial x_1^*} = -k \frac{\partial \psi}{\partial x_1^*} \quad \cdots \quad \frac{\partial \mathrm{IP}}{\partial x_N^*} = -k \frac{\partial \psi}{\partial x_N^*} \qquad k > 0 \qquad (10)$$

However, since Eq. (9) includes the equality sign to allow consideration of a neighboring point on the boundary, so Eq. (8) will also if Eq. (10) is valid. Thus, as we have just ensured that $\Delta\mathrm{IP} \geqslant 0$ by use of Eq. (10), only a necessary condition for a local minimum on the boundary has been found. As before,

Eq. (10) can be interpreted geometrically as requiring the rate of change of the index of performance to be largest normal to the boundary and positively directed toward the interior of the admissible region. If Eq. (10) does not hold, then the only other way a local minimum might exist at the point would be if $d\mathrm{IP}(x_1^*, \cdots, x_N^*) = 0$.

To summarize: $\mathrm{IP} = \mathrm{IP}(x_1, x_2, \cdots, x_N)$, $\psi(x_1, x_2, \cdots, x_N) \leqslant 0$.

INTERIOR POINT: $\psi(x_1^*, \cdots, x_N^*) < 0$. A necessary condition for local minima in the interior is that

$$d\mathrm{IP} = \frac{\partial \mathrm{IP}}{\partial x_1^*}\, dx_1 + \cdots + \frac{\partial \mathrm{IP}}{\partial x_N^*}\, dx_N = 0 \tag{11}$$

which requires

$$\frac{\partial \mathrm{IP}}{\partial x_1^*} = 0 \quad \cdots \quad \frac{\partial \mathrm{IP}}{\partial x_N^*} = 0 \tag{12}$$

The sufficient conditions for a local minimum would be determined by the requirement

$$d^2\mathrm{IP} > 0 \tag{13}$$

BOUNDARY POINT: $\psi(x_1^*, \cdots, x_N^*) = 0$. A necessary condition for local minima is that

$$\frac{d\mathrm{IP}}{dn} \geqslant 0 \tag{14}$$

where dn is normal to the surface $\psi(x_1^*, \cdots, x_N^*) = 0$ and positively directed toward the interior of the admissible region. Equivalent to (14) are the two conditions

$$\frac{\partial \mathrm{IP}}{\partial x_1^*} = -k\,\frac{\partial \psi}{\partial x_1^*} \quad \cdots \quad \frac{\partial \mathrm{IP}}{\partial x_N^*} = -k\,\frac{\partial \psi}{\partial x_N^*} \qquad k > 0 \tag{15a}$$

or

$$\frac{\partial \mathrm{IP}}{\partial x_1^*} = 0 \quad \cdots \quad \frac{\partial \mathrm{IP}}{\partial x_N^*} = 0 \tag{15b}$$

The sufficiency conditions would again be found by consideration of the second-order terms in Eq. (1).

2-6 CONSTRAINT EQUATIONS

We consider a topic now that was first introduced in a different context in Section 2-4. To begin, let us say that our problem is to minimize the function $IP = IP(x_1, x_2)$, where x_1 and x_2 are not independent but related by the constraint equation.

$$\psi_1(x_1, x_2) = 0 \tag{1}$$

As a necessary condition for a minimum we have

$$dIP = \frac{\partial IP}{\partial x_1} dx_1 + \frac{\partial IP}{\partial x_2} dx_2 = 0 \tag{2}$$

However, since dx_1 and dx_2 are not arbitrary but related by

$$d\psi_1 = \frac{\partial \psi_1}{\partial x_1} dx_1 + \frac{\partial \psi_1}{\partial x_2} dx_2 = 0 \tag{3}$$

it is not possible to conclude as before that $\partial IP/\partial x_1^* = \partial IP/\partial x_2^* = 0$ in Eq. (2). This is easily seen, since if $\partial IP/\partial x_1^* = \partial IP/\partial x_2^* = 0$ we would have two equations for the determination of x_1^*, x_2^*. Although the values thus found would satisfy Eq. (2) they would not, in general, satisfy Eq. (1).

To find values that satisfy both equations we arbitrarily choose one of the variables as the independent one. Then using Eq. (3), dIP is expressed in terms of the independent differential. Let dx_2 be the independent differential. Assuming $\partial \psi_1/\partial x_1 \neq 0$, Eq. (3) yields

$$dx_1 = -\frac{\partial \psi_1/\partial x_2}{\partial \psi_1/\partial x_1} dx_2 \tag{4}$$

and

$$dIP = \left[-\frac{\partial IP}{\partial x_1} \frac{\partial \psi_1/\partial x_2}{\partial \psi_1/\partial x_1} + \frac{\partial IP}{\partial x_2} \right] dx_2 \tag{5}$$

As dx_2 can now be considered as arbitrary, we conclude that to satisfy Eq. (5) requires

$$\frac{\partial IP}{\partial x_1} \frac{\partial \psi_1}{\partial x_2} - \frac{\partial IP}{\partial x_2} \frac{\partial \psi_1}{\partial x_1} = 0 \tag{6}$$

Equation (6) and the constraint equation

$$\psi_1(x_1, x_2) = 0 \tag{7}$$

are two equations to be solved simultaneously for the solution point (x_1^*, x_2^*) satisfying the necessary conditions for a local minimum. This procedure of elimination of the dependent variables from the IP could conceptually be extended to a IP which is a function of many variables, subject to multiple constraint equations. However, in practice it would be a tedious method. With this in mind, an alternative technique is introduced which shall find extensions in our future work.

Consider, without further a priori justification, the function termed the *augmented index of performance*,

$$\mathcal{IP}(x_1, x_2) = \text{IP}(x_1, x_2) + \lambda_1 \psi_1(x_1, x_2) \tag{8}$$

where λ_1 is a parameter (multiplier) to be determined. In actuality, since by Eq. (1) $\psi_1 = 0$, we have

$$\mathcal{IP}(x_1, x_2) = \text{IP}(x_1, x_2) \tag{9}$$

and therefore a necessary condition for a local minimum is that

$$d\mathcal{IP}(x_1, x_2) = d\text{IP}(x_1, x_2) = 0 \tag{10}$$

Retaining the idea that Eq. (8) can be used to better advantage than Eq. (9), one has from Eq. (10) that

$$d\mathcal{IP} = d\text{IP} + \lambda_1 \, d\psi_1 = 0 \tag{11}$$

or

$$\left(\frac{\partial \text{IP}}{\partial x_1} + \lambda_1 \frac{\partial \psi_1}{\partial x_1}\right) dx_1 + \left(\frac{\partial \text{IP}}{\partial x_2} + \lambda_1 \frac{\partial \psi_1}{\partial x_2}\right) dx_2 = 0 \tag{12}$$

Now dx_1 and dx_2, as before, are both not independent; hence one cannot conclude immediately that each coefficient must be zero to satisfy Eq. (12). Let us choose dx_1 to be the dependent differential and dx_2 to be the independent one. Further, let us use the multiplier λ_1, which is at *our* disposal, to *make* one of the coefficients zero in Eq. (12). For example, let λ_1 take on the value that makes the coefficient of the dependent differential dx_1 equal zero; that is,

$$\frac{\partial \text{IP}}{\partial x_1} + \lambda_1 \frac{\partial \psi_1}{\partial x_1} = 0 \tag{13}$$

Equation (12) now reduces to

$$\left(\frac{\partial \text{IP}}{\partial x_2} + \lambda_1 \frac{\partial \psi_1}{\partial x_2}\right) dx_2 = 0 \tag{14}$$

However, since dx_2 is the independent differential it can be varied at will. Hence for Eq. (14) to be satisfied for all dx_2 its coefficient must also vanish.

$$\frac{\partial \text{IP}}{\partial x_2} + \lambda_1 \frac{\partial \psi_1}{\partial x_2} = 0 \tag{15}$$

Combining our results

$$\frac{\partial \mathrm{IP}}{\partial x_1} + \lambda_1 \frac{\partial \psi_1}{\partial x_1} = 0 \tag{16a}$$

$$\frac{\partial \mathrm{IP}}{\partial x_2} + \lambda_1 \frac{\partial \psi_1}{\partial x_2} = 0 \tag{16b}$$

$$\psi_1(x_1, x_2) = 0 \tag{16c}$$

Eqs. (16a, b, c) are three equations to be solved simultaneously for the three quantities x_1^*, x_2^*, and the multiplier λ_1. On eliminating λ_1 between Eqs. (16a) and (16b) our equations become

$$\frac{\partial \mathrm{IP}}{\partial x_1} \frac{\partial \psi_1}{\partial x_2} - \frac{\partial \mathrm{IP}}{\partial x_2} \frac{\partial \psi_1}{\partial x_1} = 0 \tag{17}$$

$$\psi_1(x_1, x_2) = 0$$

which are the same ones found to determine (x_1^*, x_2^*) by our first procedure.

Note that Eqs. (16a) and (16b) are just those equations which would have resulted from considering dx_1 and dx_2 as though they were independent in Eq. (12). Introduction of the multiplier λ_1 has allowed us to treat all the variables in the augmented index of performance *as though* each one were independent. The extension to more variables and constraints follows directly.

To summarize: Given that $\mathrm{IP} = \mathrm{IP}(x_1, \cdots, x_N)$ and that the variables are related by the $p < N$ constraint equations,

$$\psi_1(x_1, \cdots, x_N) = 0$$
$$\psi_2(x_1, \cdots, x_N) = 0$$
$$\vdots \tag{18}$$
$$\psi_p(x_1, \cdots, x_N) = 0$$

then necessary conditions for a local minimum to occur are found by forming the augmented index of performance

$$\mathscr{IP}(x_1, \cdots, x_N) = \mathrm{IP}(x_1, \cdots, x_N) + \sum_{i=1}^{p} \lambda_i \psi_i(x_1, \cdots, x_N) \tag{19}$$

The necessary conditions are

$$\frac{\partial \mathscr{IP}}{\partial x_1} = 0 \quad \frac{\partial \mathscr{IP}}{\partial x_2} = 0 \quad \cdots \quad \frac{\partial \mathscr{IP}}{\partial x_N} = 0 \tag{20}$$

Equations (18) and (20) are $p + N$ equations for the p unknowns λ_i and the N

unknowns x_i^*. The quantities $x_1^*, \cdots, x_N^*$ determine the location of possible points of local minima as well as the magnitude of the IP $= \text{IP}(x_1^*, \cdots, x_N^*)$.

The multipliers introduced in this section are commonly called *Lagrange multipliers*. We shall use them extensively later on.

2-7 MINIMA OF DEFINITE INTEGRALS— THE CALCULUS OF VARIATIONS

On reflection, it seems clear that not all system performance indices will be of the type discussed in Sections 2-2 through 2-6. For example, one can imagine an IP which requires, not the determination of a *single* best value for each system variable as before, but rather determination of the best behavior of the system variables over an entire interval between some initial and final times. Such an IP is

$$\text{IP} = \int_{t_i}^{t_f} \bar{f}(t, x, \dot{x})\, dt \tag{1}$$

where for the present we consider t_i and $x_0 = x(t_i)$, t_f and $x_f = x(t_f)$ to be specified values.

Given the IP of Eq. (1), in line with our design philosophy, the task would be to find the function $x(t)$ in the interval between t_i and t_f which minimizes the IP, while taking on the specified end values x_0 and x_f.

$$\text{IP*} = \min_{x(t)} \int_{t_i}^{t_f} \bar{f}(t, x, \dot{x})\, dt \tag{2}$$

The IP given by Eq. (1) is thus seen to be a function of the function $x(t)$; that is, $\text{IP} = \text{IP}[x(t)]$. In this manner, we begin our consideration of the requirements for the minimization of functionals. Having gained a certain amount of insight in earlier sections we proceed, not by stating an all-encompassing theorem, even if one exists, but by digesting small pieces before going on.

To repeat, instead of finding points at which a function is minimum, we must now find an entire function $x(t)$ which makes a definite integral a minimum. Let us imagine that such a minimizing function exists. Then, in line with our earlier procedure, the value of the IP corresponding to use of the minimizing function in Eq. (1) must be smaller than the IP obtained by using *any* other function. Although this is what is desired, as before, it is easiest to start our development by first treating comparison functions which differ from the minimizing function by only a small amount.

Denoting the comparison functions by $x(t, \epsilon)$, where ϵ is a small parameter, we will assume that $x(t, \epsilon)$ lies in some ϵ neighborhood of $x(t)$, the minimizing function, that is $|x(t, \epsilon) - x(t)| \leq \epsilon$. To derive necessary conditions that the minimizing function must satisfy, let us further restrict our comparison functions by also requiring that their slope $\dot{x}(t, \epsilon)$ be in an ϵ neighborhood of the slope $\dot{x}(t)$ of the minimizing function, that is $|\dot{x}(t, \epsilon) - \dot{x}(t)| \leqslant \epsilon$.

Variations of this type about the minimizing function are termed weak

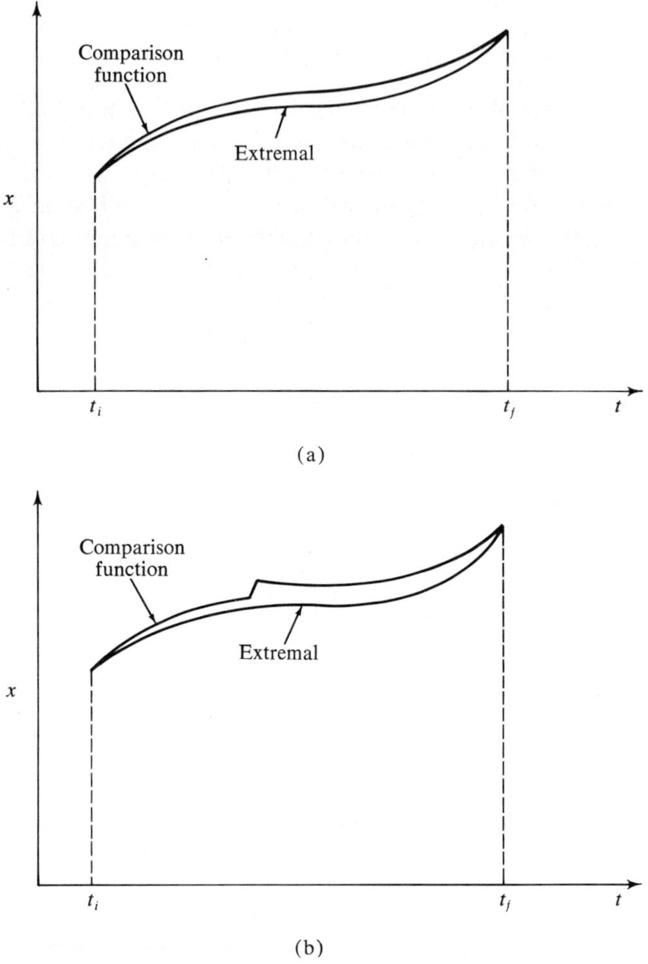

FIGURE 2-2   (a) Example of a weak variation; (b) example of a strong variation

variations. If the restriction on the slope $\dot{x}(t, \epsilon)$ is removed, the variation is termed a strong variation [see. Fig. 2-2]. Strong variations will later be employed in Section 3-6 to derive additional results. For the present, however, we restrict ourselves to consideration of comparison functions yielding weak variations.

Functions of this type have a series expansion in ϵ about $\epsilon = 0$; that is, one may write

$$x(t, \epsilon) = x(t) + \frac{\partial x}{\partial \epsilon}\bigg|_{\epsilon=0} \epsilon + O(\epsilon^2) \tag{3}$$

Since the first-order term in ϵ will obviously predominate in Eq. (3), provided its coefficient is not zero, it is convenient to represent the term by a special notation,

$$\delta x(t) \equiv \frac{\partial x}{\partial \epsilon}\bigg|_{\epsilon=0} \epsilon \tag{4}$$

Then to first order, $\delta x(t)$ represents the variation at time t between the comparison function $x(t, \epsilon)$ and the minimizing function $x(t)$,

$$\delta x(t) = x(t, \epsilon) - x(t) + O(\epsilon^2) \tag{5}$$

As the δ notation will be used a great deal in our work, it is worthwhile to restate its meaning in a more general context.

Definition   The quantity $\delta\Omega$, given by

$$\delta\Omega = \frac{d\Omega}{d\epsilon}\bigg|_{\epsilon=0} \epsilon \tag{6}$$

is termed the *first variation* of Ω and represents the differential change in Ω about $\epsilon = 0$ with respect to those variables in Ω which depend on ϵ. If all the variables in Ω depend on ϵ, then $\delta\Omega = d\Omega$.

2-8   THE EULER EQUATION

Retaining only the dominant first-order terms in Eq. 2-7(3) we begin our analysis by considering $x(t, \epsilon)$ to be of the form

$$x(t, \epsilon) = x(t) + \epsilon z(t) \tag{1}$$

where $z(t)$ is arbitrary except at t_i and t_f, where it must be zero to satisfy the end conditions. Thus

$$\delta x(t) = \epsilon z(t) \tag{2a}$$

and

$$\delta x_0 = \delta x_f = 0 \tag{2b}$$

Further, since

$$\dot{x}(t, \epsilon) = \dot{x}(t) + \epsilon\dot{z}(t) \tag{3}$$

then by our definition of Eq. 2-7(6),

$$\delta\dot{x}(t) = \epsilon\dot{z}(t) \tag{4}$$

On differentiating Eq. (2a) with respect to time and comparing the result with Eq. (4) one has that

$$\frac{d\delta x(t)}{dt} = \delta\frac{dx(t)}{dt} \tag{5}$$

Equation (5) shows that the δ operation and the derivative operation are interchangeable.

Substituting the function $x(t, \epsilon)$ from Eq. (1) into the IP of Eq. 2-7(1) the system index of performance is seen to be a function of ϵ:

$$IP(\epsilon) = \int_{t_i}^{t_f} \bar{f}[t, x(t, \epsilon), \dot{x}(t, \epsilon)]\, dt \tag{6}$$

On expanding $IP(\epsilon)$ about $\epsilon = 0$ one has

$$IP(\epsilon) = IP(0) + \frac{dIP}{d\epsilon}\bigg|_{\epsilon=0}\epsilon + \frac{d^2IP}{d\epsilon^2}\bigg|_{\epsilon=0}\frac{\epsilon^2}{2} + \cdots \tag{7}$$

Thus by Eq. 2-3(3), since the value of the IP in Eq. (7) is seen to be only a function of ϵ, a necessary condition for $IP(0)$ to be a minimum is seen to be

$$\delta IP = \frac{dIP}{d\epsilon}\bigg|_{\epsilon=0}\epsilon = 0 \tag{8}$$

Utilizing our definition of δ, the quantity δIP, the first variation of IP, is formed by simply treating the δ as if it were a differential and varying those terms which depend on ϵ. As t_i and t_f are fixed and hence not functions of ϵ, one finds

$$\delta IP = \int_{t_i}^{t_f} \delta\bar{f}[t, x(t, \epsilon), \dot{x}(t, \epsilon)]\, dt$$

which on expanding $\delta\bar{f}$ yields

$$\delta IP = \int_{t_i}^{t_f}\left[\frac{\partial\bar{f}}{\partial x}[t, x(t), \dot{x}(t)]\delta x + \frac{\partial\bar{f}}{\partial\dot{x}}[t, x(t), \dot{x}(t)]\,\delta\dot{x}\right] dt \tag{9}$$

The convenient δ notation will be used throughout our work in this manner.†

---

† Note in Eq. (9) that since in $x(t, \epsilon)$ and $\dot{x}(t, \epsilon)$, the variable t is not a function of ϵ, we do not write dx or $d\dot{x}$ in forming δIP but rather δx and $\delta\dot{x}$. The quantities δx and $\delta\dot{x}$ are only that part of dx and $d\dot{x}$ resulting from changes in ϵ. For example, about $\epsilon = 0$, $dx = \partial x/\partial t|_{\epsilon=0}\, dt + \partial x/\partial\epsilon|_{\epsilon=0}\epsilon$, or $dx = \dot{x}(t)\, dt + \delta x$.

The reader should satisfy himself that this procedure is completely equivalent to formally obtaining $\delta \text{IP} = d\text{IP}/d\epsilon|_{\epsilon=0}\,\epsilon$ by carrying out the latter.

Using Eq. (5) in Eq. (9) to interchange the δ and d/dt operations in $\delta \dot{x}$ one obtains

$$\delta \text{IP} = \int_{t_i}^{t_f} \left[\frac{\partial \bar{f}}{\partial x}\,\delta x + \frac{d\bar{f}}{\partial \dot{x}}\,\frac{d}{dt}\,\delta x \right] dt$$

The second term under the integral may now be integrated by parts to yield

$$\delta \text{IP} = \frac{\partial \bar{f}}{\partial \dot{x}}\,\delta x \bigg|_{t_i}^{t_f} - \int_{t_i}^{t_f} \left[\frac{d}{dt}\,\frac{\partial \bar{f}}{\partial \dot{x}} - \frac{\partial \bar{f}}{\partial x} \right] \delta x\, dt \tag{10}$$

where since $\delta x_0 = \delta x_f = 0$ by Eq. (2b), our necessary condition for a minimum becomes

$$\delta \text{IP} = 0 = - \int_{t_i}^{t_f} \left[\frac{d}{dt}\,\frac{\partial \bar{f}}{\partial \dot{x}} - \frac{\partial \bar{f}}{\partial x} \right] \delta x\, dt \tag{11}$$

It is seen that one way in which Eq. (11) can be satisfied is if the coefficient of δx vanishes. To see if any other possibilities exist, consider the following argument. If IP(0) is to be a minimum, then Eq. (11) must hold for all possible δx within t_i and t_f. Suppose

$$\frac{d}{dt}\,\frac{\partial \bar{f}}{\partial \dot{x}} - \frac{\partial \bar{f}}{\partial x}$$

was not zero, but positive, over some interval Δt, within t_i and t_f and zero everywhere else. Then since Eq. (11) must hold for all possible $\delta x(t)$, it must hold for a $\delta x(t)$, which is also positive over the same interval Δt. But in this case the integrand in Eq. (11) would be either positive or zero and integration would yield a positive δIP in contradiction to Eq. (11). Thus a necessary and sufficient condition for $\delta \text{IP} = 0$ is that

$$\frac{d}{dt}\,\frac{\partial \bar{f}}{\partial \dot{x}} - \frac{\partial \bar{f}}{\partial x} = 0 \tag{12a}$$

or equivalently, on expansion of $(d/dt)(\partial \bar{f}/\partial \dot{x})$,

$$\frac{\partial^2 \bar{f}}{\partial t\, \partial \dot{x}} + \frac{\partial^2 \bar{f}}{\partial x\, \partial \dot{x}}\,\dot{x} + \frac{\partial^2 \bar{f}}{\partial \dot{x}^2}\,\ddot{x} - \frac{\partial \bar{f}}{\partial x} = 0 \tag{12b}$$

Equation (12a), called the *Euler equation* or *characteristic equation* for the problem, is a first necessary condition to be satisfied by any function $x(t)$ which is to minimize our IP of Eq. 2-7(1). Functions (paths) which satisfy the Euler equation are said to be *extremals*.

Example 2-8(1)   Let our problem be to minimize the IP given by

$$\mathrm{IP} = \int_0^2 (\dot{x}^2 - 1)^2 \, dt$$

where $x(0) = 1$ and $x(2) = 1$. Forming the quantities required for the Euler equation,

$$\frac{\partial \bar{f}}{\partial \dot{x}} = 4\dot{x}(\dot{x}^2 - 1) \qquad \frac{\partial \bar{f}}{\partial x} = 0$$

one finds from (12a) that

$$\frac{d}{dt}\frac{\partial \bar{f}}{\partial \dot{x}} = 0$$

which on integration yields

$$\frac{\partial \bar{f}}{\partial \dot{x}} = \text{const.} = 4\dot{x}(\dot{x}^2 - 1)$$

Since the characteristic equation reduces to a cubic algebraic equation for $\dot{x}$, one concludes that $\dot{x}$ must be a constant; that is, $\dot{x} = c_1$. On integrating, one finds the extremals $x(t) = c_1 t + c_2$, where the constants c_1 and c_2 are to be used to meet the initial and final conditions, $x(0) = x(2) = 1$. In this manner we see that the only solution is $x(t) = 1$ for all t between 0 and 2. Corresponding to this extremal, the IP yields

$$\mathrm{IP}^* = \int_0^2 (0 - 1)^2 \, dt = 2$$

Example 2-8(2)   Our problem is taken to be that of minimizing, by proper choice of $x(t)$,

$$\mathrm{IP} = \int_0^1 (\dot{x}^2 - a^2 x^2) \, dt \tag{13}$$

where the values $x(0) = 0$, $x(1) = 0$, and a^2 are specified. For use in Eq. (12a) one forms

$$\frac{\partial \bar{f}}{\partial \dot{x}} = 2\dot{x} \qquad \frac{\partial \bar{f}}{\partial x} = -2a^2 x$$

which yields the characteristic equation

$$\ddot{x} + a^2 x = 0 \tag{14}$$

Equation (14) is a second-order ordinary differential equation to be satisfied by the minimizing function. Its solution is

$$x(t) = \bar{A} \sin at + \bar{B} \cos at$$

where the constants $\bar{A}$ and $\bar{B}$ are to be used to satisfy the end conditions, $x(0) = x(1) = 0$. In this manner one finds that $\bar{B} = 0$ and $\bar{A} \sin a = 0$.

Thus, either $\bar{A} = 0$ or $\sin a = 0$, in which case the parameter a must take on one of the values $a = n\pi$, $n = 1, 2, \cdots$. The possible extremals are

$$x(t) \equiv 0 \qquad\qquad\qquad \text{for } a \neq n\pi$$

$$x(t) = \bar{A} \sin n\pi t \quad \text{or} \quad x(t) \equiv 0 \qquad \text{for } a = n\pi \tag{15}$$

Substituting each of these extremal solutions in turn into the IP of Eq. (13) one finds that it takes on the value zero in both cases. It appears that $\text{IP*} = 0$ for this problem, independent of the value of a.

As a simple check on these results, consider the function $x(t) = t(1 - t)$, which, it is seen, satisfies the required boundary conditions. Differentiating to obtain $\dot{x} = 1 - 2t$ and substituting into the IP of Eq. (13), one finds $\text{IP} = (10 - a^2)/30$. Note that for $a^2 > 10$ the IP would be negative and hence lower than the value $\text{IP*} = 0$ resulting from using the extremals $x(t) \equiv 0$ or $x(t) = \bar{A} \sin n\pi t$.

The difficulty encountered results because our extremal solution as determined by the Euler equation, Eq. (12a), has only satisfied a necessary condition on the minimizing solution. The Euler equation comes from satisfying the requirement that $\delta\text{IP} = 0$. Using the extremal solution the change with ϵ in the value of the IP about the value $\text{IP}(0)$ is given in Eq. (7) as

$$\text{IP}(\epsilon) - \text{IP}(0) = \delta^2\text{IP} + O(\epsilon^3) \tag{16}$$

where the definition

$$\delta^2\text{IP} = \left.\frac{d^2\text{IP}}{d\epsilon^2}\right|_{\epsilon=0} \frac{\epsilon^2}{2} \tag{17}$$

has been employed to define the second variation of the IP about ϵ equal to zero. Then from Eq. (16) it is seen that the question of whether $\text{IP}(0)$ is the minimum performance index can only be answered after examination of the sign of the second variation. This is done in Section 2-11, where sufficient conditions for the extremal solution to minimize the performance index under weak variations are determined.

Due to the difficulty involved in working with the second variation to determine sufficiency conditions, solutions of the Euler equation in engineering problems are often checked only by physical arguments. Except for Section 2-11, in the main it is this latter course that will be taken in continuing our work. It is believed that the needs of the engineering reader are often best served by first learning to use basic tools before complicating the subject to the extent that interest and perhaps utility are lost.

2-9   THE CASE $\bar{f} = \bar{f}(x, \dot{x})$

It will be shown here that for $\bar{f}$ not an explicit function of t, a first integral of our characteristic equation

$$\frac{d}{dt}\frac{\partial \bar{f}}{\partial \dot{x}} - \frac{\partial \bar{f}}{\partial x} = 0 \tag{1}$$

can be obtained. Multiplying Eq. (1) by $\dot{x}$ it may be rewritten as

$$\frac{d}{dt}\left(\dot{x}\frac{\partial \bar{f}}{\partial \dot{x}}\right) - \frac{\partial \bar{f}}{\partial x}\dot{x} - \frac{\partial \bar{f}}{\partial \dot{x}}\ddot{x} = 0 \tag{2}$$

In addition, consider the quantity

$$\frac{d\bar{f}}{dt} = \frac{\partial \bar{f}}{\partial t} + \frac{\partial \bar{f}}{\partial x}\dot{x} + \frac{\partial \bar{f}}{\partial \dot{x}}\ddot{x} \tag{3}$$

Elimination of $(\partial \bar{f}/\partial x)\dot{x} + (\partial \bar{f}/\partial \dot{x})\ddot{x}$ between Eq. (2) and Eq. (3) then yields

$$\frac{d}{dt}\left(\dot{x}\frac{\partial \bar{f}}{\partial \dot{x}} - \bar{f}\right) = -\frac{\partial \bar{f}}{\partial t} \tag{4}$$

Equations (1), (2), (3), and (4) are for the general case $\bar{f} = \bar{f}(t, x, \dot{x})$. However, for the case treated here $\bar{f} = \bar{f}(x, \dot{x})$. As $\bar{f}$ is not an explicit function of t, then $\partial \bar{f}/\partial t \equiv 0$. Thus Eq. (4) may be integrated to yield

$$\dot{x}\frac{\partial \bar{f}}{\partial \dot{x}} - \bar{f} = \text{const.} \tag{5}$$

giving the desired first integral.

2-10   THE WEIERSTRASS–ERDMANN CORNER CONDITIONS

In the earlier sections it has been implicitly assumed that the extremal $x(t)$ sought was a continuous function of time and had as well, a continuous first derivative $\dot{x}(t)$, Although it might be argued that the continuity requirement on the system variable $x(t)$ seems reasonable, it is possible to imagine the rate $\dot{x}(t)$ changing discontinuously. Thus the interval between t_i and t_f may be thought of as made up of subarcs along which solutions of the Euler equation hold, joined together at points of discontinuity of $\dot{x}$, called *corners* of the extremal arc, by certain continuity conditions. It is these continuity conditions that we now seek.†

---

† A more rigorous derivation than that presented here will be given in Section 4-4 after introduction of the control formulation.

Equations 2-8(12) and 2-9(4) are repeated below for this purpose:

$$\frac{d}{dt}\frac{\partial \bar{f}}{\partial \dot{x}} - \frac{\partial \bar{f}}{\partial x} = 0 \tag{1a}$$

$$\frac{d}{dt}\left(\dot{x}\frac{\partial \bar{f}}{\partial \dot{x}} - \bar{f}\right) = -\frac{\partial \bar{f}}{\partial t} \tag{1b}$$

The first is the Euler equation, which defines the extremal; the second was obtained from it in Section 2-9. Thus both relations must hold along any extremal.

Imagine that a corner occurs between t_i and t_f at $t = t_1$. Then integrating Eqs. (1a) and (1b) across the corner from $t_1 - \epsilon$ to $t_1 + \epsilon$ one finds

$$\frac{\partial \bar{f}}{\partial \dot{x}}\bigg|_{t_1 + \epsilon} = \frac{\partial \bar{f}}{\partial \dot{x}}\bigg|_{t_1 - \epsilon} + \int_{t_1 - \epsilon}^{t_1 + \epsilon} \frac{\partial \bar{f}}{\partial x}\,dt \tag{2a}$$

$$\left(\dot{x}\frac{\partial \bar{f}}{\partial \dot{x}} - \bar{f}\right)\bigg|_{t_1 + \epsilon} = \left(\dot{x}\frac{\partial \bar{f}}{\partial \dot{x}} - \bar{f}\right)\bigg|_{t_1 - \epsilon} - \int_{t_1 - \epsilon}^{t_1 + \epsilon} \frac{\partial \bar{f}}{\partial t}\,dt \tag{2b}$$

Assuming that $\bar{f}$ and its required partial derivatives are bounded but possibly discontinuous, then as ϵ approaches zero the integrals in Eqs. (2a) and (2b) will vanish. This results because under the above condition of finiteness the integrals would represent the integration of a finite quantity over an interval of length zero. Denoting the direction the limit is approached as ϵ goes to zero by $t_1^{(+)}$ and $t_1^{(-)}$, Eq. (2) becomes

$$\frac{\partial \bar{f}}{\partial \dot{x}}\bigg|_{t_1^{(-)}} = \frac{\partial \bar{f}}{\partial \dot{x}}\bigg|_{t_1^{(+)}} \tag{3a}$$

$$\left(\dot{x}\frac{\partial \bar{f}}{\partial \dot{x}} - \bar{f}\right)\bigg|_{t_1^{(-)}} = \left(\dot{x}\frac{\partial \bar{f}}{\partial \dot{x}} - \bar{f}\right)\bigg|_{t_1^{(+)}} \tag{3b}$$

Equations (3a) and (3b) are the Weierstrass–Erdmann corner conditions for this problem. They are necessary conditions to be satisfied at corners of the extremal arc.

Example 2-10(1)   Let us return to Example 2-8(1), which requires minimizing

$$\mathrm{IP} = \int_0^2 (\dot{x}^2 - 1)^2 \, dt \tag{4}$$

where $x(0) = x(2) = 1$. There it is found that the extremal solutions to the Euler equation are straight lines, $x(t) = c_1 t + c_2$. On satisfying the end conditions the solution becomes $x(t) = 1$, with the corresponding index of performance, $\mathrm{IP}^* = 2$.

Consider now the possibility that a corner exists along the minimizing arc at some time t_1. At the corner let the value of $\dot{x}$ as t_1 is approached from above and below be $\dot{x}_+$ and $\dot{x}_-$, respectively. Applying the corner conditions of Eqs. (3a) and (3b) one finds

$$\dot{x}_-(\dot{x}_-^2 - 1) = \dot{x}_+(\dot{x}_+^2 - 1) \tag{5a}$$

$$4\dot{x}_-^2(\dot{x}_-^2 - 1) - (\dot{x}_-^2 - 1)^2 = 4\dot{x}_+^2(\dot{x}_+^2 - 1) - (\dot{x}_+^2 - 1)^2 \tag{5b}$$

or, on rearranging,

$$(\dot{x}_-^3 - \dot{x}_+^3) - (\dot{x}_- - \dot{x}_+) = 0 \tag{6a}$$

$$3(\dot{x}_-^4 - \dot{x}_+^4) - 2(\dot{x}_-^2 - \dot{x}_+^2) = 0 \tag{6b}$$

Equations (6a) and (6b) are two simultaneous equations to be solved for possible values of $\dot{x}_-$ and $\dot{x}_+$. Since we are interested in determining if other solutions than the obvious one $\dot{x}_- = \dot{x}_+$ exist, let us factor $(\dot{x}_- - \dot{x}_+)$ from each equation, obtaining

$$(\dot{x}_- - \dot{x}_+)[\dot{x}_-^2 + \dot{x}_-\dot{x}_+ + \dot{x}_+^2 - 1] = 0 \tag{7a}$$

$$(\dot{x}_- - \dot{x}_+)[3(\dot{x}_- + \dot{x}_+)(\dot{x}_-^2 + \dot{x}_+^2) - 2(\dot{x}_- + \dot{x}_+)] = 0 \tag{7b}$$

Thus, by Eq. (7), either $\dot{x}_+ = \dot{x}_-$ and no corner exists at t_1 or

$$\dot{x}_-^2 + \dot{x}_-\dot{x}_+ + \dot{x}_+^2 = 1 \tag{8a}$$

$$(\dot{x}_- + \dot{x}_+)[3(\dot{x}_-^2 + x_+^2) - 2] = 0 \tag{8b}$$

Assume first that $\dot{x}_- + \dot{x}_+ \neq 0$ in Eq. (8b). One then obtains by solution of Eqs. (8a) and (8b) that $\dot{x}_+ = \dot{x}_- = \pm\frac{1}{3}$, which means there is no discontinuity in this case. The only other possibility which satisfies Eq. (8b), is $\dot{x}_+ + \dot{x}_- = 0$. Using this relation in Eq. (8a) one finds that $\dot{x}_+ = -\dot{x}_- = \pm 1$, a possible discontinuity.

Since the extremals are straight lines, the solution in this case consists of continuous-line segments with slope of ± 1. In this manner an alternative solution to $x(t) \equiv 1$ is found to be

$$\begin{aligned} x(t) &= t + 1 & 0 \leqslant t \leqslant 1 \\ x(t) &= -t + 3 & 1 \leqslant t \leqslant 2 \end{aligned} \tag{9}$$

Substituting this solution for which $\dot{x} = \pm 1$ back into the IP of Eq. (4) we find that IP* $= 0$. As zero is the lower bound of the IP since $\bar{f}$ is positive semidefinite, a minimizing solution has been found. That this solution is nonunique should be apparent (see Fig. 2-3).

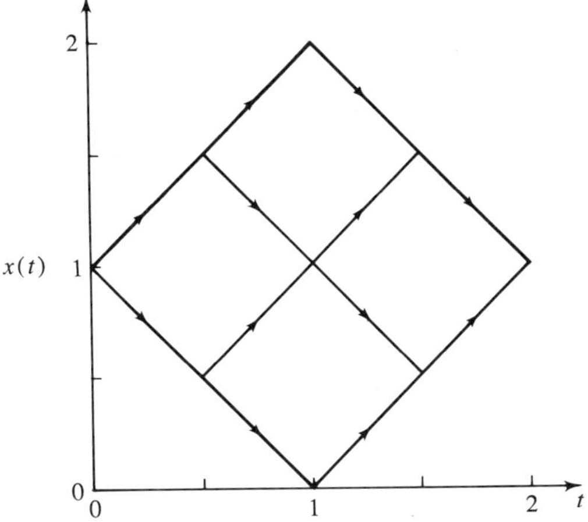

FIGURE 2-3   Several minimizing extremals for Example 2-10(1).

2-11   THE SECOND VARIATION

To review, our problem has been stated as that of minimizing

$$\text{IP} = \int_{t_i}^{t_f} \bar{f}(t, x, \dot{x})\, dt \tag{1}$$

by determination of $x(t)$ in the interval between t_i and t_f where t_i, t_f, $x(t_i)$, and $x(t_f)$ are all specified values. Assuming that $x(t)$ is the minimizing solution, we have considered weak comparison functions of the form

$$x(t, \epsilon) = x(t) + \epsilon z(t) \tag{2}$$

Then substituting the comparison function into the IP of Eq. (1) and expanding about ϵ equal to zero, there resulted

$$\Delta\text{IP} = \text{IP}(\epsilon) - \text{IP}(0) = \frac{d\text{IP}}{d\epsilon}\bigg|_{\epsilon=0}\epsilon + \frac{d^2\text{IP}}{d\epsilon^2}\bigg|_{\epsilon=0}\frac{\epsilon^2}{2} + O(\epsilon^3) \tag{3}$$

The condition that

$$\Delta\text{IP} = \text{IP}(\epsilon) - \text{IP}(0) > 0 \tag{4}$$

for arbitrary infinitesimal variations about the minimizing solution led to the requirement that the first-order terms in ϵ in Eq. (3) must vanish. Using the notation

$$\delta\text{IP} = \frac{d\text{IP}}{d\epsilon}\bigg|_{\epsilon=0}\epsilon = 0 \tag{5}$$

we found that for Eq. (5) to be satisfied, $x(t)$ must satisfy the Euler equation

$$\frac{d}{dt}\frac{\partial \bar{f}}{\partial \dot{x}} - \frac{\partial \bar{f}}{\partial x} = 0 \tag{6}$$

Examining Eq. (3), it is seen that $x(t)$ will only correspond to a minimizing solution if the second-order term in ϵ is positive semidefinite. That is, using the solution to Eq. (6), the relation

$$\left.\frac{d^2\text{IP}}{d\epsilon^2}\right|_{\epsilon=0}\frac{\epsilon^2}{2} \equiv \delta^2\text{IP} \geqslant 0 \tag{7}$$

obtained from Eqs. (3) and (4) must be satisfied for all allowable variations $\delta x(t)$ if $x(t)$ is to be the minimizing solution. If Eq. (7) is satisfied without the equality sign for all nontrivial admissible variations $\delta x(t)$, it can be concluded that the value of the IP obtained using the extremal solution will be lower than the value obtained by using a weak comparison function in the IP. In this context, the extremal solution of Eq. (6), which satisfies Eq. (7), would be said to yield a weak minimum in the IP.

Let us now determine what is involved in requiring that Eq. (7) be satisfied. The quantity $\delta^2\text{IP}$ must be formed to start. Substituting from Eq. (2) into Eq. (1) we have

$$\text{IP}(\epsilon) = \int_{t_i}^{t_f} \bar{f}[t, x(t, \epsilon), \dot{x}(t, \epsilon)]\, dt$$

from which one finds

$$\frac{d\text{IP}}{d\epsilon} = \int_{t_i}^{t_f} \left\{ \frac{\partial \bar{f}}{\partial x}[t, x(t, \epsilon), \dot{x}(t, \epsilon)]z(t) + \frac{\partial \bar{f}}{\partial \dot{x}}[t, x(t, \epsilon), \dot{x}(t, \epsilon)]\dot{z}(t) \right\} dt \tag{8}$$

where the relation of Eq. (2) has been used to express $\partial x(t, \epsilon)/\partial \epsilon$ and $\partial \dot{x}(t, \epsilon)/\partial \epsilon$ as $z(t)$ and $\dot{z}(t)$, respectively. Then taking the derivative of Eq. (8) with respect to ϵ and evaluating the result at ϵ equal to zero, one has

$$\left.\frac{d^2\text{IP}}{d\epsilon^2}\right|_{\epsilon=0} = \int_{t_i}^{t_f}\left\{ \frac{\partial^2\bar{f}}{\partial x^2}[t, x(t), \dot{x}(t)]z(t)^2 + 2\,\frac{\partial^2\bar{f}}{\partial x\,\partial\dot{x}}[t, x(t), \dot{x}(t)]z(t)\dot{z}(t) \right.$$

$$\left. + \frac{\partial^2\bar{f}}{\partial\dot{x}^2}[t, x(t), \dot{x}(t)]\dot{z}(t)^2 \right\} dt \tag{9}$$

Finally, multiplying Eq. (9) by $\epsilon^2/2$, using Eqs. 2-8(2), 2-8(4) and the definitions

$$\bar{f}_{xx}(t) = \frac{\partial^2\bar{f}}{\partial x^2}[t, x(t), \dot{x}(t)]$$

$$\bar{f}_{x\dot{x}}(t) = \frac{\partial^2\bar{f}}{\partial x\,\partial\dot{x}}[t, x(t), \dot{x}(t)] \tag{10}$$

$$\bar{f}_{\dot{x}\dot{x}}(t) = \frac{\partial^2\bar{f}}{\partial\dot{x}^2}[t, x(t), \dot{x}(t)]$$

one obtains

$$\delta^2 \text{IP} = \frac{1}{2} \int_{t_i}^{t_f} [\bar{f}_{\dot{x}\dot{x}}(t)\, \delta\dot{x}^2 + 2\bar{f}_{x\dot{x}}(t)\, \delta x\, \delta\dot{x} + \bar{f}_{xx}(t)\, \delta x^2]\, dt \tag{11}$$

The quantities $\bar{f}_{\dot{x}\dot{x}}$, $\bar{f}_{x\dot{x}}$, and $\bar{f}_{xx}$ are shown as explicit functions of time because they are evaluated along the known extremal solution to Eq. (6). All variations $\delta x(t)$ are allowable provided they satisfy the end conditions

$$\delta x(t_i) = \delta x(t_f) = 0 \tag{12}$$

resulting from the end points being fixed.

Equation (11) can also be written in a different form that will be of use. The first two terms in the integrand are integrated using Eq. (5) once, by parts, to yield

$$2\delta^2\text{IP} = [\bar{f}_{\dot{x}\dot{x}}\, \delta\dot{x}\, \delta x + \bar{f}_{x\dot{x}}\, \delta x^2]\, \Big|_{t_i}^{t_f}$$

$$- \int_{t_i}^{t_f} \left[\delta x\, \frac{d}{dt}\, (\bar{f}_{\dot{x}\dot{x}}\, \delta\dot{x}) - \left(\bar{f}_{xx} - \frac{d}{dt}\bar{f}_{x\dot{x}} \right) \delta x^2 \right] dt$$

Then with Eq. (12) one has

$$\delta^2\text{IP} = -\frac{1}{2} \int_{t_i}^{t_f} \delta x \left[\frac{d}{dt}\, (\bar{f}_{\dot{x}\dot{x}}\, \delta\dot{x}) - \left(\bar{f}_{xx} - \frac{d}{dt}\bar{f}_{x\dot{x}} \right) \delta x \right] dt \tag{13}$$

The relations of Eqs. (11) and (13) are equivalent.

Representing the second variation of the IP as given in Eq. (11) by

$$\delta^2\text{IP} = \int_{t_i}^{t_f} \omega(t,\, \delta x,\, \delta\dot{x})\, dt \tag{14a}$$

where

$$\omega(t, x, \dot{x}) = \tfrac{1}{2}[\bar{f}_{\dot{x}\dot{x}}(t)\, \delta\dot{x}^2 + 2\bar{f}_{x\dot{x}}(t)\, \delta x\, \delta\dot{x} + \bar{f}_{xx}(t)\, \delta x^2] \tag{14b}$$

it is seen on comparing Eqs. (14a) and (1) that the variations which minimize $\delta^2\text{IP}$ must satisfy a characteristic equation analogous to Eq. (6). Thus

$$\frac{d}{dt}\frac{\partial\omega}{\partial\, \delta\dot{x}} - \frac{\partial\omega}{\partial\, \delta x} = 0 \tag{15}$$

The problem of determining the minimum of $\delta^2\text{IP}$ is called the *accessory minimum problem* while the Euler equation for the problem is called the *Jacobi* or *accessory differential equation*. Let us denote a variation which satisfies the accessory equation of Eq. (15) by $\delta v(t)$. Then using Eq. (14b), after canceling two terms the accessory equation for the problem becomes

$$\frac{d}{dt}[\bar{f}_{\dot{x}\dot{x}}(t)\, \delta\dot{v}] - \left[\bar{f}_{xx}(t) - \frac{d}{dt}\bar{f}_{x\dot{x}}(t) \right] \delta v(t) = 0 \tag{16}$$

The solution $\delta v(t)$ to the accessory equation of Eq. (16) can now be used to put the second variation as given by Eq. (13) in a particularly fruitful form. In Eq. (13), after multiplying and dividing by $\delta v(t)$ under the integral, Eq. (16) can be used to replace $[\bar{f}_{xx} - (d/dt)\,\bar{f}_{x\dot{x}}]\,\delta v$ by $(d/dt)[\bar{f}_{x\dot{x}}\,\delta v]$, yielding

$$\delta^2 \mathrm{IP} = \frac{1}{2}\int_{t_i}^{t_f} \frac{\delta x}{\delta v}\left[\delta x \frac{d}{dt}(\bar{f}_{\dot{x}\dot{x}}\,\delta v) - \delta v \frac{d}{dt}(\bar{f}_{\dot{x}\dot{x}}\,\delta \dot{x})\right] dt \tag{17}$$

Then since

$$\delta x \frac{d}{dt}(\bar{f}_{\dot{x}\dot{x}}\,\delta v) - \delta v \frac{d}{dt}(\bar{f}_{\dot{x}\dot{x}}\,\delta \dot{x}) = \frac{d}{dt}[\bar{f}_{\dot{x}\dot{x}}(\delta x\,\delta \dot{v} - \delta v\,\delta \dot{x})]$$

Equation (17) becomes

$$\delta^2 \mathrm{IP} = \frac{1}{2}\int_{t_i}^{t_f} \frac{\delta x}{\delta v}\frac{d}{dt}[\bar{f}_{\dot{x}\dot{x}}(\delta x\,\delta \dot{v} - \delta v\,\delta \dot{x})]\,dt \tag{18}$$

Integrating Eq. (18) by parts one obtains

$$\delta^2 \mathrm{IP} = \frac{1}{2}\left(\frac{\delta x^2\,\delta \dot{v}}{\delta v} - \delta x\,\delta \dot{x}\right)\bar{f}_{\dot{x}\dot{x}}\bigg|_{t_i}^{t_f} - \frac{1}{2}\int_{t_i}^{t_f}\bar{f}_{\dot{x}\dot{x}}[\delta x\,\delta \dot{v} - \delta v\,\delta \dot{x}]\frac{d}{dt}\left(\frac{\delta x}{\delta v}\right)dt \tag{19}$$

which, on using Eq. (12) to eliminate the first term, yields the desired relation

$$\delta^2 \mathrm{IP} = \frac{1}{2}\int_{t_i}^{t_f}\bar{f}_{\dot{x}\dot{x}}(t)\left[\delta \dot{x} - \frac{\delta x\,\delta \dot{v}}{\delta v}\right]^2 dt \tag{20}$$

Recall that we wished to determine the conditions under which $\delta^2 \mathrm{IP} > 0$ for $\delta x(t)$ not identically zero. From Eq. (20) it is seen that this will occur whenever

$$\bar{f}_{\dot{x}\dot{x}}(t) > 0 \tag{21}$$

provided the term in the brackets in the integrand does not vanish identically for any admissible variation. Examination of the bracketed quantity reveals that it cannot be zero throughout the interval t_i to t_f if $\delta x(t)$ is not identically equal to $\delta v(t)$.

$$\delta x(t) \not\equiv \delta v(t) \tag{22}$$

The variation $\delta v(t)$ is a nontrivial (not identically zero) solution of Eq. (16), a linear second-order differential equation. Let the two linearly independent solutions to Eq. (16) be $\delta v_1(t)$ and $\delta v_2(t)$. Then the general solution may be written

$$\delta v(t) = A\,\delta v_1(t) + B\,\delta v_2(t)$$

As there are two arbitrary constants, the variation $\delta v(t)$ can be made to vanish at t_i by choosing A and B so that they satisfy

$$0 = A\,\delta v_1(t_i) + B\,\delta v_2(t_i) \tag{23}$$

Then since the admissible variations $\delta x(t)$ are required by Eq. (12) to vanish at t_i and t_f the question of whether $\delta x(t)$ can identically equal $\delta v(t)$ depends on whether the next zero of $\delta v(t)$ after t_i is inside or outside the interval $t_i \leqslant t \leqslant t_f$.

Let us introduce the following definition.

Conjugate Point Definition   A point t_1 is said to be conjugate to the initial point t_i if there exists a nontrivial solution $\delta v(t)$ to the accessory equation which vanishes at t_i and again at t_1.

If the first conjugate point occurs after t_f, the relation of Eq. (22) is satisfied since $\delta v(t)$ does not vanish again after t_i until $t_1 > t_f$, whereas the admissible variation $\delta x(t)$ must vanish again at t_f. Thus it can be concluded that if the first conjugate point occurs at $t_1 > t_f$, and Eq. (21) is satisfied, the extremal solution will yield a weak minimum in the IP.

On the other hand, suppose the first conjugate point occurs inside the interval at $t_1 < t_f$. Then since the relation of Eq. (20) was obtained by multiplying and dividing by $\delta v(t)$, which is zero at t_1, its validity is doubtful. For this case we prove the following theorem.

CONJUGATE POINT (JACOBI) CONDITION: Let $x^*(t)$ be the function which minimizes our IP. Then using $x^*(t)$ in the accessory problem, there exists no conjugate point to t_i in the interval $t_i < t < t_f$.

PROOF BY CONTRADICTION: To prove the theorem let us start by assuming that on using $x^*(t)$ there is a conjugate point at $t_1 < t_f$. This means there is a *nontrivial* solution $\delta v(t)$ of Eq. (16) which vanishes at t_i and again at $t_1 < t_f$, as shown in Fig. 2-4a. Since $\delta x(t)$ must vanish at t_i and t_f, let us construct an admissible variation $\delta x(t)$ by letting

$$\delta x(t) = \begin{cases} \delta v(t) & t_i \leqslant t \leqslant t_1 \\ 0 & t_1 \leqslant t \leqslant t_f \end{cases} \tag{24}$$

as shown in Fig. 2-4b.

For $\delta x(t)$ as given by Eq. (24) let us determine the value of $\delta^2 \text{IP}$. Breaking the interval of integration in Eq. (11) into the two subintervals $t_i \leqslant t \leqslant t_1$ and $t_1 \leqslant t \leqslant t_f$, it is seen that the integral over the latter portion vanishes since $\delta x(t)$ is identically zero there. Thus

$$\delta^2 \text{IP} = \int_{t_i}^{t_1} \omega(t, \delta x, \delta \dot x)\, dt$$

$$\omega(t, \delta x, \delta \dot x) = \tfrac{1}{2}[\bar f_{\dot x \dot x}(t)\, \delta \dot x^2 + 2\bar f_{x \dot x}\, \delta x\, \delta \dot x + \bar f_{xx}(t)\, \delta x^2] \tag{25}$$

Then following the procedure that led to Eq. (13) from Eq. (11), realizing that

$$\delta x(t_i) = \delta x(t_1) = 0 \tag{26}$$

δ^2IP can be written

$$\delta^2\text{IP} = -\frac{1}{2} \int_{t_i}^{t_1} \delta x \left[\frac{d}{dt} (\bar{f}_{\dot{x}\dot{x}} \, \delta\dot{x}) - \left(\bar{f}_{xx} - \frac{d}{dt}\bar{f}_{x\dot{x}} \right) \delta x \right] dt \tag{27}$$

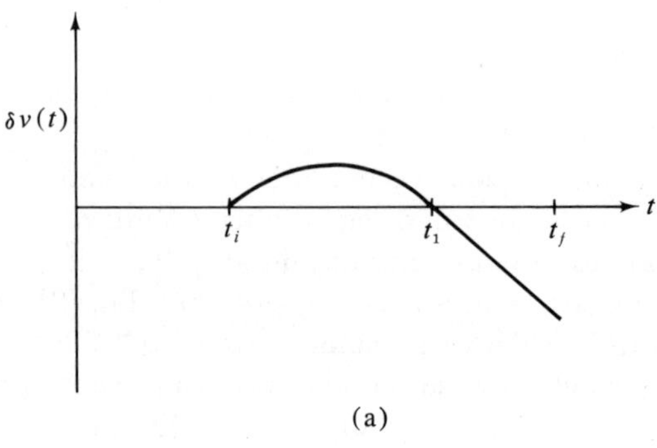

(a)

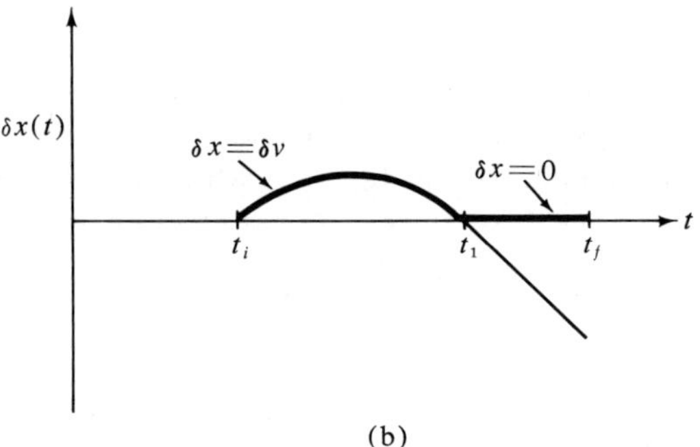

(b)

FIGURE 2-4   (a) Accessory equation solution—
conjugate point at t_1; (b) assumed variation $\delta x(t)$.

Finally, since in the interval $t_i \leqslant t \leqslant t_1$, $\delta x(t)$ has been taken equal to $\delta v(t)$, which satisfies the accessory equation of Eq. (16), it is seen that Eq. (27) gives

$$\delta^2\text{IP} = 0 \qquad \text{using } \delta x(t) \text{ from Eq. (24)} \tag{28}$$

Recall now that we are considering variations $\delta x(t)$ about the minimizing solution $x^*(t)$. In this context, one sees from Eq. (7) that the minimum value of δ^2IP is zero. Thus $\delta x(t)$ as given by Eq. (24) minimizes δ^2IP. Since this is true, the variation $\delta x(t)$ must satisfy the corner conditions for the accessory problem. Analogous to Eq. 2-10(3a) we require for the corner at t_1 that

$$\left.\frac{\partial \omega}{\partial\,\delta\dot{x}}\right|_{t_1^{(-)}} = \left.\frac{\partial \omega}{\partial\,\delta\dot{x}}\right|_{t_1^{(+)}} \tag{29}$$

yielding

$$\left.[\bar{f}_{\dot{x}\dot{x}}\,\delta\dot{x} + \bar{f}_{x\dot{x}}\,\delta x]\right|_{t_1^{(-)}} = \left.[\bar{f}_{\dot{x}\dot{x}}\,\delta\dot{x} + \bar{f}_{x\dot{x}}\,\delta x]\right|_{t_1^{(+)}} \tag{30}$$

Moreover, as $\delta x(t) \equiv 0$ for $t \geqslant t_1$, Eq. (30) reduces to

$$\bar{f}_{\dot{x}\dot{x}}(t_1)\,\delta\dot{x}(t_1^{(-)}) = 0 \tag{31}$$

Assume, consistent with our earlier work of Eq. (21), that $\bar{f}_{\dot{x}\dot{x}}(t_1) \neq 0$. Equation (31) then requires that

$$\delta\dot{x}(t_1^{(-)}) = 0 \tag{32}$$

Since $\delta x(t) = \delta v(t)$ for $t \leqslant t_1$, Eqs. (26) and (32) requires that at t_1 the solution $\delta v(t)$ to the accessory equation of Eq. (16) satisfy the conditions

$$\delta v(t_1) = 0$$
$$\delta\dot{v}(t_1) = 0 \tag{33}$$

Equation (16) is a homogeneous linear second-order ordinary differential equation for $\delta v(t)$. The conditions of Eq. (33) then require† the solution to be

$$\delta v(t) \equiv 0 \tag{34}$$

The result that $\delta v(t)$ must be identically zero contradicts the initial assumption that there be a conjugate point in the interval completing the proof.

Example 2-11(1)   Let us examine the solution given to the problem stated in Example 2-8(2). The task undertaken was that of minimizing

$$\text{IP} = \int_0^1 (\dot{x}^2 - a^2 x^2)\,dt \tag{35}$$

---

† Let the two linearly independent solutions to Eq. (16) be $\delta v_1(t)$ and $\delta v_2(t)$. Then the general solution is $\delta v(t) = A\,\delta v_1(t) + B\,\delta v_2(t)$. For the conditions of Eq. (33) to hold requires

$$0 = A\,\delta v_1(t_1) + B\,\delta v_2(t_1)$$
$$0 = A\,\delta\dot{v}_1(t_1) + B\,\delta\dot{v}_2(t_1)$$

Thus either $A = B = 0$ or the quantity $[\delta v_1\,\delta\dot{v}_2 - \delta\dot{v}_1\,\delta v_2]|_{t_1} = 0$. However, by the definition of linear independence, the latter quantity cannot vanish; hence $A = B = 0$ and the result of Eq. (34) is obtained.

where the values $x(0) = 0$, $x(1) = 0$, and a^2 are specified. The Euler equation for the problem gave the extremal solutions

$$x(t) \equiv 0 \qquad\qquad \text{for } a \neq n\pi$$

$$x(t) = \bar{A} \sin n\pi t \quad \text{or} \quad x(t) \equiv 0 \qquad \text{for } a = n\pi \tag{36}$$

Employing Eq. (11) the second variation of the IP of Eq. (35) is obtained as

$$\delta^2 \text{IP} = \int_0^1 (\delta \dot{x}^2 - a^2 \, \delta x^2) \, dt \tag{37}$$

from which the accessory equation is found to be

$$\frac{d^2}{dt^2} \delta x + a^2 \, \delta x = 0 \tag{38}$$

The general solution of the accessory equation is then

$$\delta x(t) = A \sin at + B \cos at \tag{39}$$

To determine if a conjugate point to $t_i = 0$ exists in the interval $0 < t < 1$ we first require $\delta x(t_i) = 0$, yielding

$$\delta x(t_i) = 0 = B$$

Equation (39) then reduces to

$$\delta x(t) = A \sin at \tag{40}$$

Now a point t_i is said to be conjugate to the initial point t_i if there exists a non-trivial solution to the accessory equation which vanishes at t_i and again at t_1. Examining $\delta x(t)$ as given by Eq. (40) it is seen that after $t_i = 0$ the variation $\delta x(t)$ next vanishes at $t_1 = \pi/a$. Thus if $a > \pi$ there is a conjugate point in the interval, whereas if $a < \pi$ there is not. In this manner it is seen by reference to the theorem just proved that if $a > \pi$, the extremal solutions found are not minimizing solutions.

2-12　THE EXTENSION TO N SYSTEM VARIABLES

As the real world is complicated, so also are the problems with which the engineer must deal. There are few interesting systems to be described by a single variable. With system-behavior dependent on many variables, optimal system response, defined in terms of a system index of performance, may require the determination of the best path in time to be followed by *each* of the system variables.

Let us treat the problem of minimizing by proper choice of the N variables $x_1(t), x_2(t), \cdots, x_N(t)$, the quantity

$$\text{IP} = \int_{t_i}^{t_f} \bar{f}[t, x_1(t), \cdots, x_N(t), \dot{x}_1(t), \cdots, \dot{x}_N(t)] \, dt \tag{1}$$

where t_i and $x_{1_0}, x_{2_0}, \cdots, x_{N_0}$; t_f and $x_{1_f}, x_{2_f}, \cdots, x_{N_f}$ are specified values. The procedure to be followed is identical in form to that already used in the case of one dependent variable. Denoting the functions that minimize the IP simply by $x_1(t), x_2(t), \cdots, x_N(t)$, we form the comparison functions $x_1(t, \epsilon), \cdots, x_N(t, \epsilon)$ so as to consider weak variations about the minimizing functions. Then

$$\begin{aligned}
x_1(t, \epsilon) &= x_1(t) + \epsilon z_1(t) \\
x_2(t, \epsilon) &= x_2(t) + \epsilon z_2(t) \\
&\vdots \\
x_N(t, \epsilon) &= x_N(t) + \epsilon z_N(t)
\end{aligned} \tag{2}$$

where since $x_1(t), \cdots, x_N(t)$ by themselves take on the correct end values at t_i and t_f one has that $z_1(t_i) = z_1(t_f) = \cdots = z_N(t_i) = z_N(t_f) = 0$ or, equivalently, $\delta x_1(t_i) = \delta x_1(t_f) = \cdots = \delta x_N(t_i) = \delta x_N(t_f) = 0$. Substituting relations (2) into the IP of Eq. (1), then $\text{IP} = \text{IP}(\epsilon)$. On expansion about $\epsilon = 0$ there results

$$\text{IP}(\epsilon) = \text{IP}(0) + \left. \frac{d\text{IP}}{d\epsilon} \right|_{\epsilon = 0} + O(\epsilon^2) \tag{3}$$

Our problem corresponds to that treated in Section 2-5, leading to the conclusion that all the first-order terms in ϵ must vanish for a minimum of the IP at $\epsilon = 0$. In the δ notation this means that a necessary condition for a minimum is

$$\delta\text{IP} = 0 \tag{4}$$

Forming δIP, using Eq. (1) and the fact that t_i and t_f are fixed, then

$$\delta\text{IP} = \int_{t_i}^{t_f} \delta\bar{f} \, dt$$

Hence Eq. (4) requires

$$\delta\text{IP} = 0 = \int_{t_i}^{t_f} \left[\frac{\partial \bar{f}}{\partial x_1} \delta x_1 + \cdots + \frac{\partial \bar{f}}{\partial x_N} \delta x_N + \frac{\partial \bar{f}}{\partial \dot{x}_1} \delta \dot{x}_1 + \cdots + \frac{\partial \bar{f}}{\partial \dot{x}_N} \delta \dot{x}_N \right] dt \tag{5}$$

Interchanging the time derivative and δ operations in the $\delta \dot{x}_k$ terms and integrating these quantities once, by parts, one finds

$$\delta\text{IP} = 0 = \left. \frac{\partial \bar{f}}{\partial x_1} \delta x_1 \right|_{t_i}^{t_f} + \cdots + \left. \frac{\partial \bar{f}}{\partial x_N} \delta x_N \right|_{t_i}^{t_f} - \int_{t_i}^{t_f} \left[\frac{d}{dt} \frac{\partial \bar{f}}{\partial \dot{x}_1} - \frac{\partial \bar{f}}{\partial x_1} \right] \delta x_1 \, dt$$

$$- \cdots - \int_{t_i}^{t_f} \left[\frac{d}{dt} \frac{\partial \bar{f}}{\partial \dot{x}_N} - \frac{\partial \bar{f}}{\partial x_N} \right] \delta x_N \, dt \tag{6}$$

But since $x_k(t_i)$ and $x_k(t_f)$ are fixed values, then $\delta x_k(t_i) = \delta x_k(t_f) = 0$ and Eq. (6) reduces to

$$\delta IP = 0 = \int_{t_i}^{t_f} \left[\frac{d}{dt} \frac{\partial \bar{f}}{\partial \dot{x}_1} - \frac{\partial \bar{f}}{\partial x_1} \right] \delta x_1 \, dt + \cdots + \int_{t_i}^{t_f} \left[\frac{d}{dt} \frac{\partial \bar{f}}{\partial \dot{x}_N} - \frac{\partial \bar{f}}{\partial x_N} \right] \delta x_N \, dt \quad (7)$$

Remember that each of the x_k may be varied independently. Therefore, to start, let us set each of the $\delta x_k = 0$ except δx_j, which is left arbitrary. In this case the condition $\delta IP = 0$ becomes

$$\delta IP = 0 = \int_{t_i}^{t_f} \left[\frac{d}{dt} \frac{\partial \bar{f}}{\partial \dot{x}_j} - \frac{\partial \bar{f}}{\partial x_j} \right] \delta x_j \, dt \quad (8)$$

Equation (8) is similar in form to Eq. 2-8(11) and the argument to be followed is identical to that carried out there. The only way Eq. (8) can be satisfied for all allowable δx_j is for

$$\frac{d}{dt} \frac{\partial \bar{f}}{\partial \dot{x}_j} - \frac{\partial \bar{f}}{\partial x_j} = 0 \quad (9)$$

As the choice of the variable x_j for first consideration was arbitrary, it is apparent that Eq. (9) must hold for all $j, j = 1, 2, \cdots, N$. Thus

$$\frac{d}{dt} \frac{\partial \bar{f}}{\partial \dot{x}_1} - \frac{\partial \bar{f}}{\partial x_1} = 0$$
$$\vdots \qquad (N \text{ equations}) \qquad (10)$$
$$\frac{d}{dt} \frac{\partial \bar{f}}{\partial \dot{x}_N} - \frac{\partial \bar{f}}{\partial x_N} = 0$$

These N relations, the Euler equations for the problem, are in general second-order, ordinary, differential equations [see Eq. 2-8(12)]. They are necessary conditions to be satisfied by the N minimizing functions $x_1(t), \cdots, x_N(t)$.

To allow consideration of problems where the rates $\dot{x}_1, \cdots, \dot{x}_N$ might change discontinuously, we develop now the corner conditions which must hold at such points. Multiplying the jth equation in Eq. (10) by $\dot{x}_j$ and rearranging as in Eq. 2-9(2) yields

$$\frac{d}{dt} \left(\dot{x}_j \frac{\partial \bar{f}}{\partial \dot{x}_j} \right) - \frac{\partial \bar{f}}{\partial x_j} \dot{x}_j - \frac{\partial \bar{f}}{\partial \dot{x}_j} \ddot{x}_j = 0$$

On doing the same to each of the other relations in Eq. (10) and adding the resulting equations together one obtains

$$\frac{d}{dt} \left(\sum_{i=1}^{N} \dot{x}_i \frac{\partial \bar{f}}{\partial \dot{x}_i} \right) - \sum_{i=1}^{N} \left(\frac{\partial \bar{f}}{\partial x_i} \dot{x}_i + \frac{\partial \bar{f}}{\partial \dot{x}_i} \ddot{x}_i \right) = 0 \quad (11)$$

But since

$$\frac{d\bar{f}}{dt} = \frac{\partial \bar{f}}{\partial t} + \sum_{i=1}^{N} \left(\frac{\partial \bar{f}}{\partial x_i} \dot{x}_i + \frac{\partial \bar{f}}{\partial \dot{x}_i} \ddot{x}_i \right) \tag{12}$$

then Eq. (12) can be used to rewrite Eq. (11) in the form

$$\frac{d}{dt} \left(\sum_{i=1}^{N} \dot{x}_i \frac{\partial \bar{f}}{\partial \dot{x}_i} - \bar{f} \right) = -\frac{\partial \bar{f}}{\partial t} \tag{13}$$

Suppose t_1 is a point at which a corner of the extremal arc is to exist. Integrate Eqs. (10) and (13) from $t_1 - \epsilon$ to $t_1 + \epsilon$ across the corner as in Section 2-10. Then on the assumption that $\partial \bar{f}/\partial t|_{t_1}$ and $\partial \bar{f}/\partial \dot{x}_i|_{t_1}$ are finite one has

$$\left. \frac{\partial \bar{f}}{\partial \dot{x}_1} \right|_{t_1^{(+)}} = \left. \frac{\partial \bar{f}}{\partial \dot{x}_1} \right|_{t_1^{(-)}}$$

$$\vdots \qquad\qquad (N \text{ conditions}) \tag{14}$$

$$\left. \frac{\partial \bar{f}}{\partial \dot{x}_N} \right|_{t_1^{(+)}} = \left. \frac{\partial \bar{f}}{\partial \dot{x}_N} \right|_{t_1^{(-)}}$$

and

$$\left. \left(\sum_{i=1}^{N} \dot{x}_i \frac{\partial \bar{f}}{\partial \dot{x}_i} - \bar{f} \right) \right|_{t_1^{(+)}} = \left. \left(\sum_{i=1}^{N} \dot{x}_i \frac{\partial \bar{f}}{\partial \dot{x}_i} - \bar{f} \right) \right|_{t_1^{(-)}} \tag{15}$$

Equations (14) and (15) provide $N + 1$ conditions to be satisfied at any corner of the extremal arc.

Note that if our problem is one in which $\bar{f}$ is not an explicit function of time, that is, $\partial \bar{f}/\partial t \equiv 0$, then a first integral of Eq. (10) follows by direct integration of Eq. (13). The result is

$$\sum_{i=1}^{N} \dot{x}_i \frac{\partial \bar{f}}{\partial \dot{x}_i} - \bar{f} = \text{const.} \tag{16}$$

When $\partial \bar{f}/\partial t \equiv 0$, Eq. (16) may be used to replace any one of Eqs. (10), should it be desirable.

Example 2-12(1)   Minimize by proper choice of $x_1(t)$ and $x_2(t)$ the quantity

$$\text{IP} = \int_0^1 \left(\frac{\dot{x}_1^2 + \dot{x}_2^2}{2} + x_1 x_2 \right) dt$$

where $x_1(0) = x_{1_0}$, $x_2(0) = x_{2_0}$, and $x_1(1) = x_2(1) = 0$.

By Eq. (10), the characteristic equations for the problems are

$$\ddot{x}_1 - x_2 = 0 \tag{17a}$$

$$\ddot{x}_2 - x_1 = 0 \tag{17b}$$

Equations (17a) and (17b) must be satisfied by the minimizing functions $x_1(t)$ and $x_2(t)$; thus they may be used to determine them. Eliminating x_2 between the two equations one has $\ddddot{x}_1 - x_1 = 0$, which has as its solution

$$x_1(t) = A \sinh t + B \cosh t + C \sin t + D \cos t \tag{18}$$

Since by Eq. (17a) $x_2 = \ddot{x}_1$, then

$$x_2(t) = A \sinh t + B \cosh t - C \sin t - D \cos t \tag{19}$$

The solutions for x_1 and x_2 as given by Eqs. (18) and (19) contain four arbitrary constants to be determined by the total of four initial and terminal conditions. At $t = 0$,

$$x_{1_0} = B + D$$

$$x_{2_0} = B - D$$

yielding

$$B = \frac{x_{1_0} + x_{2_0}}{2} \qquad D = \frac{x_{1_0} - x_{2_0}}{2} \tag{20}$$

while at $t = 1$,

$$0 = A \sinh 1 + B \cosh 1 + C \sin 1 + D \cos 1$$

$$0 = A \sinh 1 + B \cosh 1 - C \sin 1 - D \cos 1$$

yielding

$$A = -B \operatorname{ctnh} 1 = -\frac{x_{1_0} + x_{2_0}}{2} \operatorname{ctnh} 1 \tag{21}$$

$$C = -D \operatorname{ctn} 1 = \frac{x_{2_0} - x_{1_0}}{2} \operatorname{ctn} 1$$

Equations (18) and (19) with the constants as determined by Eqs. (20) and (21) determine the extremals which satisfy the end conditions. As

$$\bar{f} = \frac{\dot{x}_1^2 + \dot{x}_2^2}{2} + x_1 x_2$$

is not an explicit function of t, it must also be true by Eq. (16) that

$$\dot{x}_1 \frac{\partial \bar{f}}{\partial \dot{x}_1} + \dot{x}_2 \frac{\partial \bar{f}}{\partial \dot{x}_2} - \bar{f} = \frac{\dot{x}_1^2 + \dot{x}_2^2}{2} - x_1 x_2 = \text{const.} \tag{22}$$

Although it has not been found useful to employ Eq. (22) in our solution, the reader should satisfy himself as to its validity.

The possible existence of corners remains to be checked. But by Eq. (14),

$$\dot{x}_{1-} = \dot{x}_{1+} \qquad \dot{x}_{2-} = \dot{x}_{2+}$$

Hence no corners of the extremal arc exist.

2-13   DIFFERENTIAL CONSTRAINTS AMONG THE SYSTEM VARIABLES

In the preceding sections the designer has been free to choose the system variables $x_i(t)$ in any manner desired so as to minimize the system IP. However, in actuality it is quite likely that the variables will not be completely free but be required to satisfy certain differential equations governing system changes. These differential constraints then limit the possible paths from among which the minimizing functions can be chosen. It is the purpose of this section to determine the necessary conditions analogous to Eqs. 2-12(10) of the un-restricted case, which must hold in the presence of differential constraints among the system variables.

To start, let us imagine that our problem is to minimize by proper choice of the two variables $x_1(t)$ and $x_2(t)$ the quantity

$$\text{IP} = \int_{t_i}^{t_f} \bar{f}(t, x_1, x_2, \dot{x}_1, \dot{x}_2) \, dt \tag{1}$$

where t_i and $x_1(t_i)$, $x_2(t_i)$; t_f and $x_1(t_f)$, $x_2(t_f)$ are given. Further, it will be assumed that $x_1(t)$ and $x_2(t)$ are related by the differential constraint

$$\varphi(t, x_1, x_2, \dot{x}_1, \dot{x}_2) = 0 \tag{2}$$

Forming δIP, which for the minimizing functions must be zero, one finds after the usual integration by parts of the terms $\delta\dot{x}_1$, $\delta\dot{x}_2$ that [see Eq. 2-12(7)]

$$\delta\text{IP} = 0 = -\int_{t_i}^{t_f} \left[\frac{d}{dt}\frac{\partial\bar{f}}{\partial\dot{x}_1} - \frac{\partial\bar{f}}{\partial x_1}\right] \delta x_1 \, dt - \int_{t_i}^{t_f} \left[\frac{d}{dt}\frac{\partial\bar{f}}{\partial\dot{x}_2} - \frac{\partial\bar{f}}{\partial x_2}\right] \delta x_2 \, dt \tag{3}$$

where use has also been made of the end conditions requiring $\delta x_1 = \delta x_2 = 0$ at t_i and t_f. The argument in Section 2-12 proceeded by stating that since the δx_j could all be varied independently, the only way for Eq. 2-12(7) to be satisfied is for the coefficient of each δx_j to be identically zero. If one tried to apply this argument to Eq. (3) there would result two ordinary, second-order equations of the form of Eqs. 2-12(10) for the determination of $x_1(t)$ and $x_2(t)$. Although solutions of these equations have the capability of satisfying the end conditions, there would be no way of ensuring that the differential constraint of Eq. (2) would be satisfied. The fallacy of the procedure is simply that δx_1 and δx_2 are not independent but rather are related through the additional condition that $\varphi(t, x_1, x_2, \dot{x}_1, \dot{x}_2) = 0$ and hence $\delta\varphi = 0$, where

$$\delta\varphi = \frac{\partial\varphi}{\partial x_1}\,\delta x_1 + \frac{\partial\varphi}{\partial x_2}\,\delta x_2 + \frac{\partial\varphi}{\partial\dot{x}_1}\,\delta\dot{x}_1 + \frac{\partial\varphi}{\partial\dot{x}_2}\,\delta\dot{x}_2 = 0 \tag{4}$$

To overcome this difficulty we make use of the concept of the undetermined (Lagrange) multiplier first used in Section 2-6. To the integrand of our IP of Eq. (1) is added the product of an undetermined function $\lambda(t)$ and the constraint relation $\varphi(t, x_1, x_2, \dot{x}_1, \dot{x}_2) = 0$. Giving this *augmented integrand* the symbol $F(t, x_1, x_2, \dot{x}_1, \dot{x}_2, \lambda)$ one has

$$F = \bar{f}(t, x_1, x_2, \dot{x}_1, \dot{x}_2) + \lambda(t)\varphi(t, x_1, x_2, \dot{x}_1, \dot{x}_2) \tag{5}$$

The augmented index of performance, $\mathscr{IP}$, is then defined by

$$\mathscr{IP} = \int_{t_i}^{t_f} F(t, x_1, x_2, \dot{x}_1, \dot{x}_2, \lambda)\,dt \tag{6}$$

where since $\varphi = 0$ it is apparent that $\bar{f} = F$ and $\mathrm{IP} = \mathscr{IP}$. Notwithstanding this apparent lack of gain by our procedure, let us go on as in Section 2-6 with the hope that our difficulty may be overcome. Forming $\delta\mathscr{IP} = \delta\mathrm{IP} = 0$ one has, comparable to Eq. (3), that

$$\delta\mathscr{IP} = 0 = -\int_{t_i}^{t_f}\left[\frac{d}{dt}\frac{\partial F}{\partial\dot{x}_1} - \frac{\partial F}{\partial x_1}\right]\delta x_1\,dt - \int_{t_i}^{t_f}\left[\frac{d}{dt}\frac{\partial F}{\partial\dot{x}_2} - \frac{\partial F}{\partial x_2}\right]\delta x_2\,dt \tag{7}$$

Remembering that δx_1 and δx_2 are both not independent, we choose δx_2 as the independent variation and δx_1 as the dependent one. Now let us use the multiplier $\lambda(t)$, which was arbitrarily introduced and is at our disposal. The quantity

$$\frac{d}{dt}\frac{\partial F}{\partial\dot{x}_1} - \frac{\partial F}{\partial x_1}$$

which is the coefficient of the dependent variation δx_1 is seen to be a function of $\lambda(t)$, $x_1(t)$, $x_2(t)$, and some of their derivatives. We shall choose the function $\lambda(t)$ to be just that function of time which makes this quantity vanish. Therefore,

$$\frac{d}{dt}\frac{\partial F}{\partial\dot{x}_1} - \frac{\partial F}{\partial x_1} = 0$$

With this choice, Eq. (7) then reduces to

$$\delta\mathscr{IP} = 0 = -\int_{t_i}^{t_f}\left[\frac{d}{dt}\frac{\partial F}{\partial\dot{x}_2} - \frac{\partial F}{\partial x_2}\right]\delta x_2\,dt \tag{8}$$

But since δx_2 is the independent variation and hence arbitrary, the only way Eq. (8) can be satisfied in general is if the coefficient of δx_2 also vanishes, yielding

$$\frac{d}{dt}\frac{\partial F}{\partial \dot{x}_2} - \frac{\partial F}{\partial x_2} = 0$$

Thus two necessary conditions to be satisfied by the minimizing functions $x_1(t)$ and $x_2(t)$ are that

$$\frac{d}{dt}\frac{\partial F}{\partial \dot{x}_1} - \frac{\partial F}{\partial x_1} = 0 \tag{9a}$$

$$\frac{d}{dt}\frac{\partial F}{\partial \dot{x}_2} - \frac{\partial F}{\partial x_2} = 0 \tag{9b}$$

Note that these relations are just the ones that would have followed from Eq. (7) if δx_1 and δx_2 had both been independent. Introduction of the multiplier $\lambda(t)$ has enabled us to treat $x_1(t)$ and $x_2(t)$ *as though* they were independent. The two second-order, ordinary differential equations (9a, b), with the given end conditions at t_i and t_f, along with the differential constraint of Eq. (2), provide the three equations required for the determination of $x_1(t)$, $x_2(t)$, and $\lambda(t)$.

Further, since the form of Eq. (9) is identical to that of Eq. 2-12(10), with the augmented integrand $F = \bar{f} + \lambda\varphi$ simply replacing $\bar{f}$, the additional results obtained there can be transposed directly. Thus if a corner of the extremal arc is to exist at some time t_1, the necessary conditions corresponding to Eqs. 2-12(14) and 2-12(15) are

$$\left.\frac{\partial F}{\partial \dot{x}_1}\right|_{t_1^{(+)}} = \left.\frac{\partial F}{\partial \dot{x}_1}\right|_{t_1^{(-)}}$$

$$\left.\frac{\partial F}{\partial \dot{x}_2}\right|_{t_1^{(+)}} = \left.\frac{\partial F}{\partial \dot{x}_2}\right|_{t_1^{(-)}} \tag{10}$$

and

$$\left.\left(\sum_{i=1}^{2} \dot{x}_i \frac{\partial F}{\partial \dot{x}_i} - F\right)\right|_{t_1^{(+)}} = \left.\left(\sum_{i=1}^{2} \dot{x}_i \frac{\partial F}{\partial \dot{x}_i} - F\right)\right|_{t_1^{(-)}} \tag{11}$$

If the problem is such that the augmented integrand is not an explicit function of time, that is, if $\partial F/\partial t \equiv 0$, then a first integral of Eq. (9) exists, corresponding to Eq. 2-12(16) and is given by

$$\sum_{i=1}^{2} \dot{x}_i \frac{\partial F}{\partial \dot{x}_i} - F = \text{const.} \tag{12}$$

The generalization of these results to systems whose description requires more than two variables and one differential constraint follows immediately.

General Statement   The problem is to minimize the performance index given by

$$\text{IP} = \int_{t_i}^{t_f} \bar{f}(t, x_1, \cdots, x_N, \dot{x}_1, \cdots, \dot{x}_N)\, dt \tag{13}$$

where t_i and $x_1(t_i) \cdots x_N(t_i)$, t_f, and $x_1(t_f) \cdots x_N(t_f)$ are specified, by proper choice of $x_1(t) \cdots x_N(t)$ between t_i and t_f subject to the $M < N$ differential constraints†

$$\varphi_1(t, x_1, \cdots, x_N, \dot{x}_1, \cdots, \dot{x}_N) = 0$$
$$\vdots \tag{14}$$
$$\varphi_M(t, x_1, \cdots, x_N, \dot{x}_1, \cdots, \dot{x}_N) = 0$$

Forming the augmented integrand given by

$$F = \bar{f}(t, x_1, \cdots, \dot{x}_N) + \sum_{i=1}^{M} \lambda_i(t)\varphi_i(t, x_1, \cdots, \dot{x}_N) \tag{15}$$

the functions $x_1(t) \cdots x_N(t)$ must satisfy the N characteristic equations for the problem

$$\frac{d}{dt}\frac{\partial F}{\partial \dot{x}_1} - \frac{\partial F}{\partial x_1} = 0$$

$$\vdots \qquad (N \text{ equations}) \tag{16}$$

$$\frac{d}{dt}\frac{\partial F}{\partial \dot{x}_N} - \frac{\partial F}{\partial x_N} = 0$$

The N equations of Eq. (16) along with the M differential constraint equations of (14) are $N + M$ conditions to be satisfied by the N system variables $x_i(t)$ and the M multipliers $\lambda_i(t)$.

If a corner is to exist in the problem at $t = t_1$ then the additional conditions to be satisfied are

$$\left.\frac{\partial F}{\partial \dot{x}_1}\right|_{t_1^{(+)}} = \left.\frac{\partial F}{\partial \dot{x}_1}\right|_{t_1^{(-)}}$$

$$\vdots \qquad (N \text{ conditions}) \tag{17}$$

$$\left.\frac{\partial F}{\partial \dot{x}_N}\right|_{t_1^{(+)}} = \left.\frac{\partial F}{\partial \dot{x}_N}\right|_{t_1^{(-)}}$$

† Since all the end conditions are specified, if the number of independent differential constraints (M) were equal to the number of system variables (N), the variables would be completely determined. If this were the case, there would be nothing left free with which to minimize the IP.

and

$$\left(\sum_{i=1}^{N} \dot{x}_i \frac{\partial F}{\partial \dot{x}_i} - F\right)\Bigg|_{t_1^{(+)}} = \left(\sum_{i=1}^{N} \dot{x}_i \frac{\partial F}{\partial \dot{x}_i} - F\right)\Bigg|_{t_1^{(-)}} \tag{18}$$

If the augmented integrand F is not an explicit function of time, that is, if $\partial F/\partial t \equiv 0$, then a first integral of Eq. (16) is

$$\sum_{i=1}^{N} \dot{x}_i \frac{\partial F}{\partial \dot{x}_i} - F = \text{const.} \tag{19}$$

Should it be desirable, when $\partial F/\partial t \equiv 0$, Eq. (19) may be used in place of any one of Eqs. (16).

Example 2-13(1)† Minimize by proper choice of $x_1(t)$ and $x_2(t)$ the quantity

$$\text{IP} = \int_0^1 \left(\frac{\dot{x}_1^2 + \dot{x}_2^2}{2} + x_1 x_2\right) dt \tag{20}$$

where $x_1(0) = x_{1_0}$, $x_2(0) = x_{2_0}$, and $x_1(1) = x_2(1) = 0$. The variables are to satisfy the system constraint that $\dot{x}_2 = x_1$; thus

$$\varphi = \dot{x}_2 - x_1 = 0 \tag{21}$$

Forming first the augmented integrand

$$F = \frac{\dot{x}_1^2 + \dot{x}_2^2}{2} + x_1 x_2 + \lambda(\dot{x}_2 - x_1)$$

one finds that the characteristic equations, Eq. (16), are

$$\ddot{x}_1 - x_2 + \lambda = 0 \tag{22a}$$

$$\ddot{x}_2 - x_1 + \dot{\lambda} = 0 \tag{22b}$$

Differentiating Eq. (22a) and eliminating λ between the resulting equation and Eq. (22b) one finds on using the system constraint, $\dot{x}_2 = x_1$, that

$$\dddot{x}_1 - \dot{x}_1 = 0 \tag{23}$$

The solution of Eq. (23) for the extremal $x_1(t)$ is

$$x_1(t) = C_1 \sinh t + C_2 \cosh t + C_3 \tag{24}$$

Having $x_1(t)$, the extremal $x_2(t)$ is found by direct integration of the differential constraint of Eq. (21), yielding

$$x_2(t) = C_1 \cosh t + C_2 \sinh t + C_3 t + C_4 \tag{25}$$

† Compare this example with Example 2-12(1).

The constants in Eqs. (24) and (25) are to be determined so as to satisfy the end conditions at $t = 0$ and $t = 1$. Thus at $t = 0$,

$$x_{1_0} = C_2 + C_3$$
$$x_{2_0} = C_1 + C_4$$

(26)

while at $t = 1$,

$$0 = C_1 \sinh 1 + C_2 \cosh 1 + C_3$$
$$0 = C_1 \cosh 1 + C_2 \sinh 1 + C_3 + C_4$$

(27)

Using Eq. (26) to eliminate C_3 and C_4 from Eq. (27) and then solving for C_1 and C_2 one obtains

$$C_1 = \frac{x_{1_0}(\cosh 1 - \sinh 1) + x_{2_0}(\cosh 1 - 1)}{2(\cosh 1 - 1) - \sinh 1}$$

$$C_2 = \frac{x_{1_0}(\cosh 1 - \sinh 1 - 1) - x_{2_0} \sinh 1}{2(\cosh 1 - 1) - \sinh 1}$$

(28)

which along with Eq. (26) determines C_1, C_2, C_3, and C_4.

The functions† $x_1(t)$ and $x_2(t)$ given by Eqs. (24) and (25) with the constants as given by Eqs. (26) and (28) satisfy the necessary conditions thus far developed for minimization of the IP. The corresponding value of the IP would be found by substitution into Eq. (20). As the IP does not involve the function $\lambda(t)$, there is no need to actually determine it here. If desired it could be obtained directly from Eq. (22a).

---

† Remembering that the state variables $x_1(t)$ and $x_2(t)$ are taken to be continuous, the reader should verify using the corner conditions of Eqs. (17) and (18) that no corners exist, that is, that $\dot{x}_1$ and $\dot{x}_2$ must also be continuous along the extremal in this problem.

3
DEVELOPMENT THROUGH
THE CONTROL FORMULATION

In Chapter 2 some formative concepts and techniques for the minimization of the system index of performance were developed. Conditions were found for minimization of the IP when it is a function of several independent variables as well as when some of the variables are interrelated by constraint equations, the latter case being treated by introduction of the technique of the undetermined (Lagrange) multiplier. For situations in which system design considerations require the IP to be an integral over the process lifetime of an integrand involving the process variables, the basic techniques of the calculus of variations were derived and employed.

Although this work provides a basis for hope that solutions to system optimization problems may be found, the problems that have been treated so far are still quite restrictive. Thus, in this chapter the theory is extended to provide a tool more capable of handling engineering problems. Within the chapter the point is reached where the theory's scope is sufficient to attempt the solution of realistic problems.

3-1 A MORE REALISTIC PROBLEM

The most general statement of the problem as yet formulated is repeated here. It is assumed that the designer is faced with a system defined by N variables $x_1(t), \cdots, x_N(t)$ which are governed by a number of differential constraints (system equations of motion) of the form† $\varphi_i(t, \mathbf{x}, \dot{\mathbf{x}}) = 0$. Further, it is assumed that the process to be optimized is such that the duration of the process is fixed as well as the values of each of the system variables at both the initial and final times. Thus t_i and $x_1(t_i), \cdots, x_N(t_i)$; t_f and $x_1(t_f), \cdots, x_N(t_f)$ are given values. The designer, after careful consideration of the system purpose, then states: The best system will minimize the system index of performance given by

$$\text{IP} = \int_{t_i}^{t_f} \bar{f}(t, x_1, \cdots, x_N, \dot{x}_1, \cdots, \dot{x}_N)\, dt \tag{1}$$

With this all-important decision made, the task of the designer is clear. The paths in time which the functions $x_i(t)$, $i = 1, \cdots, N$, follow between their fixed initial and final states must be chosen from among those satisfying the system equations of motion so as to minimize the IP of Eq. (1). As the theory to achieve this goal was developed in Chapter 2, all is thus well.

However, it is possible that the problem as presented is not the one which the designer may wish to solve. One may imagine a process whose end point is not determined by reaching a specified time at which the variables have given values but, alternatively, is reached when some terminal relationship among the variables is satisfied. Furthermore, one may suppose that the system purpose is such that the desired IP to be minimized is not of the form of Eq. (1) but rather is given by

$$\text{IP} = g[t_i, x_1(t_i), \cdots, x_N(t_i), t_f, x_1(t_f), \cdots, x_N(t_f)] + \int_{t_i}^{t_f} \bar{f}(t, x_1, \cdots, \dot{x}_N)\, dt \tag{2}$$

The minimization of Eq. (2) under the associated differential constraints and end conditions is known as the *problem of Bolza*. If in Eq. (2) g is identically zero, the *problem* is called that *of Lagrange*, whereas if $\bar{f}$ is identically zero and not g, the minimization is termed the *problem of Mayer* (*1*). In the framework of this more general problem the necessary conditions for a local minimum to exist are obtained next.

3-2 THE FIRST VARIATION

To initiate the work, the development first centers about a system defined by two state variables, $x_1(t)$ and $x_2(t)$, interrelated by a single differential constraint (system equation of motion). The extension to a more complex process will then follow easily.

---

† The notation $\varphi_i(t, \mathbf{x}, \dot{\mathbf{x}})$ is used as shorthand to indicate that φ_i may be a function of all x_j and $\dot{x}_j$. Thus the quantity $\varphi_i(t, \mathbf{x}, \dot{\mathbf{x}})$ is equivalent to $\varphi_i(t, x_1 \cdots x_N, \dot{x}_1 \cdots \dot{x}_N)$. Vectors and matrices will be indicated by boldface type in this manner throughout the book.

It is assumed that the designer has decided to minimize the IP given by

$$\text{IP} = g[t_i, x_1(t_i), x_2(t_i), t_f, x_1(t_f), x_2(t_f)]$$

$$+ \int_{t_i}^{t_f} \bar{f}[t, x_1(t), x_2(t), \dot{x}_1(t), \dot{x}_2(t)]\, dt \quad (1)$$

where the functions g and $\bar{f}$ have been chosen so as to satisfy system design requirements. Although the initiation and termination times of the process are not specified, it is known that certain relations between the end-point variables must be met. These are the equations

$$\psi_j[t_i, x_1(t_i), x_2(t_i), t_f, x_1(t_f), x_2(t_f)] = 0 \quad j = 1, 2, \cdots, p \leqslant 6 \quad (2)$$

That is, it is assumed that there may be as many as six relations among the six end-point variables t_i, $x_1(t_i)$, $x_2(t_i)$, t_f, $x_1(t_f)$, and $x_2(t_f)$. It is obvious that there can be no more than six equations, because it would then not, in general, be possible for the six variables to satisfy the additional equations. If N is the number of system variables, there must be no more than $2N + 2$ relations: two for each state variable to be used to determine initial and terminal values and two to be used to determine initial and terminal times.

Moreover, if we digress for a moment to imagine the case in which $\bar{f}$ is identically zero in our IP of Eq. (1), it is seen that in this case the number of end-point relations of the type of Eq. (2) must be less than or equal to five, or less than or equal to $2N + 1$ in the general case. This follows since if $\bar{f}$ is identically zero, the quantity to be minimized in Eq. (1) is only a function of the end-point values. If the number of relations were to equal six, then six equations would exist for the determination of the six end variables. Thus in this case there would be no free quantities at the disposal of the designer with which to minimize the IP.

Finally, the changes in x_1 and x_2 over the process lifetime are assumed to be governed by the single system equation of motion,

$$\varphi_1(t, x_1, x_2, \dot{x}_1, \dot{x}_2) = 0 \quad (3)$$

As in Section 2-13 an augmented integrand (F) is formed by taking the product of the differential constraint of Eq. (3) with an as yet undetermined multiplier $\lambda_1(t)$, yielding

$$F(t, \mathbf{x}, \dot{\mathbf{x}}) = \bar{f}(t, \mathbf{x}, \dot{\mathbf{x}}) + \lambda_1(t)\varphi_1(t, \mathbf{x}, \dot{\mathbf{x}}) \quad (4)$$

The reader should remember that introduction of the Lagrange multiplier enables us to treat the changes δx_1 and δx_2 as though they were independent during the process lifetime. The augmented performance index is then

$$\mathscr{IP} = \text{IP} = g[t_i, \mathbf{x}(t_i), t_f, \mathbf{x}(t_f)] + \int_{t_i}^{t_f} F(t, \mathbf{x}, \dot{\mathbf{x}}, \lambda_1)\, dt \quad (5)$$

Now, as in Section 2-13, let us determine the conditions that must hold for the first variation of our $\mathscr{IP}$ to vanish. Assuming that $x_1(t)$ and $x_2(t)$ represent the minimizing functions, then during the process the comparison functions are given by

$$x_1(t, \epsilon) = x_1(t) + \epsilon z_1(t)$$

$$x_2(t, \epsilon) = x_2(t) + \epsilon z_2(t)$$

(6)

Further, as in the present case neither the initial nor the terminal times or the values of the system variables at these times is assumed known, each of the end quantities must be varied to determine its effect on the value of the IP.

Thus the initial and final times will also be functions of ϵ; that is,

$$t_i = t_i(\epsilon)$$

$$t_f = t_f(\epsilon)$$

(7)

Changes in these values about their optimum, for which $\epsilon = 0$, are given by

$$dt_i(0) = \frac{dt_i}{d\epsilon}\bigg|_{\epsilon = 0} \epsilon = \delta t_i \qquad dt_f(0) = \frac{dt_f}{d\epsilon}\bigg|_{\epsilon = 0} \epsilon = \delta t_f \qquad (8)$$

In addition, the initial and final values of x_1 and x_2 are functions of ϵ;

$$x_{1_0}(\epsilon) = x_1(t_i, \epsilon) = x_1(t_i(\epsilon), \epsilon)$$

$$x_{2_0}(\epsilon) = x_2(t_i, \epsilon) = x_2(t_i(\epsilon), \epsilon)$$

$$x_{1_f}(\epsilon) = x_1(t_f, \epsilon) = x_1(t_f(\epsilon), \epsilon)$$

$$x_{2_f}(\epsilon) = x_2(t_f, \epsilon) = x_2(t_f(\epsilon), \epsilon)$$

(9)

Note that changing ϵ affects the end values of x_1 and x_2 in two ways. For example, consider $x_{1_0}(\epsilon) = x_1(t_i(\epsilon), \epsilon)$. Its value can change with ϵ due to the change in t_i with ϵ or it can be varied, holding t_i fixed, through the explicit dependence of x_{1_0} on ϵ. Therefore, the total change in x_{1_0} with ϵ may be written

$$dx_{1_0}(\epsilon) = \frac{\partial x_{1_0}}{\partial t_i}\frac{dt_i}{d\epsilon}d\epsilon + \frac{\partial x_{1_0}}{\partial \epsilon}d\epsilon$$

or about the minimizing value $\epsilon = 0$, using the δ notation and Eq. (8) one has

$$dx_1(t_i) = \dot{x}_1(t_i)\,dt_i + \delta x_1(t_i)$$

In the same manner the differential change of all the end variables about their minimizing values is obtained:

$$dx_1(t_i) = \dot{x}_1(t_i)\,dt_i + \delta x_1(t_i)$$

$$dx_2(t_i) = \dot{x}_2(t_i)\,dt_i + \delta x_2(t_i)$$

$$dx_1(t_f) = \dot{x}_1(t_f)\,dt_f + \delta x_1(t_f)$$

$$dx_2(t_f) = \dot{x}_2(t_f)\,dt_f + \delta x_2(t_f)$$

(10)

With the realization that the end-point quantities given in Eqs. (7) and (9) may all be varied, that is, that they are functions of ϵ, let us return to the minimization of the IP.

To review: For the IP to be a minimum, or equivalently the augmented $\mathscr{IP}$, its value when the comparison functions of Eqs. (6), (7), and (9) are inserted must be greater than its value when the minimizing functions are inserted. Thus $\mathscr{IP}(\epsilon) - \mathscr{IP}(0) > 0$. Expanding $\mathscr{IP}(\epsilon)$ about $\epsilon = 0$ there results

$$\mathscr{IP}(\epsilon) - \mathscr{IP}(0) = \left.\frac{d\mathscr{IP}}{d\epsilon}\right|_{\epsilon=0} \epsilon + \left.\frac{d^2\mathscr{IP}}{d\epsilon^2}\right|_{\epsilon=0} \frac{\epsilon^2}{2} + \cdots$$

One sees that if the dominant term $\delta\mathscr{IP} = d\mathscr{IP}/d\epsilon|_{\epsilon=0}\,\epsilon$ does not vanish, since ϵ can be either positive or negative, it will not be possible to satisfy the condition that $\mathscr{IP}(\epsilon) - \mathscr{IP}(0) > 0$. This argument leads us to conclude that a necessary condition for the $\mathscr{IP}$ to be a minimum is that its first variation vanish, $\delta\mathscr{IP} = 0$.

Before applying this condition to our problem let us substitute the comparison functions of Eqs. (6), (7), and (9) into the $\mathscr{IP}$ of Eq. (5) so that the dependence on ϵ is clearly shown.

$$\mathscr{IP}(\epsilon) = g[t_i(\epsilon), \mathbf{x}(t_i(\epsilon), \epsilon), t_f(\epsilon), \mathbf{x}(t_f(\epsilon), \epsilon)]$$
$$+ \int_{t_i(\epsilon)}^{t_f(\epsilon)} F[t, \mathbf{x}(t, \epsilon), \dot{\mathbf{x}}(t, \epsilon), \lambda_1(t)]\, dt \quad (11)$$

It is the first variation of the $\mathscr{IP}$ of Eq. (11) which must equal zero as a necessary condition for the $\mathscr{IP}(0)$ (IP(0)) to be a minimum.

In forming $\delta\mathscr{IP} = 0$ from Eq. (11), the reader should recall that the δ notation means that one is to take the variation of those quantities which are functions of ϵ. Thus, since *all* the quantities in the first term (g) on the right of Eq. (11) depend on ϵ, the variation of this term is equivalent to taking its total differential. Then, using the fact that by Eq. (8), $\delta t_i = dt_i$ and $\delta t_f = dt_f$, one finds† for the first variation of the $\mathscr{IP}$ that

$$\delta\mathscr{IP} = dg + \int_{t_i}^{t_f} \delta F\, dt + F\bigg|_{t_f} dt_f - F\bigg|_{t_i} dt_i = 0 \quad (12)$$

† Given an integral where the integrand as well as the upper and lower limits may be functions of a parameter, such as in the integral

$$I(\alpha) = \int_{a(\alpha)}^{b(\alpha)} f(y, \alpha)\, dy$$

the rate of change of $I(\alpha)$ with respect to α is found to be

$$\frac{dI(\alpha)}{d\alpha} = \int_{a(\alpha)}^{b(\alpha)} \frac{\partial f}{\partial \alpha}(y, \alpha)\, dy + f(b(\alpha), \alpha)\frac{db(\alpha)}{d\alpha} - f(a(\alpha), \alpha)\frac{da(\alpha)}{d\alpha}$$

This result is derived in most texts on advanced calculus. (See F. B. Hildebrand, *Advanced Calculus for Applications*, Prentice-Hall, Inc., Englewood Cliffs, N.J., 1962, Chap. 7.) It has been used in obtaining Eq. (12) from Eq. (11), where t_i, t_f, and the integrand through $\mathbf{x}$ and $\dot{\mathbf{x}}$ are all functions of the parameter ϵ.

On expansion, the first term on the right is given by

$$dg = \frac{\partial g}{\partial t_i}\,dt_i + \frac{\partial g}{\partial x_1(t_i)}\,dx_1(t_i) + \frac{\partial g}{\partial x_2(t_i)}\,dx_2(t_i) + \frac{\partial g}{\partial t_f}\,dt_f$$

$$+ \frac{\partial g}{\partial x_1(t_f)}\,dx_1(t_f) + \frac{\partial g}{\partial x_2(t_f)}\,dx_2(t_f) \quad (13)$$

Further, since only the quantities $\mathbf{x}$ and $\dot{\mathbf{x}}$ depend on ϵ in the integrand F of Eq. (11) its first variation is given, as in Eq. 2-8(9), by

$$\delta F = \sum_{i=1}^{2}\left(\frac{\partial F}{\partial x_i}\,\delta x_i + \frac{\partial F}{\partial \dot{x}_i}\,\delta \dot{x}_i\right) \quad (14)$$

Recall that by Eq. 2-8(5) the δ and derivative operations may be interchanged. In this manner Eq. (12) may be rewritten†

$$\delta \mathscr{IP} = 0 = dg + F\,dt\,\Big|_{t_i}^{t_f} + \int_{t_i}^{t_f}\sum_{i=1}^{2}\left[\frac{\partial F}{\partial x_i}\,\delta x_i + \frac{\partial F}{\partial \dot{x}_i}\frac{d}{dt}\,\delta x_i\right]dt \quad (15)$$

The second term under the integral is now integrated once by parts, as in Eq. 2-8(10), to yield

$$\delta \mathscr{IP} = 0 = dg + F\,dt\,\Big|_{t_i}^{t_f} + \sum_{j=1}^{2}\frac{\partial F}{\partial \dot{x}_j}\,\delta x_j\,\Big|_{t_i}^{t_f} - \int_{t_i}^{t_f}\sum_{i=1}^{2}\left[\frac{d}{dt}\frac{\partial F}{\partial \dot{x}_i} - \frac{\partial F}{\partial x_i}\right]\delta x_i\,dt \quad (16)$$

Note that by Eq. (13) the quantity dg is seen to involve $dx_j(t_i)$ and $dx_j(t_f)$, while the third term in Eq. (16) involves $\delta x_j(t_i)$ and $\delta x_j(t_f)$. Since by Eq. (10) dx_j and δx_j at the end points are related; that is,

$$dx_j = \dot{x}_j\,dt + \delta x_j \qquad \text{(at } t_i \text{ and } t_f) \quad (17)$$

then $\delta x_j(t_i)$ and $\delta x_j(t_f)$ may be eliminated from Eq. (16) through use of Eq. (17). Equation (16) then becomes

$$\delta \mathscr{IP} = 0 = dg - \left[\sum_{i=1}^{2}\frac{\partial F}{\partial \dot{x}_i}\,\dot{x}_i - F\right]dt\,\Big|_{t_i}^{t_f} + \sum_{i=1}^{2}\frac{\partial F}{\partial \dot{x}_i}\,dx_i\,\Big|_{t_i}^{t_f}$$

$$- \int_{t_i}^{t_f}\sum_{i=1}^{2}\left[\frac{d}{dt}\frac{\partial F}{\partial \dot{x}_i} - \frac{\partial F}{\partial x_i}\right]\delta x_i\,dt \quad (18)$$

---

† By the notation $F\,dt\,\big|_{t_i}^{t_f}$ is meant the difference between the value of the quantity at the upper limit and its value at the lower limit. Thus

$$F\,dt\,\Big|_{t_i}^{t_f} = F\big|_{t_f}\,dt_f - F\big|_{t_i}\,dt_i$$

Equation (18) is made up of two distinct types of terms. First occur those terms giving the change in the $\mathscr{IP}$ caused by varying the end-point values. They are followed by the integral terms, which express the change in the $\mathscr{IP}$ caused by varying x_1 and x_2 between the end points. From Eq. (18) the basic necessary conditions for the minimization of our IP are to be obtained.

3-3 THE EULER (CHARACTERISTIC) EQUATIONS

Imagine for the moment that the problem posed in Section 3-2 has been solved and that the optimum values for the end-point variables $t_i, x_1(t_i), x_2(t_i), t_f, x_1(t_f)$, and $x_2(t_f)$ have been found. With this assumed knowledge let us form a new problem.

Let each of the end-point variables be fixed at its optimum value for the original problem. Then the optimum functions $x_1(t)$ and $x_2(t)$ are to be found between these fixed end points. Note that if the end-point values are given this requires $dt_i = dx_1(t_i) = dx_2(t_i) = dt_f = dx_1(t_f) = dx_2(t_f) = 0$. Thus the condition of Eq. 3-2(18) for the first variation to vanish becomes

$$\delta \mathscr{IP} = 0 = \int_{t_i}^{t_f} \left[\frac{d}{dt}\frac{\partial F}{\partial \dot{x}_1} - \frac{\partial F}{\partial x_1}\right] \delta x_1 \, dt + \int_{t_i}^{t_f} \left[\frac{d}{dt}\frac{\partial F}{\partial \dot{x}_2} - \frac{\partial F}{\partial x_2}\right] \delta x_2 \, dt \quad (1)$$

Equation (1) is of the form already studied in Section 2-13. Although δx_1 and δx_2 are not both independent, owing to the differential constraint $\varphi_1(t, x_1, x_2, \dot{x}_1, \dot{x}_2) = 0$, the introduction of the Lagrange multiplier $\lambda_1(t)$ enables us to treat δx_1 and δx_2 as though they were independent variations. Therefore, the coefficients of δx_1 and δx_2 in Eq. (1) must each be zero, yielding the Euler equations

$$\frac{d}{dt}\frac{\partial F}{\partial \dot{x}_1} - \frac{\partial F}{\partial x_1} = 0$$

$$\frac{d}{dt}\frac{\partial F}{\partial \dot{x}_2} - \frac{\partial F}{\partial x_2} = 0$$

$$(2)$$

These differential equations, along with the system equation $\varphi_1 = 0$, are necessary conditions to be satisfied for optimum behavior of $x_1(t)$, $x_2(t)$, and $\lambda_1(t)$ between the initial and terminal times t_i and t_f. Now the requirement that the paths $x_1(t)$ and $x_2(t)$ satisfy Eq. (2) has been established here for the case where the end-point variables are given values. However, since the fixed values chosen were assumed to be the optimum values obtained from the solution of the original problem, the optimum paths $x_1(t)$, $x_2(t)$ obtained by satisfying Eq. (2) yield the optimum paths for the original problem. In this manner it is seen that the Euler or characteristic equations for the problem with variable end points are the same as those for the problem with fixed end points.

3-4 THE TRANSVERSALITY CONDITION

For the problem with variable end points it is seen from Eq. 3-2(18) that satisfying Eq. 3-3(2) is not a sufficient condition to ensure the vanishing of the first variation. Not only must Eq. 3-3(2) be satisfied, but in addition the remaining terms in Eq. 3-2(18) must also equal zero. Thus we must require

$$0 = dg - \left[\sum_{i=1}^{2} \frac{\partial F}{\partial \dot{x}_i} \dot{x}_i - F \right] dt \Big|_{t_i}^{t_f} + \sum_{i=1}^{2} \frac{\partial F}{\partial \dot{x}_i} dx_i \Big|_{t_i}^{t_f} \tag{1}$$

This condition is termed the *transversality condition*. It is the role of the transversality condition to provide the additional relations required, along with the p end conditions of Eq. 3-2(2), to determine the six end values t_i, $x_1(t_i)$, $x_2(t_i)$, t_f, $x_1(t_f)$, and $x_2(t_f)$.

Equation (1) must be satisfied for all possible values of the end differentials. However, in obtaining these relations from the transversality condition it is necessary to remember that not all dt_i, $dx_1(t_i)$, $dx_2(t_i)$, dt_f, $dx_1(t_f)$, and $dx_2(t_f)$ are allowable. This follows as there are p relations between these six differentials from the p end conditions of Eq. 3-2(2), since on taking the differential of $\psi_j = 0$ one has

$$d\psi_j = 0 = \frac{\partial \psi_j}{\partial t_i} dt_i + \frac{\partial \psi_j}{\partial x_1(t_i)} dx_1(t_i) + \frac{\partial \psi_j}{\partial x_2(t_i)} dx_2(t_i) + \frac{\partial \psi_j}{\partial t_f} dt_f$$

$$+ \frac{\partial \psi_j}{\partial x_1(t_f)} dx_1(t_f) + \frac{\partial \psi_j}{\partial x_2(t_f)} dx_2(t_f) \qquad j = 1, 2, \cdots, p \tag{2}$$

Thus there are only $6 - p$ independent end differentials. In this light it is seen that the transversality condition, Eq. (1), must be satisfied for only those differentials dt_i, $dx_1(t_i)$, $dx_2(t_i)$, dt_f, $dx_1(t_f)$, and $dx_2(t_f)$ consistent with the p constraints of Eq. (2).

As an example of how one might proceed in a particular problem suppose that there were three end conditions; that is, $p = 3$. In this case there are only three independent differentials which we shall arbitrarily choose to be $dx_1(t_i)$, dt_f, and $dx_2(t_f)$. Now using the three equations of (2) one can solve for dt_i, $dx_2(t_i)$, and $dx_1(t_f)$ in terms of the independent differentials. Having done this, dt_i, $dx_2(t_i)$, and $dx_1(t_f)$ can be eliminated from the transversality condition, Eq. (1), by substituting their values in terms of $dx_1(t_i)$, dt_f, and $dx_2(t_f)$. The transversality condition then reduces to the form

$$A\, dx_1(t_i) + B\, dt_f + C\, dx_2(t_f) = 0 \tag{3}$$

where A, B, and C will in general be functions of all the end variables, including $\lambda_1(t_i)$ and $\lambda_1(t_f)$.

Since $dx_1(t_i)$, dt_f, and $dx_2(t_f)$ are independent differentials, we can set dt_f and $dx_2(t_f)$ to zero. Equation (3) then becomes

$$A\,dx_1(t_i) = 0$$

As this relation must be satisfied for all $dx_1(t_i)$, it follows that $A = 0$. In a similar manner one finds that Eq. (3) also requires $B = 0$, $C = 0$. These three equations,

$$A[t_i, \cdots, x_2(t_f)] = 0$$
$$B[t_i, \cdots, x_2(t_f)] = 0$$
$$C[t_i, \cdots, x_2(t_f)] = 0$$

are the three additional equations required which along with the three specified end constraints and the one differential constraint $\varphi_1 = 0$ evaluated at t_i or t_f specify the six terminal values t_i, $x_1(t_i)$, $x_2(t_i)$, t_f, $x_1(t_f)$, and $x_2(t_f)$ and the value of either $\lambda_1(t_i)$ or $\lambda_1(t_f)$.†

To summarize, in the procedure just outlined the dependent differentials are eliminated from the transversality condition through use of the end constraints of Eq. (2). Once this has been done, the transversality condition requires that the coefficient of each independent differential be set equal to zero. This then provides the additional number of end conditions needed for the solution of the problem.

The procedure just given is usually the one followed in the solution of problems. It does not, however, provide an explicit set of equations with which to work until the problem of interest has been defined. Since it is sometimes useful to have such explicit equations for the general case and since the method of obtaining them would again illustrate the use of the undetermined (Lagrange) multipliers, let us proceed with their development.

Each of the p end constraints of Eq. 3-2(2) is multiplied by an undetermined constant ν_j and the sum of these p products added to the function g of our IP of Eq. 3-2(1) to form

$$G[t_i, x_1(t_i), x_2(t_i), t_f, x_1(t_f), x_2(t_f)]$$
$$= g[t_i, x_1(t_i), x_2(t_i), t_f, x_1(t_f), x_2(t_f)]$$
$$+ \sum_{j=1}^{p} \nu_j \psi_j[t_i, x_1(t_i), x_2(t_i), t_f, x_1(t_f), x_2(t_f)] \quad (4)$$

---

† After differentiating the constraint $\varphi_1 = 0$ it is possible to solve for $\ddot{x}_1$, $\ddot{x}_2$, and λ_1 in terms of the other variables by using Eq. 3-3(2) along with $\dot{\varphi}_1 = 0$; that is,

$$\ddot{x}_1 = f(t, x_1, x_2, \dot{x}_1, \dot{x}_2, \lambda_1)$$
$$\ddot{x}_2 = g(t, x_1, x_2, \dot{x}_1, \dot{x}_2, \lambda_1)$$
$$\dot{\lambda}_1 = h(t, x_1, x_2, \dot{x}_1, \dot{x}_2, \lambda_1)$$

Thus it is seen that it is necessary to specify two end conditions for each of $x_1(t)$ and $x_2(t)$, but only one end value determines $\lambda_1(t)$.

In the augmented $\mathscr{IP}$ the function g is then replaced by G, yielding

$$\mathscr{IP} = G[t_i, x_1(t_i), x_2(t_i), t_f, x_1(t_f), x_2(t_f)]$$

$$+ \int_{t_i}^{t_f} F[t, x_1(t), x_2(t), \dot{x}_1(t), \dot{x}_2(t), \lambda_1(t)] \, dt \quad (5)$$

where since $G = g$ and $F = \bar{f}$ we have, as before, that the augmented $\mathscr{IP}$ of Eq. (5) equals the system IP of Eq. 3-2(1).

Following the development of Sections 3-2 and 3-3 it is seen that the analogue of the transversality condition of Eq. (1) is

$$0 = dG - \left[\sum_{i=1}^{2} \frac{\partial F}{\partial \dot{x}_i} \dot{x}_i - F \right] dt \Big|_{t_i}^{t_f} + \sum_{i=1}^{2} \frac{\partial F}{\partial \dot{x}_i} dx_i \Big|_{t_i}^{t_f} \quad (6)$$

where dG has simply replaced dg in Eq. (1).†

Grouping together the coefficients of each of the differentials, Eq. (6) can be written in the form

$$A \, dt_i + B \, dx_1(t_i) + C \, dx_2(t_i) + D \, dt_f + E \, dx_1(t_f) + F \, dx_2(t_f) = 0 \quad (7)$$

The coefficients $A, B, \cdots, F$ will in general be functions of all the end variables as well as the p constants ν_j. If all the differentials $dt_i, dx_1(t_i), \cdots, dx_2(t_f)$ were independent, it would be necessary for each of their coefficients to vanish to satisfy Eq. (7) for all possible values of $dt_i, \cdots, dx_2(t_f)$. Although they are not all independent because of the p constraints of Eqs. (2), it shall be seen that the p multipliers ν_j enable us to treat them as though they were.

Let us make use of the ν_j which we introduced and which are at our disposal so as to achieve this goal. For the sake of illustration let $p = 3$. As before, there are three dependent differentials and three independent ones. The three multipliers ν_1, ν_2, and ν_3 are determined by the requirement that they take on those values which make three of the coefficients $A, B, \cdots, F$ in which they appear equal zero. Suppose then that ν_1, ν_2, and ν_3 are chosen so that they make $A = 0$, $B = 0$, and $C = 0$. Equation (7) then becomes

$$D \, dt_f + E \, dx_1(t_f) + F \, dx_2(t_f) = 0 \quad (8)$$

But as only three differentials remain they may be taken to be the independent ones for the problem. Then since they are independent differentials, the only way Eq. (8) can be satisfied for all possible dt_f, $dx_1(t_f)$, and $dx_2(t_f)$ is for $D = 0$, $E = 0$, and $F = 0$. Thus the introduction of the multipliers ν_j has allowed us

† This substitution could have been done directly in Eq. (1) as

$$dG = dg + \sum_{i=1}^{p} \nu_i \, d\psi_i = dg$$

since $d\psi_i = 0$ by Eq. (2).

to set the coefficients of each of the differentials to zero, $A = B = \cdots = F = 0$ just as would be the case if the differentials were independent.

In terms of the problem variables setting the coefficients in Eq. (7) to zero then yields the six equations

$$A = \frac{\partial G}{\partial t_i} + \left[\sum_{i=1}^{2} \frac{\partial F}{\partial \dot{x}_i}\,\dot{x}_i - F\right]\Bigg|_{t_i} = 0 \qquad D = \frac{\partial G}{\partial t_f} - \left[\sum_{i=1}^{2} \frac{F}{\dot{x}_i}\,\dot{x}_i - F\right]\Bigg|_{t_f} = 0$$

$$B = \frac{\partial G}{\partial x_1(t_i)} - \frac{\partial F}{\partial \dot{x}_1}\Bigg|_{t_i} = 0 \qquad\qquad E = \frac{\partial G}{\partial x_1(t_f)} + \frac{\partial F}{\partial \dot{x}_1}\Bigg|_{t_f} = 0 \qquad (9)$$

$$C = \frac{\partial G}{\partial x_2(t_i)} - \frac{\partial F}{\partial \dot{x}_2}\Bigg|_{t_i} = 0 \qquad\qquad F = \frac{\partial G}{\partial x_2(t_f)} + \frac{\partial F}{\partial \dot{x}_2}\Bigg|_{t_f} = 0$$

Available to determine the end variables and the multipliers v_j are the six equations of (9) along with the p end conditions of Eq. 3-2(2) and the single differential constraint of Eq. 3-2(3) evaluated at either t_i or t_f. These $7 + p$ equations determine the $7 + p$ quantities made up of the six end values t_i, $x_1(t_i)$, $x_2(t_i)$, t_f, $x_1(t_f)$, and $x_2(t_f)$; the value of either $\lambda_1(t_i)$ or $\lambda_1(t_f)$; and the p multipliers $v_1, v_2, \cdots, v_p$.

3-5 THE RESULTS FOR N SYSTEM VARIABLES

The system is now taken to be defined by N variables $x_1(t), x_2(t), \cdots, x_N(t)$ interrelated by M system equations of the form

$$\varphi_i(t, \mathbf{x}, \dot{\mathbf{x}}) = 0 \qquad i = 1, 2, \cdots, M \qquad (1)$$

It is assumed that optimum design requires minimization of a system IP given by

$$\text{IP} = g[t_i, \mathbf{x}(t_i), t_f, \mathbf{x}(t_f)] + \int_{t_i}^{t_f} \bar{f}(t, \mathbf{x}, \dot{\mathbf{x}})\, dt \qquad (2)$$

The initial and final times t_i and t_f may be unspecified, but the end variables will be required to satisfy the p terminal constraints

$$\psi_j[t_i, \mathbf{x}(t_i), t_f, \mathbf{x}(t_f)] = 0 \qquad j = 1, 2, \cdots, p \qquad (3)$$

where if in Eq. (2) $g = 0$, then $p \leqslant 2N + 2$. If $\bar{f} = 0$ and not g, then $p \leqslant 2N + 1$.

Introducing the M undetermined multipliers $\lambda_1(t), \cdots, \lambda_M(t)$ and forming the augmented integrand

$$F = \bar{f} + \sum_{i=1}^{M} \lambda_i \varphi_i \qquad (4)$$

one has by comparison with Eq. 3-3(2), that the minimizing functions $x_1(t)$, $\cdots$, $x_N(t)$ must satisfy the N characteristic equations given by

$$\frac{d}{dt}\frac{\partial F}{\partial \dot{x}_1} - \frac{\partial F}{\partial x_1} = 0$$

$$\vdots \qquad (N \text{ equations}) \qquad (5)$$

$$\frac{d}{dt}\frac{\partial F}{\partial \dot{x}_N} - \frac{\partial F}{\partial x_N} = 0$$

Equation (5) is the same as Eq. 2-13(16) thus the results derived in Chapter 2 from these equations also apply here. Therefore, by Eqs. 2-13(17, 18) one has that for a corner to exist in the problem at t_1 the additional conditions to be satisfied are

$$\left.\frac{\partial F}{\partial \dot{x}_1}\right|_{t_1^{(+)}} = \left.\frac{\partial F}{\partial \dot{x}_1}\right|_{t_1^{(-)}}$$

$$\vdots \qquad (N \text{ conditions}) \qquad (6a)$$

$$\left.\frac{\partial F}{\partial \dot{x}_N}\right|_{t_1^{(+)}} = \left.\frac{\partial F}{\partial \dot{x}_N}\right|_{t_1^{(-)}}$$

and

$$\left.\left[\sum_{i=1}^{N}\frac{\partial F}{\partial \dot{x}_i}\dot{x}_i - F\right]\right|_{t_1^{(+)}} = \left.\left[\sum_{i=1}^{N}\frac{\partial F}{\partial \dot{x}_i}\dot{x}_i - F\right]\right|_{t_1^{(-)}} \qquad (6b)$$

Further, if the augmented integrand F is not an explicit function of time, that is, if $\partial F/\partial t \equiv 0$, then a first integral of Eq. (5) is

$$\sum_{i=1}^{N}\frac{\partial F}{\partial \dot{x}_i}\dot{x}_i - F = C = \text{const.} \qquad (7)$$

Along with the characteristic equations of Eq. (5), the condition that the first variation of the IP vanish requires that the transversality condition be satisfied. Referring to Eq. 3-4(1) this condition for our problem is seen to be

$$0 = dg - \left[\sum_{i=1}^{N}\frac{\partial F}{\partial \dot{x}_i}\dot{x}_i - F\right] dt \Big|_{t_i}^{t_f} + \sum_{i=1}^{N}\frac{\partial F}{\partial \dot{x}_i} dx_i \Big|_{t_i}^{t_f} \qquad (8)$$

The transversality condition of Eq. (8) must be satisfied for all end differentials consistent with the p relations obtained from Eq. (3) and given by

$$d\psi_j[t_i, \mathbf{x}(t_i), t_f, \mathbf{x}(t_f)] = 0 \qquad j = 1, 2, \cdots, p \qquad (9)$$

Alternatively, introducing the p undetermined constants v_j and forming

$$G = g + \sum_{j=1}^{p} v_j \psi_j \tag{10}$$

the transversality condition can be expressed as

$$0 = dG - \left[\sum_{i=1}^{N} \frac{\partial F}{\partial \dot{x}_i} \dot{x}_i - F \right] dt \bigg|_{t_i}^{t_f} + \sum_{i=1}^{N} \frac{\partial F}{\partial \dot{x}_i} dx_i \bigg|_{t_i}^{t_f} \tag{11}$$

which, analogous to Eq. 3-4(9) yields the set of $2N + 2$ equations

$$\frac{\partial G}{\partial t_i} + \left[\sum_{i=1}^{N} \frac{\partial F}{\partial \dot{x}_i} \dot{x}_i - F \right] \bigg|_{t_i} = 0 \tag{12a}$$

$$\frac{\partial G}{\partial t_f} - \left[\sum_{i=1}^{N} \frac{\partial F}{\partial \dot{x}_i} \dot{x}_i - F \right] \bigg|_{t_f} = 0 \tag{12b}$$

$$\frac{\partial G}{\partial x_j(t_i)} - \frac{\partial F}{\partial \dot{x}_j} \bigg|_{t_i} = 0 \quad j = 1, \ldots, N \tag{12c}$$

$$\frac{\partial G}{\partial x_j(t_f)} + \frac{\partial F}{\partial \dot{x}_j} \bigg|_{t_f} = 0 \quad j = 1, \ldots, N \tag{12d}$$

These conditions are now applied to several examples to illustrate the method of solution.

Example 3-5(1)   Minimize the IP given by

$$\text{IP} = \frac{x_1^2(t_f)}{2} + \int_0^{t_f} \left[1 + \frac{(\dot{x}_1^2 + \dot{x}_2^2)}{2} \right] dt \tag{13}$$

where the system is governed by the equation of motion

$$\dot{x}_1 = x_2 \tag{14}$$

and x_1 and x_2 are subject to end conditions such that

$$\text{at } t = t_i = 0 \quad \text{at } t = t_f \text{ (unspecified)}$$

$$x_1(0) = 0 \quad x_1(t_f) + x_2(t_f) = a$$

$$x_2(0) = 0$$

The end conditions may be expressed in the form of Eq. 3-2(2) by writing

$$\psi_1 = t_i = 0 \tag{15a}$$

$$\psi_2 = x_1(t_i) = 0 \tag{15b}$$

$$\psi_3 = x_2(t_i) = 0 \tag{15c}$$

$$\psi_4 = x_1(t_f) + x_2(t_f) - a = 0 \tag{15d}$$

while the differential constraint of Eq. (14) in the form of Eq. 3-2(3) is given by

$$\varphi_1 = \dot{x}_1 - x_2 = 0 \tag{16}$$

On obtaining the augmented integrand

$$F = 1 + \frac{\dot{x}_1^2 + \dot{x}_2^2}{2} + \lambda_1(\dot{x}_1 - x_2)$$

one finds from Eq. (5) that the characteristic equations for the problem are

$$\ddot{x}_1 + \dot{\lambda}_1 = 0 \tag{17a}$$

$$\ddot{x}_2 + \lambda_1 = 0 \tag{17b}$$

Differentiating Eq. (17b) and eliminating $\dot{\lambda}_1$ between Eqs. (17a) and (17b) one has on using Eq. (16) that

$$\dddot{x}_2 - \dot{x}_2 = 0 \tag{18}$$

Solving Eq. (18) for $x_2(t)$ then yields

$$x_2 = C_1 \sinh t + C_2 \cosh t + C_3 \tag{19}$$

the use of which in Eq. (17b) gives

$$\lambda_1 = -C_1 \sinh t - C_2 \cosh t \tag{20}$$

Finally, on employing Eq. (19) in Eq. (16) and integrating there results

$$x_1 = C_1 \cosh t + C_2 \sinh t + C_3 t + C_4 \tag{21}$$

From Eqs. (19), (20), and (21) it is seen that to fix $x_1(t)$, $x_2(t)$, and $\lambda_1(t)$ six constants must be determined, t_i, t_f, C_1, C_2, C_3, and C_4, whereas thus far there are available only the four end conditions of Eqs. (15). The additional relations are found from the transversality condition of Eq. (8). For this problem Eq. (8) becomes

$$0 = x_1(t_f)\, dx_1(t_f) - \left[-1 + \frac{\dot{x}_1^2 + \dot{x}_2^2}{2} + \lambda_1 \dot{x}_1 - \lambda_1(\dot{x}_1 - x_2) \right] dt \bigg|_{t_i}^{t_f}$$

$$+ \left[(\dot{x}_1 + \lambda_1)\, dx_1 + \dot{x}_2\, dx_2 \right] \bigg|_{t_i}^{t_f} \tag{22}$$

Using Eq. (16) and the result from Eqs. (15) that $dt_i = dx_1(t_i) = dx_2(t_i) = 0$, Eq. (22) can be simplified to

$$0 = x_1(t_f)\,dx_1(t_f) - \left[-1 + \frac{\dot{x}_1^2(t_f) + \dot{x}_2^2(t_f)}{2} + \lambda_1(t_f)\dot{x}_1(t_f)\right] dt_f$$

$$+ \left[[\dot{x}_1(t_f) + \lambda_1(t_f)]\,dx_1(t_f) + \dot{x}_2(t_f)\,dx_2(t_f)\right] \quad (23)$$

However, from Eqs. (15) it is seen that $dx_1(t_f)$ and $dx_2(t_f)$ are not both independent since the relation $d\psi_4 = 0$ gives

$$dx_1(t_f) + dx_2(t_f) = 0 \quad (24)$$

Employing Eq. (24) to express $dx_2(t_f)$ in terms of $dx_1(t_f)$ and then using the result to eliminate $dx_2(t_f)$ from Eq. (23) one obtains
$$0 = [x_1(t_f) + \dot{x}_1(t_f) + \lambda_1(t_f) - \dot{x}_2(t_f)]\,dx_1(t_f)$$

$$- \left[-1 + \frac{\dot{x}_1^2(t_f) + \dot{x}_2^2(t_f)}{2} + \lambda_1(t_f)\dot{x}_1(t_f)\right] dt_f \quad (25)$$

Since in Eq. (25) $dx_1(t_f)$ and dt_f are independent differentials, the only way Eq. (25) can be satisfied for all allowable values of $dx_1(t_f)$, dt_f is for each of their coefficients to vanish. Thus Eq. (25) yields two additional end conditions,

$$x_1(t_f) + \dot{x}_1(t_f) + \lambda_1(t_f) - \dot{x}_2(t_f) = 0 \quad (26a)$$

$$-1 + \frac{\dot{x}_1^2(t_f) + \dot{x}_2^2(t_f)}{2} + \lambda_1(t_f)\dot{x}_1(t_f) = 0 \quad (26b)$$

The two conditions of Eq. (26) along with the four end conditions of Eq. (15) provide the requisite number of conditions for determining the six unknown quantities t_i, t_f, C_1, C_2, C_3, and C_4.

Before finding the solution let us illustrate the alternative method of obtaining the necessary end conditions from the transversality condition. Forming G as in Eq. (10) by introduction of the four undetermined constant multipliers v_1, v_2, v_3, and v_4 one has

$$G = \frac{x_1^2(t_f)}{2} + v_1 t_i + v_2 x_1(t_i) + v_3 x_2(t_i) + v_4[x_1(t_f) + x_2(t_f) - a] \quad (27)$$

For this problem Eq. (12) then yields

$$v_1 + \left[\frac{\dot{x}_1^2(t_i) + \dot{x}_2^2(t_i)}{2} + \lambda_1(t_i)\dot{x}_1(t_i)\right] = 0 \quad (28a)$$

$$-1 + \frac{\dot{x}_1^2(t_f) + \dot{x}_2^2(t_f)}{2} + \lambda_1(t_f)\dot{x}_1(t_f) = 0 \quad (28b)$$

$$\nu_2 - [\dot{x}_1(t_i) + \lambda_1(t_i)] = 0 \tag{28c}$$

$$\nu_3 - \dot{x}_2(t_i) = 0 \tag{28d}$$

$$x_1(t_f) + \nu_4 + \dot{x}_1(t_f) + \lambda_1(t_f) = 0 \tag{28e}$$

$$\nu_4 + \dot{x}_2(t_f) = 0 \tag{28f}$$

On eliminating ν_4 between Eq. (28e) and Eq. (28f) one obtains

$$x_1(t_f) + \dot{x}_1(t_f) + \lambda_1(t_f) - \dot{x}_2(t_f) = 0 \tag{29}$$

Equations (29) and (28b) are the same two additional end conditions previously found in Eq. (26). Thus the equivalence of the two methods of employing the transversality condition has been demonstrated. It is up to the user to decide which method better suits his needs in any particular problem.†

Proceeding with the solution of the problem, on utilizing the initial conditions in Eq. (15) one finds

$$t_i = 0$$

$$C_1 + C_4 = 0 \tag{30}$$

$$C_2 + C_3 = 0$$

with which the solutions for $x_1(t)$, $x_2(t)$, and $\lambda_1(t)$ become

$$x_1(t) = C_1(\cosh t - 1) + C_2(\sinh t - t)$$

$$x_2(t) = C_1 \sinh t + C_2(\cosh t - 1) \tag{31}$$

$$\lambda_1(t) = -C_1 \sinh t - C_2 \cosh t$$

Substituting from Eq. (31) into the end condition of Eq. (15d) one has

$$a = (C_1 + C_2)(\cosh t_f + \sinh t_f - 1) - C_2 t_f \tag{32}$$

Before using the end condition of Eq. (26a) note that it can be simplified considerably by employing the differential constraint of Eq. (16) and then the constraint of Eq. (15d) to yield

$$a = \dot{x}_2(t_f) - \lambda_1(t_f) \tag{33}$$

Then using Eq. (31) in Eq. (33) one obtains

$$C_1 + C_2 = \frac{a}{\cosh t_f + \sinh t_f} \tag{34}$$

With Eq. (34), Eq. (32) can be rewritten

$$-C_2 t_f = \frac{a}{\cosh t_f + \sinh t_f} = C_1 + C_2 \tag{35}$$

---

† As previously stated, Eqs. (28b) and (29) along with the four end conditions of Eq. (15) are to be used to determine the six unknowns t_i, t_f, C_1, C_2, C_3, and C_4. Once this has been done, Eqs. (28a), (28c), (28d), and either Eq. (28e) or (28f) can be used to obtain values for the constants ν_1, ν_2, ν_3, and ν_4.

Finally, on using Eq. (33) to eliminate λ_1, the remaining constraint, Eq. (26b), can be written

$$-1 + \left[\frac{\dot{x}_1(t_f) + \dot{x}_2(t_f)}{2}\right]^2 - a\dot{x}_1(t_f) = 0 \tag{36}$$

On substituting from Eq. (31) into Eq. (36), and employing Eq. (35) to express C_1 and C_2 in terms of t_f, Eq. (36) becomes

$$\left(1 + \frac{1}{t_f}\right)^2 = \frac{2}{a^2}(\sinh t_f + \cosh t_f)^2 \tag{37}$$

The left side of Eq. (37) is a monotonically decreasing function of t_f, while the right side monotonically increases with t_f. Thus there is only one positive value of t_f which satisfies Eq. (37) for a given a.

To illustrate a useful iterative numerical procedure for solving Eq. (37) let us pick the value $a = \sqrt{2}$, and attempt to find the t_f which then satisfies Eq. (37). Setting $a = \sqrt{2}$, Eq. (37) can be written

$$t_f = \frac{t_f + 1}{\sinh t_f + \cosh t_f} \tag{38}$$

The method of solution is quite simple. An initial guess is made for t_f, termed $t_f(0)$, and inserted into the right side of Eq. (38). The left side then gives a first approximation to t_f termed $t_f(1)$. Continuing in this manner $t_f(1)$ is inserted into the right side of Eq. (38) and $t_f(2)$ is generated. The procedure is continued until the difference between the mth approximation, $t_f(m)$, and the $(m + 1)$ approximation, $t_f(m + 1)$, is as small as desired.

Thus since in our problem $0 \leqslant t_f \leqslant \infty$, let us choose as our initial guess $t_f(0) = 0$. Substituting into Eq. (38) one finds $t_f(1) = 1$, which when inserted back into Eq. (38) gives $t_f(2) = 0.736$. Putting $t_f(2)$ into Eq. (38) yields $t_f(3) = 0.832$. Continuing in this manner the difference between $t_f(6)$ and $t_f(7)$ is small enough for our purposes, giving us as the solution of Eq. (38) the value $t_f = 0.806$.

Then using Eqs. (34) and (35) one obtains $C_1 = 1.414$ and $C_2 = -0.783$ which with Eqs. (30) and (31) completes the solution to the problem.

Example 3-5(2)   The problem worked in the previous example was stated in the form of the problem of Bolza. In the present example the problem will be transformed into a problem of Mayer and solved to illustrate how such a problem is treated.

The equivalence of the two formulations will also be shown. After having once done this, it can be seen that the problem of Bolza (or of Lagrange) is no more general than the problem of Mayer. Thus, should it be convenient, there is no reason why the general formulation of our physical problems should not be given in the form of a problem of Mayer.

The problem in Example 3-5(1) required the minimization of the system performance index:

$$\text{IP} = \frac{x_1^2(t_f)}{2} + \int_0^{t_f} \left(1 + \frac{\dot{x}_1^2 + \dot{x}_2^2}{2} \right) dt \tag{39}$$

under the differential constraint of Eq. (14) and the end conditions of Eq. (15). Now, as discussed in Section 3-1, for the problem to be in the form of a problem of Mayer the quantity to be minimized must only be a function of the end-point values of the variables. To achieve this, an additional variable $x_3(t)$ is introduced such that its initial value is zero and its time derivative equals the integrand in the IP of Eq. (39). Thus

$$\dot{x}_3 = 1 + \frac{\dot{x}_1^2 + \dot{x}_2^2}{2}$$

$$x_3(0) = 0 \tag{40}$$

and it is seen that

$$x_3(t_f) = \int_0^{t_f} \left(1 + \frac{\dot{x}_1^2 + \dot{x}_2^2}{2} \right) dt$$

The original problem in Example 3-5(1) is therefore equivalent to the following problem.

Minimize the IP given by

$$\text{IP} = \frac{x_1^2(t_f)}{2} + x_3(t_f) \tag{41}$$

where the system is governed by the equations of motion

$$\dot{x}_1 = x_2$$

$$\dot{x}_3 = 1 + \frac{\dot{x}_1^2 + \dot{x}_2^2}{2}$$

and x_1, x_2, and x_3 are subject to end conditions such that

$$\text{at } t = t_i = 0 \qquad \text{at } t = t_f \text{ (unspecified)}$$

$$x_1(0) = 0$$

$$x_2(0) = 0 \qquad x_1(t_f) + x_2(t_f) - a = 0$$

$$x_3(0) = 0$$

The end conditions may be expressed in the form of Eq. (3) by writing

$$\psi_1 = t_i = 0$$
$$\psi_2 = x_1(t_i) = 0$$
$$\psi_3 = x_2(t_i) = 0$$
$$\psi_4 = x_1(t_f) + x_2(t_f) - a = 0$$
$$\psi_5 = x_3(t_i) = 0$$

(42)

while the differential constraints put in the form of Eq. (1) are given by

$$\varphi_1 = \dot{x}_1 - x_2 = 0$$
$$\varphi_2 = \dot{x}_3 - 1 - \frac{\dot{x}_1^2 + \dot{x}_2^2}{2} = 0$$

(43)

Since there is no integrand $\bar{f}$ in our IP, the augmented integrand F is simply made up of the portion resulting from the two differential constraints. Introducing the multipliers $\lambda_1(t)$ and $\lambda_2(t)$, by Eq. (4),

$$F = \lambda_1(\dot{x}_1 - x_2) + \lambda_2\left(\dot{x}_3 - 1 - \frac{\dot{x}_1^2 + \dot{x}_2^2}{2}\right)$$

Thus, from Eq. (5), the characteristic equations for the problem are seen to be

$$\dot{\lambda}_1 - \frac{d}{dt}(\lambda_2 \dot{x}_1) = 0 \tag{44a}$$

$$\frac{d}{dt}(\lambda_2 \dot{x}_2) + \lambda_1 = 0 \tag{44b}$$

$$\dot{\lambda}_2 = 0 \tag{44c}$$

As $\dot{\lambda}_2 = 0$ by Eq. (44c), the multiplier λ_2 is a constant and is denoted in what follows by $\bar{\lambda}_2$. With this knowledge, Eq. (44) becomes

$$-\bar{\lambda}_2 \ddot{x}_1 + \dot{\lambda}_1 = 0$$
$$-\bar{\lambda}_2 \ddot{x}_2 + \lambda_1 = 0$$

(45)

In addition to Eq. (45), the minimizing functions must satisfy the transversality condition of Eq. (8). For this problem the condition requires

$$0 = x_1(t_f)\, dx_1(t_f) + dx_3(t_f)$$

$$- \left[(\lambda_1 - \lambda_2 \dot{x}_1)\dot{x}_1 - \lambda_2 \dot{x}_2^2 + \lambda_2 \dot{x}_3\right] dt \,\Big|_{t_i}^{t_f}$$

$$+ \left[(\lambda_1 - \lambda_2 \dot{x}_1)\, dx_1 - \lambda_2 \dot{x}_2\, dx_2 + \lambda_2\, dx_3\right]\Big|_{t_i}^{t_f} \tag{46}$$

where the fact that $F = \sum_{i=1}^{3} \lambda_i \varphi_i = 0$ for this problem has been used in Eq. (46) to eliminate F from the coefficient of dt.

Using the knowledge that by Eq. (42) the initial conditions are all fixed; the relation $d\psi_4 = 0$ from Eq. (42) to express $dx_2(t_f)$ in terms of $dx_1(t_f)$; and the fact that λ_2 is a constant, hence $\lambda_2(t_f) = \bar{\lambda}_2$, Eq. (46) becomes

$$
\begin{aligned}
0 = {}& [x_1(t_f) + \lambda_1(t_f) - \bar{\lambda}_2 \dot{x}_1(t_f) + \bar{\lambda}_2 \dot{x}_2(t_f)]\, dx_1(t_f) \\
& - [\lambda_1(t_f)\dot{x}_1(t_f) - \bar{\lambda}_2(\dot{x}_1^2(t_f) + \dot{x}_2^2(t_f)) + \bar{\lambda}_2 \dot{x}_3(t_f)]\, dt_f \\
& + [1 + \bar{\lambda}_2]\, dx_3(t_f)
\end{aligned} \tag{47}
$$

As $dx_1(t_f)$, dt_f, and $dx_3(t_f)$ are independent differentials, Eq. (47) can only be satisfied if each of their coefficients vanishes. Thus

$$x_1(t_f) + \lambda_1(t_f) - \bar{\lambda}_2 \dot{x}_1(t_f) + \bar{\lambda}_2 \dot{x}_2(t_f) = 0 \tag{48a}$$

$$\lambda_1(t_f)\dot{x}_1(t_f) - \bar{\lambda}_2[\dot{x}_1^2(t_f) + \dot{x}_2^2(t_f)] + \bar{\lambda}_2 \dot{x}_3(t_f) = 0 \tag{48b}$$

$$\bar{\lambda}_2 + 1 = 0 \tag{48c}$$

On using Eq. (48c), from which one finds $\bar{\lambda}_2 = -1$, and Eq. (43b), which expresses $\dot{x}_3$ in terms of $\dot{x}_1$ and $\dot{x}_2$, Eqs. (48a) and (48b) can be written

$$x_1(t_f) + \dot{x}_1(t_f) + \lambda_1(t_f) - \dot{x}_2(t_f) = 0$$

$$-1 + \frac{\dot{x}_1^2(t_f) + \dot{x}_2^2(t_f)}{2} + \lambda_1(t_f)\dot{x}_1(t_f) = 0 \tag{49}$$

Further, with $\bar{\lambda}_2 = -1$, the characteristic equations of Eq. (45) are

$$\ddot{x}_1 + \dot{\lambda}_1 = 0$$

$$\ddot{x}_2 + \dot{\lambda}_1 = 0 \tag{50}$$

Equation (50), along with the two end conditions of Eq. (49) and the first four end conditions of Eq. (42) determine $x_1(t)$, $x_2(t)$, and $\lambda_1(t)$. Comparing the transversality conditions and characteristic equations of (49) and (50) with those obtained in the previous formulation of the problem and given in Eqs. (26) and (17), it is seen that they are identical. Thus from this point the solution proceeds as in the first example.

3-6 THE WEIERSTRASS AND LEGENDRE–CLEBSCH CONDITIONS

It has proved fruitful thus far to work with weak variations in the development of necessary conditions to be satisfied by the minimizing functions. As has been seen in Section 3-5, these conditions, along with the problem constraints, determine a set of functions $x_1(t), x_2(t), \cdots, x_N(t)$, called *extremals*, which render the IP of Eq. 3-5(2) stationary. That is for the extremal functions, $\delta \text{IP} = 0$, under weak variations.

However, the comparison functions used in weak variations are rather special. Consider the extremal $x_1(t)$ and the comparison function $x_1(t, \epsilon)$. For the difference between the two to correspond to a weak variation, recall (see Section 2-7) that this requires $|x_1(t, \epsilon) - x_1(t)| \leqslant \epsilon$ and $|\dot{x}_1(t, \epsilon) - \dot{x}_1(t)| \leqslant \epsilon$. Thus for a weak variation the comparison function and its time rate of change can differ only slightly from the corresponding values on the extremal.

In this light the reader could ask if the value of the IP might indeed not be lowered by using comparison functions in the IP on which the restrictions were less severe. Let us test this hypothesis by now considering strong variations. For a strong variation $|x_1(t, \epsilon) - x_1(t)| \leqslant \epsilon$, while $|\dot{x}_1(t, \epsilon) - \dot{x}_1(t)|$ need not be less than ϵ. Therefore, as ϵ approaches zero, while the value of the comparison function approaches that of the extremal, its slope can differ greatly from that of the extremal.

The problem on which the test is to be applied is that given in Section 2-13. This is the problem stated in Section 3-5 with the additional restriction that the terminal times and the values of the system variables at these times be fixed. The condition that will be obtained here by treating the simpler problem does not differ from that which would be found for the more general problem of Section 3-5, but the procedure is much easier to follow.

To review, in Section 2-13 the minimization of the IP of Eq. 2-13(13) was to be achieved by proper determination of the N system variables $x_1(t) \cdots x_N(t)$ between the specified end points subject to the differential constraints of Eq. 2-13(14). By considering weak variations, the characteristic equations of Eq. 2-13(16) and the corner conditions of Eqs. 2-13(17) and 2-13(18) were found. Then through application of these conditions, the extremals $x_1(t) \cdots x_N(t)$ for which $\delta\text{IP} = 0$ can be determined. Assume that this has been done.

To ascertain if the value of the IP can be lowered by using comparison functions that differ strongly from the extremal solution, we consider functions of the type shown in Fig. 3-1. As shown, the comparison function equals the extremal solution everywhere in $t_i \leqslant t \leqslant t_f$ except in the subinterval $t_1 \leqslant t \leqslant t_2$ between corners of the extremal arc. In the interval $t_1 \leqslant t \leqslant t_1 + e$ a strong variation is allowed, while in $t_1 + e \leqslant t \leqslant t_2$ only a weak variation is permitted in returning to the extremal. Now let $\dot{X}_i(t_1)$ be a set of values not necessarily close to $\dot{x}_i(t_1)$ such that the values $(t_1, x(t_1), \dot{X}(t_1))$ satisfy the differential constraints of Eq. 2-13(14). Then for each extremal $x_i(t)$, $i = 1, 2, \cdots, N$, the comparison function will be given by

$$
\begin{array}{lll}
x_i(t): & t_i \leqslant t \leqslant t_1 & \text{(region I)} \\[4pt]
X_i(t): & t_1 \leqslant t \leqslant t_1 + e & \text{(region II)} \\[4pt]
x_i(t, e): & t_1 + e \leqslant t \leqslant t_2 & \text{(region III)} \\[4pt]
x_i(t): & t_2 \leqslant t \leqslant t_f & \text{(region IV)}
\end{array}
\tag{1}
$$

where $e \geqslant 0$.

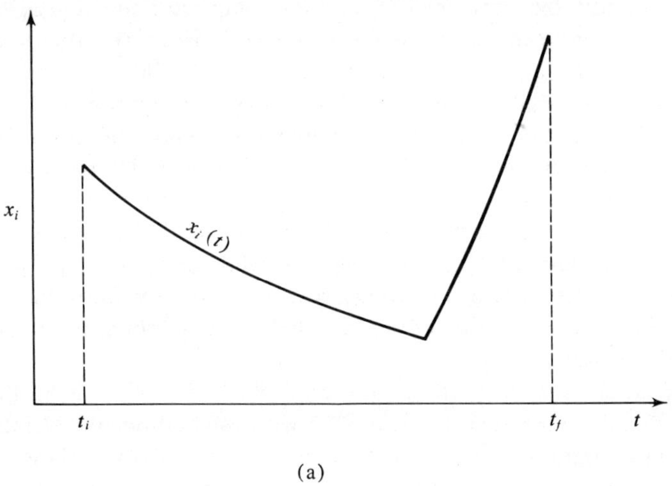

(a)

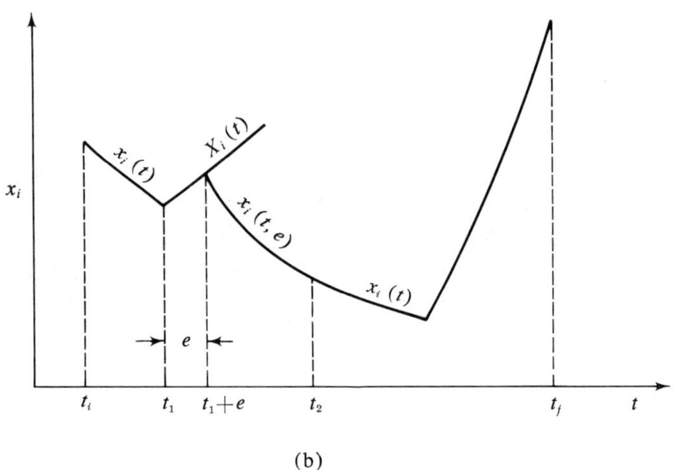

(b)

FIGURE 3-1   (a) Extremal solution to the problem; (b) comparison
function used in the derivation of the Weierstrass E condition

As the x_i are to be continuous, although their rate of change need not be,
this requires at t_1, $t_1 + e$, and t_2 that

$$x_i(t_1) = X_i(t_1) \tag{2a}$$

$$X_i(t_1 + e) = x_i(t_1 + e, e) \tag{2b}$$

$$x_i(t_2, e) = x_i(t_2) \tag{2c}$$

Note that when e equals zero the middle regions are eliminated and the comparison function of Eq. (1) becomes the extremal solution $x_i(t)$.

Recall that the question posed was whether the value of the IP could be lowered by considering strong variations. To find the answer with respect to the comparison function of Eq. (1), the augmented integrand is formed and the comparison functions substituted into the augmented performance index based on the IP of Eq. 2-13(13). The $\mathscr{IP}$ as a function of e is therefore

$$\mathscr{IP}(e) = \text{IP}(e) = \int_{t_i}^{t_1} F[t, \mathbf{x}(t), \dot{\mathbf{x}}(t)]\, dt + \int_{t_1}^{t_1+e} F[t, \mathbf{X}(t), \dot{\mathbf{X}}(t)]\, dt$$

$$+ \int_{t_1+e}^{t_2} F[t, \mathbf{x}(t, e), \dot{\mathbf{x}}(t, e)]\, dt + \int_{t_2}^{t_f} F[t, \mathbf{x}(t), \dot{\mathbf{x}}(t)]\, dt \quad (3)$$

By hypothesis, the extremals $x_i(t)$ are the functions that minimize $\mathscr{IP}(e)$. For this to be true requires that $\mathscr{IP}(e)$ be greater than $\mathscr{IP}(0)$, since the latter value corresponds to that obtained for the $\mathscr{IP}$ by inserting the extremals $x_i(t)$. In this manner it is seen that the required condition can be found by considering the change in the $\mathscr{IP}$ about its value when e equals zero; that is, $d\mathscr{IP}(0)$. If the $x_i(t)$ are to be the minimizing functions, then

$$d\mathscr{IP}(0) = \frac{d\mathscr{IP}}{de}\bigg|_{e=0} e \geqslant 0 \quad (4)$$

Before going on to form Eq. (4) from Eq. (3) let us obtain a result which will be of use later. From Eq. (2b), since

$$dX_i(t_1 + e) = dx_i(t_1 + e, e)$$

it follows that at e equal to zero one has

$$\frac{dX_i}{dt}(t_1) = \frac{\partial x_i}{\partial t}(t_1, 0) + \frac{\partial x_i}{\partial e}(t_1, 0)$$

or

$$\frac{\partial x_i}{\partial e}(t_1, 0) = \dot{X}_i(t_1) - \dot{x}_i(t_1) \quad (5)$$

In addition, since $x_i(t_2)$ is fixed, it follows from Eq. (2c) that

$$\frac{\partial x_i}{\partial e}(t_2, 0) = 0 \quad (6)$$

Returning to our task let us obtain $d\mathscr{IP}(0)/de$ from Eq. (3), so as to form Eq. (4). From Eq. (3) one finds

$$\frac{d\mathscr{IP}(e)}{de} = \{F[t, \mathbf{X}(t), \dot{\mathbf{X}}(t)] - F[t, \mathbf{x}(t), \dot{\mathbf{x}}(t)]\}\bigg|_{t_1+e}$$

$$+ \int_{t_1+e}^{t_2} \sum_{j=1}^{N}\left[\frac{\partial F}{\partial x_j}\frac{\partial x_j}{\partial e}(t, e) + \frac{\partial F}{\partial \dot{x}_j}\frac{\partial \dot{x}_j}{\partial e}(t, e)\right] dt \quad (7)$$

which on integrating the second term in the integral once by parts becomes

$$\frac{d\mathscr{IP}(e)}{de} = \{F[t, \mathbf{X}(t), \dot{\mathbf{X}}(t)] - F[t, \mathbf{x}(t), \dot{\mathbf{x}}(t)]\}\Big|_{t_1 + e}$$

$$+ \sum_{j=1}^{N} \frac{\partial F}{\partial \dot{x}_j}[t, \mathbf{x}(t, e), \dot{\mathbf{x}}(t, e)]\frac{\partial x_j}{\partial e}(t, e)\Big|_{t_1 + e}^{t_2}$$

$$- \int_{t_1 + e}^{t_2} \sum_{j=1}^{N} \left[\frac{d}{dt}\frac{\partial F}{\partial \dot{x}_j} - \frac{\partial F}{\partial x_j}\right]\frac{\partial x_j}{\partial e}(t, e)\, dt \qquad (8)$$

Evaluating Eq. (8) at $e = 0$, the functions $x_j(t, e)$ and $\dot{x}_j(t, e)$ go over into the extremal functions $x_j(t)$ and $\dot{x}_j(t)$. Then since along the extremal

$$\frac{d}{dt}\frac{\partial F}{\partial \dot{x}_j} - \frac{\partial F}{\partial x_j} = 0$$

by Eq. 2-13(16), the integral in Eq. (8) is eliminated, yielding

$$\frac{d\mathscr{IP}(0)}{de} = \{F[t, \mathbf{x}(t), \dot{\mathbf{X}}(t)] - F[t, \mathbf{x}(t), \dot{\mathbf{x}}(t)]\}\Big|_{t_1}$$

$$+ \sum_{j=1}^{N} \frac{\partial F}{\partial \dot{x}_j}[t, \mathbf{x}(t), \dot{\mathbf{x}}(t)]\frac{\partial x_j}{\partial e}(t, 0)\Big|_{t_1}^{t_2} \qquad (9)$$

where Eq. (2a) has been used to replace $\mathbf{X}(t_1)$ by $\mathbf{x}(t_1)$ in the first term. Further, as $\partial x_j(t_2, 0)/\partial e$ is zero by Eq. (6), on using Eq. (5), Eq. (9) becomes

$$\frac{d\mathscr{IP}(0)}{de} = \left\{F[t, \mathbf{x}(t), \dot{\mathbf{X}}(t)] - F[t, \mathbf{x}(t), \dot{\mathbf{x}}(t)]\right.$$

$$\left. - \sum_{j=1}^{N} \frac{\partial F}{\partial \dot{x}_j}[t, \mathbf{x}(t), \dot{\mathbf{x}}(t)][\dot{X}_j(t) - \dot{x}_j(t)]\right\}\Big|_{t_1} \qquad (10)$$

Having $d\mathscr{IP}(0)/de$, the change $d\mathscr{IP}(0)$ about the value $\mathscr{IP}(0)$ corresponding to use of the extremal functions $x_i(t)$ can now be found by multiplying Eq. (10) by e [see Eq. (4)].

Recall that for the extremals to be the minimizing functions, $d\mathscr{IP}(0) \geqslant 0$. But since $e > 0$ this means that its coefficient $d\mathscr{IP}(0)/de$ must be greater than or equal to zero. Define the coefficient to be E. For the extremals $x_i(t)$ to be the minimizing functions requires that

$$E = F[t, \mathbf{x}(t), \dot{\mathbf{X}}(t)] - F[t, \mathbf{x}(t), \dot{\mathbf{x}}(t)]$$

$$- \sum_{j=1}^{N} \frac{\partial F}{\partial \dot{x}_j}[t, \mathbf{x}(t), \dot{\mathbf{x}}(t)][\dot{X}_j(t) - \dot{x}_j(t)] \geqslant 0 \qquad (11)$$

where since t_1 was arbitrary the designation has been dropped.

The function E is known as the Weierstrass E function. Equation (11), called the *necessary condition* of Weierstrass, must be satisfied at each t for all slope functions $\dot{\mathbf{X}}(t)$ consistent with the differential constraints of Eq. 2-13(14), $\varphi_i(t, \mathbf{x}, \dot{\mathbf{X}}) = 0$, if the extremals $x_i(t)$ are to be the minimizing functions for our IP under strong variations. Although the condition is developed here for the problem with fixed end points, as stated earlier, a more general study would show it to be valid also for problems in which the end points are variable.

Consider now the form that Eq. (11) takes if the variations are not strong but rather are weak. Doing this a condition will be found which must be satisfied by the extremals if they are to minimize the IP under weak variations.

By definition, for a weak variation $\dot{X}_i$ can only differ from $\dot{x}_i$ by a small amount. Thus for weak variations, the quantity $F(t, \mathbf{x}, \dot{\mathbf{X}})$ in Eq. (11) can be expanded about its value when $\dot{X}_i = \dot{x}_i$. One finds

$$F(t, \mathbf{x}, \dot{\mathbf{X}}) = F(t, \mathbf{x}, \dot{\mathbf{x}}) + \sum_{j=1}^{N} \frac{\partial F}{\partial \dot{x}_j} (t, \mathbf{x}, \dot{\mathbf{x}}) \, \delta \dot{x}_j$$

$$+ \frac{1}{2} \sum_{j=1}^{N} \sum_{k=1}^{N} \frac{\partial^2 F}{\partial \dot{x}_j \, \partial \dot{x}_k} (t, \mathbf{x}, \dot{\mathbf{x}}) \, \delta \dot{x}_j \, \delta \dot{x}_k + O(\delta \dot{x}^3) \quad (12)$$

where the notation $\delta \dot{x}_j = \dot{X}_j - \dot{x}_j$ has been used as in our earlier work with weak variations.

Substituting Eq. (12) into Eq. (11), assuming that the second-order terms do not all vanish, there results

$$\sum_{j=1}^{N} \sum_{k=1}^{N} \frac{\partial^2 F}{\partial \dot{x}_j \, \partial \dot{x}_k} (t, \mathbf{x}, \dot{\mathbf{x}}) \, \delta \dot{x}_j \, \delta \dot{x}_k \geqslant 0 \quad (13)$$

The relation of Eq. (13) is called the *Legendre–Clebsch condition*. It must be satisfied by the extremals $x_i(t)$ for all $\delta \dot{x}$ consistent with the constraints of Eq. 2-13(14) in the form

$$\delta \varphi = \sum_{i=1}^{N} \frac{\partial \varphi}{\partial \dot{x}_i} \, \delta \dot{x}_i = 0$$

if the IP is to be minimized by the extremals under weak variations.

Example 3-6(1)   Return to the problem first presented in Section 2-10, which required the minimization of

$$\text{IP} = \int_0^2 (\dot{x}^2 - 1)^2 \, dt$$

under the condition that $x(0) = x(2) = 1$. Recall that the characteristic equations yielded the result $\dot{x} = \text{const.}$, while from the Weierstrass–Erdmann corner conditions one found that at a corner $\dot{x}_+ = -\dot{x}_- = \pm 1$.

The solution is thus either a continuous straight line of zero slope, or a series of continuous straight-line segments of slope ± 1. Denoting the first solution by sol. 1 and a solution of the second type by sol. 2 one has

$$\text{Sol. 1:}\quad x(t) = 1 \qquad\qquad 0 \leqslant t \leqslant 2$$

$$\text{Sol. 2:}\quad x(t) = t + 1 \qquad 0 \leqslant t \leqslant 1$$

$$x(t) = -t + 3 \qquad 1 \leqslant t \leqslant 2$$

with the values $\text{IP} = 2$, $\text{IP} = 0$ corresponding to use of sol. 1 or sol. 2, respectively. Note that since the integrand of our IP is positive semidefinite the value $\text{IP} = 0$, found by using sol. 2, must be the minimum value.

Let us now apply the Weierstrass and Legendre–Clebsch conditions to the solutions found. For this problem $F = \bar{f} = (\dot{x}^2 - 1)^2$; thus the Weierstrass and Legendre–Clebsch conditions of Eqs. (11) and (13) become, respectively,

$$E = (\dot{X}^2 - 1)^2 - (\dot{x}^2 - 1)^2 - 4\dot{x}(\dot{x}^2 - 1)(\dot{X} - \dot{x}) \geqslant 0 \qquad (14)$$

and

$$(3\dot{x}^2 - 1)\,\delta\dot{x}^2 \geqslant 0 \qquad (15)$$

Testing sol. 1, for which $x(t) = 1$, $\dot{x}(t) \equiv 0$, Eq. (14) becomes

$$E = (\dot{X}^2 - 1)^2 - 1 \geqslant 0 \qquad (16)$$

As Eq. (16) cannot be satisfied for all $\dot{X}$, it is seen that sol. 1 does not satisfy the Weierstrass condition and hence cannot be the minimizing solution under strong variations. Further, on testing sol. 1 under the weaker Legendre–Clebsch condition of Eq. (15), it is seen that even this is not satisfied, since it requires $-\delta\dot{x}^2 \geqslant 0$. Thus the solution $x(t) = 1$ is not a minimizing solution under either strong or weak variations.

On the other hand, for sol. 2, in which $\dot{x}(t) = \pm 1$, the Weierstrass condition of Eq. (14) requires

$$E = (\dot{X}^2 - 1)^2 \geqslant 0$$

which is satisfied for all $\dot{X}$. Sol. 2 therefore satisfies this condition. The Legendre–Clebsch condition is then automatically satisfied.

3-7   PROBLEMS NOT INVOLVING RATES

The development of the Weierstrass and Legendre–Clebsch conditions has provided a means of testing solutions if the augmented integrand F is an explicit function of the rates $\dot{x}_i$. As this will not be the case in all problems of interest, let us see how to modify the results to obtain useful information when it is not true.

Consider the problem of minimizing

$$\text{IP} = \int_{t_i}^{t_f} \bar{f}(t, x)\, dt \tag{1}$$

by proper choice of $x(t)$, where t_i and t_f are fixed but not the values of $x(t)$ at these times. By Eq. 3-5(5) the characteristic equation for the problem is

$$\frac{\partial \bar{f}}{\partial x}(t, x) = 0 \tag{2}$$

from which the extremal $x(t)$ is determined.

On attempting to check the Weierstrass and Legendre–Clebsch conditions one finds from Eqs. 3-6(11) and 3-6(13) that $E \equiv 0$ and $\partial^2 \bar{f}/\partial \dot{x}^2 \equiv 0$. Therefore, no useful test is obtained from either condition.

To see if further information can be obtained, let us artificially introduce rates into the problem by the change of variables

$$\dot{z} = x \tag{3}$$

where the value of z at t_i and t_f is not specified.

The problem is then to minimize

$$\text{IP} = \int_{t_i}^{t_f} \bar{f}(t, \dot{z})\, dt \tag{4}$$

by proper choice of $z(t)$, where t_i and t_f are specified but not the values of $z(t)$ at these times. The characteristic equation for the problem in this form yields $\partial \bar{f}/\partial \dot{z} = \text{const.}$, while the transversality condition of Eqs. 3-5(12c, d) gives

$$\left.\frac{\partial \bar{f}}{\partial \dot{z}}\right|_{t_i} = \left.\frac{\partial \bar{f}}{\partial \dot{z}}\right|_{t_f} = 0$$

Thus since $\partial \bar{f}/\partial \dot{z}$ is a constant there results the equation

$$\frac{\partial \bar{f}}{\partial \dot{z}}(t, \dot{z}) = 0 \tag{5}$$

for determination of $\dot{z}(t)$. On substituting the symbol x for $\dot{z}$, Eq. (5) is seen to be identical to Eq. (2).

Continuing, let us test the solution found by Eq. (5) in the Weierstrass and Legendre–Clebsch conditions of Eqs. 3-6(11) and 3-6(13). Applying Eq. (5), the Weierstrass condition gives

$$E = \bar{f}(t, \dot{Z}) - \bar{f}(t, \dot{z}) \geqslant 0 \tag{6}$$

while the Legendre–Clebsch condition becomes

$$\frac{\partial^2 \bar{f}}{\partial \dot{z}^2} \, \delta\dot{z}^2 \geqslant 0 \tag{7}$$

These conditions must be satisfied by the extremal $z(t)$ if it is to be a minimizing solution.

However, as $z(t)$ never appears, the above statement is equivalent to stating that Eqs. (6) and (7) must be satisfied for $\dot{z}(t)$ as found from Eq. (5) if that function is to correspond to a minimizing solution. But a test on $\dot{z}(t)$, in terms of the original problem variables, is, by Eq. (3), a test on $x(t)$. Thus in the original problem variables the Weierstrass condition is seen to be

$$E = f(t, X) - f(t, x) \geqslant 0 \tag{8}$$

while the Legendre–Clebsch condition yields

$$\frac{\partial^2 \bar{f}}{\partial x^2} (t, x) \, \delta x^2 \geqslant 0 \tag{9}$$

Equations (8) and (9) provide the tests sought on the extremal solutions obtained from Eq. (2).

In conclusion, in any problem in which the rate of change of a variable does not appear explicitly in the augmented integrand F, the Weierstrass and Legendre–Clebsch conditions are to be employed with the variable itself used in place of its rate of change.

3-8   THE RESULTS EXPRESSED FOR CONTROL APPLICATIONS

Necessary conditions to be satisfied by the minimizing functions have been obtained in the preceding sections for a system expressed in a very general manner. Now, without restricting the system to be considered, it is useful to reexpress these conditions in terms of a somewhat more standard representation of the system to be controlled.

Following the discussion in Chapter 1, the plant is assumed to be defined by $N = n + m$ system variables, the n state variables $x_1, \cdots, x_n$ whose behavior in time is governed by the n equations of motion (system equations)

$$\begin{aligned} \dot{x}_1 &= f_1(t, x_1, \cdots, x_n, u_1, \cdots, u_m) \\ &\vdots \\ \dot{x}_n &= f_n(t, x_1, \cdots, x_n, u_1, \cdots, u_m) \end{aligned} \tag{1}$$

and the m control variables $u_1, \cdots, u_m$ appearing in the equations of motion (1).

The latter are to be determined so as to minimize the system performance index expressed in terms of the initial and final states of the system.†

$$\text{IP} = g[t_i, x_1(t_i), \cdots, x_n(t_i), t_f, x_1(t_f), \cdots, x_n(t_f)] \tag{2}$$

At t_i and t_f it is assumed that system design requires terminal conditions to be met such that

$$\psi_j[t_i, x_1(t_i), \cdots, x_n(t_i), t_f, x_1(t_f), \cdots, x_n(t_f)] = 0 \qquad j = 1, \cdots, p \leqslant 2n + 1 \tag{3}$$

Introducing the n undetermined multipliers $\lambda_1(t), \cdots, \lambda_n(t)$ the augmented integrand of Eq. 3-5(4) becomes

$$F = \sum_{i=1}^{n} \lambda_i(t)[\dot{x}_i - f_i(t, \mathbf{x}, \mathbf{u})] \tag{4}$$

Through use of Eq. (4) the characteristic equations of 3-5(5) for the n state variables and the m control variables of the problem are seen to be

$$\dot{\lambda}_1 = -\sum_{i=1}^{n} \lambda_i \frac{\partial f_i}{\partial x_1} (t, \mathbf{x}, \mathbf{u})$$
$$\vdots \qquad\qquad (n \text{ equations}) \tag{5}$$
$$\dot{\lambda}_n = -\sum_{i=1}^{n} \lambda_i \frac{\partial f_i}{\partial x_n} (t, \mathbf{x}, \mathbf{u})$$

and

$$\sum_{i=1}^{n} \lambda_i \frac{\partial f_i}{\partial u_1} (t, \mathbf{x}, \mathbf{u}) = 0$$
$$\vdots \qquad\qquad (m \text{ equations}) \tag{6}$$
$$\sum_{i=1}^{n} \lambda_i \frac{\partial f_i}{\partial u_m} (t, \mathbf{x}, \mathbf{u}) = 0$$

The equations of (5) are often termed the *adjoint equations*. The terminology follows from the work in Section 6-7. Equations (6) are referred to as the *control equations*.

---

† Although the problem has been given as a problem of Mayer, recall [see Example 3-5(2)] that had the IP been of the form

$$\text{IP} = \bar{g}[t_i, \mathbf{x}(t_i), t_f, \mathbf{x}(t_f)] + \int_{t_i}^{t_f} \bar{f}(t, \mathbf{x}, \dot{\mathbf{x}}) \, dt,$$

it could easily be put in the form of Eq. (2). For by adding the additional constraints $\dot{x}_{n+1} = \bar{f}(t, x, u)$, $x_{n+1}(t_i) = 0$, which define $x_{n+1}(t)$, the IP to be minimized is seen to be $\text{IP} = \bar{g} + x_{n+1}(t_f)$, which is of the form of Eq. (2).

Note that by Eq. (1) $F \equiv 0$. Hence it follows from Eq. 3-5(7) that if $\partial f_i / \partial t \equiv 0$ for $i = 1, \cdots, n$, a first integral of Eq. (5) is given by

$$\sum_{i=1}^{n} \lambda_i \dot{x}_i = \sum_{i=1}^{n} \lambda_i(t) f_i(t, \mathbf{x}, \mathbf{u}) = C = \text{const.} \tag{7}$$

Should a corner exist at some time t_1 because of discontinuous changes in the control variables $u_1, \cdots, u_m$, the Weierstrass–Erdmann corner conditions of Eq. 3.5(6) require

$$\lambda_1 \Big|_{t_1^{(+)}} = \lambda_1 \Big|_{t_1^{(-)}}$$
$$\vdots \qquad \qquad (n \text{ conditions}) \tag{8a}$$
$$\lambda_n \Big|_{t_1^{(+)}} = \lambda_n \Big|_{t_1^{(-)}}$$

and

$$\sum_{i=1}^{n} \lambda_i f_i \Big|_{t_1^{(+)}} = \sum_{i=1}^{n} \lambda_i f_i \Big|_{t_1^{(-)}} \tag{8b}$$

Finally, introducing the p undetermined constants $\nu_1, \cdots, \nu_p$ and forming

$$G = g[t_i, \mathbf{x}(t_i), t_f, \mathbf{x}(t_f)] + \sum_{j=1}^{p} \nu_j \psi_j[t_i, \mathbf{x}(t_i), t_f, \mathbf{x}(t_f)] \tag{9}$$

the transversality condition of Eq. 3-5(12) becomes

$$\sum_{i=1}^{n} \lambda_i f_i \Big|_{t_i} + \frac{\partial G}{\partial t_i} = 0 \tag{10a}$$

$$\sum_{i=1}^{n} \lambda_i f_i \Big|_{t_f} - \frac{\partial G}{\partial t_f} = 0 \tag{10b}$$

$$\lambda_j(t_i) = \frac{\partial G}{\partial x_j(t_i)} \qquad j = 1, \cdots, n \tag{10c}$$

$$\lambda_j(t_f) = -\frac{\partial G}{\partial x_j(t_f)} \qquad j = 1, \cdots, n \tag{10d}$$

An equivalent statement of the transversality condition as given by Eq. 3-5(8) requires that

$$0 = dg - \sum_{i=1}^{n} \lambda_i f_i \, dt \Big|_{t_i}^{t_f} + \sum_{i=1}^{n} \lambda_i \, dx_i \Big|_{t_i}^{t_f} \tag{11a}$$

for all end differentials consistent with the p constraints

$$d\psi_j[t_i, \mathbf{x}(t_i), t_f, \mathbf{x}(t_f)] = 0 \qquad j = 1, \cdots, p \qquad (11b)$$

Depending on which is most convenient, the transversality condition may be used in the form given by either Eq. (10) or (11).

The solution to the motion equations of Eq. (1) and the characteristic equations of (5) and (6) which satisfies the end conditions of Eq. (3) and the transversality condition of Eq. (10) or (11) is the extremal solution to our problem. That is, for this solution the first variation of our IP vanishes under weak variations. The solution must, in addition, also satisfy either the Weierstrass condition of Eq. 3-6(11) or the Legendre–Clebsch condition of Eq. 3-6(13) for it to be a minimizing solution. The former tests the solution under strong variations; the latter tests it under weak variations.

There are $N = n + m$ system variables in our problem, the n state variables x_i and the m control variables u_i. On examining the augmented integrand F of Eq. (4), however, it is seen that the control variable rates do not appear. Since this is the case, using the results of Section 3-7, the control variables themselves are used in the E function.

Thus remembering that $F \equiv 0$, the Weierstrass condition of Eq. 3-6(11) gives

$$-\sum_{j=1}^{n} \lambda_j(t)[\dot{X}_j(t) - \dot{x}_j(t)] - \sum_{j=1}^{m}\left[-\sum_{i=1}^{n} \lambda_i(t)\frac{\partial f_i}{\partial u_j}\right][U_j(t) - u_j(t)] \geqslant 0 \qquad (12)$$

which on employing (6) to eliminate the second term reduces to

$$\sum_{j=1}^{n} \lambda_j(t)\dot{x}_j(t) \geqslant \sum_{j=1}^{n} \lambda_j(t)\dot{X}_j(t) \qquad (13)$$

Now by Eq. (1), $\dot{x}_j = f_j(t, \mathbf{x}, \mathbf{u})$, where the values of x_i and u_i are the values on the extremal at time t. Similarly, any other admissible rate must also satisfy Eq. (1). Thus $\dot{X}_j(t) = f_j(t, \mathbf{x}, \mathbf{U})$, where the x_i are again the extremal values while U_i may be any allowable control value. Substituting for $\dot{x}_j$ and $\dot{X}_j$ in Eq. (13) one has as the condition to be satisfied that

$$\sum_{j=1}^{n} \lambda_j(t)f_j(t, \mathbf{x}, \mathbf{u}) \geqslant \sum_{j=1}^{n} \lambda_j(t)f_j(t, \mathbf{x}, \mathbf{U}) \qquad (14)$$

Equation (14) states that the left side, which contains the extremal values of the control variables, must be greater than or equal to the right side, in which any allowable control values may be inserted. Obviously then, this can only be satisfied if the extremal values of $u_1, \cdots, u_m$ maximize the left side of Eq. (14).

Thus defining

$$H = \sum_{i=1}^{n} \lambda_i^*(t) f_i(t, x_1^*, \cdots, x_n^*, u_1, \cdots, u_m) \tag{15}$$

the extremal control variables $u_1^*, \cdots, u_m^*$ are to be determined so as to maximize H at each t during the process.

$$H^* = H(t, \mathbf{x}^*, \mathbf{u}^*, \boldsymbol{\lambda}^*) \geqslant H(t, \mathbf{x}^*, \mathbf{u}, \boldsymbol{\lambda}^*) \tag{16}$$

In this way the Weierstrass condition for the minimization of our IP is found to be equivalent to Pontryagin's maximum principle (2, 3), the use of which also requires the maximization of H in the above manner for the minimization of the IP.

Note that a necessary condition for maximizing H as given by Eq. (15) is that

$$\left. \frac{\partial H}{\partial u_k} \right|_{u*} = 0 \qquad k = 1, \cdots, m \tag{17}$$

which on using Eq. (15) yields,

$$\sum_{i=1}^{n} \lambda_i(t) \frac{\partial f_i}{\partial u_k} (t, x, u) \bigg|_{*} = 0 \qquad k = 1, \cdots, m \tag{18}$$

The relations of Eq. (18) are seen to be the same as the previously derived control equations of Eq. (6). To ensure that the controls found from Eq. (18) actually maximize H requires, in addition, that the second order terms in the expansion of H about H^* be negative definite, that is

$$\sum_{j=1}^{m} \sum_{k=1}^{m} \frac{\partial^2 H}{\partial u_j \, \partial u_k} \bigg|_{u*} \delta u_j \, \delta u_k < 0 \tag{19}$$

3-9   THE LUNAR ASCENT PROBLEM

To illustrate the use of the conditions developed in Section 3-8 a somewhat simplified lunar ascent problem is treated.

A spaceship on the lunar surface is to be placed in orbit at a specified altitude with a given velocity by appropriately applying its thrust T so as to minimize the time required to achieve the maneuver. To simplify the initial analysis let us assume that the down-range distance that will be covered, although unspecified, is expected to be small enough so that the lunar curvature can be neglected.

The lunar gravity, g_m, can then be considered constant in direction.

Assuming further that the altitude to be achieved is not large compared with the moon's radius, the magnitude of the gravitational acceleration may also be considered constant.

Now defining a coordinate system with the origin fixed on the moon's surface, the x axis along the surface, and the y axis in the vertical direction, as in Fig. 3-2, the equations of motion for the spacecraft are

$$m\ddot{x} = T \cos \theta$$

$$m\ddot{y} = T \sin \theta - mg_m \tag{1}$$

The angle θ is the thrust angle and $m(t)$ is the vehicle's mass.

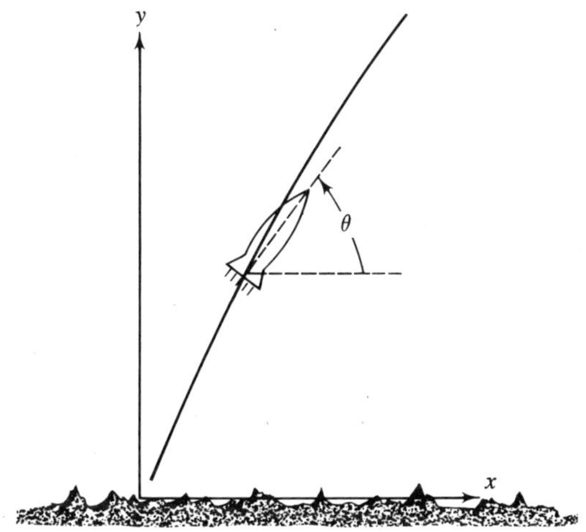

FIGURE 3-2   Geometry used in the lunar ascent problem

As a final restriction let us assume that the thrust acceleration $A \equiv T/m$ is maintained constant. Equation (1) then becomes

$$\ddot{x} = A \cos \theta$$

$$\ddot{y} = A \sin \theta - g_m \tag{2}$$

where θ, the thrust angle, is the only control variable appearing.

The problem is now taken to be that of minimizing the time t_f,

$$\text{IP} = t_f \tag{3}$$

required to go between the surface at time $t = t_i = 0$, where the velocity is zero, to the final altitude h, where the velocity is horizontal and has magnitude U. Thus

$$
\begin{array}{ll}
\text{at } t = t_i = 0 & \text{at } t = t_f \text{ (unspecified)} \\
x = 0 & \\
y = 0 & y = h \\
\dot{x} = 0 & \dot{x} = U \\
\dot{y} = 0 & \dot{y} = 0
\end{array}
\tag{4}
$$

The system equations as given by Eq. (2) are of second-order rather than first-order form, as in Eq. 3-8(1), for which the results were developed. To put them in the proper form let us set $x_1 = x$ and $x_2 = y$. Defining the velocity components $\dot{x}$ and $\dot{y}$ to be x_3 and x_4, respectively, Eq. (2) may be rewritten in terms of x_1, x_2, x_3, and x_4. Thus

$$
\begin{aligned}
\dot{x}_1 &= x_3 &&= f_1 \\
\dot{x}_2 &= x_4 &&= f_2 \\
\dot{x}_3 &= A \cos \theta &&= f_3 \\
\dot{x}_4 &= A \sin \theta - g_m &&= f_4
\end{aligned}
\tag{5}
$$

while the end conditions of Eq. (4) become

$$
\begin{array}{ll}
\text{at } t = t_i = 0 & \text{at } t = t_f \\
x_1 = 0 & \\
x_2 = 0 & x_2 = h \\
x_3 = 0 & x_3 = U \\
x_4 = 0 & x_4 = 0
\end{array}
\tag{6}
$$

Having set the problem in standard form let us find the optimum behavior of the control variable $\theta(t)$, so as to minimize Eq. (1), the time required to achieve the terminal conditions.

Introducing the four multipliers $\lambda_1(t), \cdots, \lambda_4(t)$, the characteristic equations of 3-8(5) give

$$
\begin{aligned}
\dot{\lambda}_1 &= 0 \\
\dot{\lambda}_2 &= 0 \\
\dot{\lambda}_3 &= -\lambda_1 \\
\dot{\lambda}_4 &= -\lambda_2
\end{aligned}
\tag{7}
$$

which may be integrated directly to yield

$$\lambda_1(t) = \bar{\lambda}_1$$

$$\lambda_2(t) = \bar{\lambda}_2$$

$$\lambda_3(t) = -\bar{\lambda}_1 t + \bar{\lambda}_3 \tag{8}$$

$$\lambda_4(t) = -\bar{\lambda}_2 t + \bar{\lambda}_4$$

where $\bar{\lambda}_1$, $\bar{\lambda}_2$, $\bar{\lambda}_3$, and $\bar{\lambda}_4$ are integration constants.

Further, from the control equation of 3-8(6) for the single control variable $\theta(t)$, one finds

$$\lambda_3 \sin \theta - \lambda_4 \cos \theta = 0$$

which, assuming λ_3 and $\cos \theta$ are not zero, gives

$$\tan \theta = \frac{\lambda_4(t)}{\lambda_3(t)} \tag{9}$$

Equation (9) does not completely determine $\theta(t)$ when λ_3 and λ_4 are known, because Eq. (9) only requires

$$\sin \theta = \frac{\pm \lambda_4(t)}{\lambda_3^2(t) + \lambda_4^2(t)} \qquad \cos \theta = \frac{\pm \lambda_3(t)}{\lambda_3^2(t) + \lambda_4^2(t)} \tag{10}$$

As knowledge of $\sin \theta$ and $\cos \theta$ is required for integration of Eq. (5) the determination of the proper sign to be used in Eq. (10) still remains.

The sign is found by applying the maximum principle (Weierstrass condition). Forming H as defined by Eq. 3-8(15), one has

$$H = \lambda_1 x_3 + \lambda_2 x_4 + \lambda_3 A \cos \theta + \lambda_4 (A \sin \theta - g_m) \tag{11}$$

The quantity H is to be maximized at each t by the extremal control $\theta^*(t)$. As not all the terms in H involve the control, let us collect those terms which do and give to them the symbol $\tilde{H}$.

Thus the condition that the control θ^* maximize H at each instant is equivalent to the condition that it maximize $\tilde{H}$ at each instant. From Eq. (11) one sees that

$$\tilde{H} = A(\lambda_3 \cos \theta + \lambda_4 \sin \theta) \tag{12}$$

Using the extremal values for $\sin \theta$ and $\cos \theta$ from Eq. (10), Eq. (12) gives

$$\tilde{H} = A \left[\frac{\lambda_3(\pm \lambda_3)}{\lambda_3^2 + \lambda_4^2} + \frac{\lambda_4(\pm \lambda_4)}{\lambda_3^2 + \lambda_4^2} \right] = \pm A[\lambda_3^2 + \lambda_4^2]^{\frac{1}{2}} \tag{13}$$

Since $\tilde{H}$ is to be maximized, it is apparent that the plus sign must be chosen, from which we conclude

$$\sin \theta^*(t) = \frac{\lambda_4(t)}{\lambda_3^2(t) + \lambda_4^2(t)} \qquad \cos \theta^*(t) = \frac{\lambda_3(t)}{\lambda_3^2(t) + \lambda_4^2(t)} \tag{14}$$

There are nine constants yet to be determined; the four constants λ_1, λ_2, λ_3, and λ_4; the four integration constants that will appear on integrating Eq. (5); and the final time t_f. Available to fix these nine values are the seven conditions of Eq. (6) and the transversality condition.

Let us apply transversality condition as given in the form of Eq. 3-8(11). From Eq. (6), the constraints of Eq. 3-8(11b) are seen to require

$$dt_i = 0$$

$$dx_1(t_i) = 0$$

$$dx_2(t_i) = 0 \qquad dx_2(t_f) = 0 \tag{15}$$

$$dx_3(t_i) = 0 \qquad dx_3(t_f) = 0$$

$$dx_4(t_i) = 0 \qquad dx_4(t_f) = 0$$

with which the transversality condition of Eq. 3-8(11a) reduces to

$$0 = dt_f - \sum_{i=1}^{4} \lambda_i f_i \bigg|_{t_f} dt_f + \lambda_1(t_f)\, dx_1(t_f) \tag{16}$$

Since dt_f and $dx_1(t_f)$ are independent differentials their coefficients must vanish to satisfy Eq. (16) for all possible dt_f, $dx_1(t_f)$. Thus

$$\lambda_1(t_f) = 0 \tag{17a}$$

$$[\lambda_1 x_3 + \lambda_2 x_4 + \lambda_3 A \cos \theta + \lambda_4(A \sin \theta - g_m)]\big|_{t_f} = 1 \tag{17b}$$

The two conditions of Eq. (17) along with the seven conditions of Eq. (6) provide the required nine conditions for determining the nine unknown constants remaining.

By Eq. (8) λ_1 is a constant, $\lambda_1(t) = \bar{\lambda}_1$. Use of Eq. (17a) then shows that $\lambda_1(t) = \bar{\lambda}_1 = 0$. With this result the expression for $\lambda_3(t)$ in Eq. (8) yields $\lambda_3(t) = \bar{\lambda}_3$. Thus only λ_4 remains a function of time, $\lambda_4(t) = -\bar{\lambda}_2 t + \bar{\lambda}_4$.

Going back a moment, note that in Eq. (14), by dividing numerator and denominator by $\bar{\lambda}_3$, the expressions for $\sin \theta^*$ and $\cos \theta^*$ can be seen to only depend on the ratio λ_4/λ_3. Defining the constants α and β by

$$\tan \theta^* = \frac{\lambda_4(t)}{\lambda_3(t)} = \frac{-\bar{\lambda}_2 t + \bar{\lambda}_4}{\bar{\lambda}_3} = \alpha t + \beta \tag{18}$$

then Eq. (14) can be written

$$\sin \theta^*(t) = \frac{\alpha t + \beta}{[1 + (\alpha t + \beta)^2]^{1/2}} \qquad \cos \theta^*(t) = \frac{1}{[1 + (\alpha t + \beta)^2]^{1/2}} \qquad (19)$$

Finally, the relations for $\sin \theta^*$ and $\cos \theta^*$ from Eq. (19) are inserted into the motion equations of Eq. (5), yielding

$$\dot{x}_1 = x_3$$

$$\dot{x}_2 = x_4$$

$$\dot{x}_3 = \frac{A}{[1 + (\alpha t + \beta)^2]^{1/2}} \qquad (20)$$

$$\dot{x}_4 = \frac{A(\alpha t + \beta)}{[1 + (\alpha t + \beta)^2]^{1/2}} - g_m$$

Integration of these four first-order equations introduces four integration constants which with α, β, and t_f are determined by the seven conditions of Eq. (6).†

Integration of the motion equations in the form given by Eq. (20), subject to the end conditions of Eqs. (6) and (17), would complete the solution. It is actually easier, however, to integrate the equations of motion by changing the independent variable from time to the control angle θ. To do this the control program of Eq. (18) is differentiated with respect to time, yielding

$$\dot{\theta}^* = \alpha \cos^2 \theta^* \qquad (21)$$

Dividing the left and right sides of Eq. (5) by Eq. (21), the desired equations are obtained:

$$\frac{dx_1}{d\theta^*} = \frac{x_3}{\alpha \cos^2 \theta^*} \qquad (22\text{a})$$

$$\frac{dx_2}{d\theta^*} = \frac{x_4}{\alpha \cos^2 \theta^*} \qquad (22\text{b})$$

$$\frac{dx_3}{d\theta^*} = \frac{A}{\alpha \cos \theta^*} \qquad (22\text{c})$$

$$\frac{dx_4}{d\theta^*} = \frac{A \sin \theta^*}{\alpha \cos^2 \theta^*} - \frac{g_m}{\alpha \cos^2 \theta^*} \qquad (22\text{d})$$

† Once $\alpha = -\lambda_2/\lambda_3$ and $\beta = \lambda_4/\lambda_3$ have been found, the separate values of λ_2, λ_3, and λ_4 could be found by employing the remaining transversality condition Eq. (17b). There is, however, no reason to do this.

At $t = t_i = 0$ and $t = t_f$, θ^* takes on the values θ_i^* and θ_f^*, respectively, which are to be found. The conditions of Eq. (6) then become

$$
\begin{array}{ll}
\text{at} \quad t = t_i = 0 & \text{at} \quad t = t_f \ (\text{unspecified}) \\
\theta^* = \theta_i^* \ (\text{unspecified}) & \theta^* = \theta_f^* \ (\text{unspecified})
\end{array}
$$

$$
\begin{array}{ll}
x_1 = 0 & \\
x_2 = 0 & x_2 = h \\
x_3 = 0 & x_3 = U \\
x_4 = 0 & x_4 = 0
\end{array}
\tag{23}
$$

where from Eq. (18) t and θ^* are related by

$$
\tan \theta^*(t) = \alpha t + \beta
\tag{24}
$$

Integrating Eq. (22c, d), one finds on applying the initial conditions on x_3 and x_4 that

$$
\alpha x_3 = A \ln \left[\frac{\tan \theta^* + \sec \theta^*}{\tan \theta_i^* + \sec \theta_i^*} \right]
\tag{25a}
$$

$$
\alpha x_4 = -g_m(\tan \theta^* - \tan \theta_i^*) + A(\sec \theta^* - \sec \theta_i^*)
\tag{25b}
$$

Using Eq. (25a, b) in Eqs. (22a, b) and integrating, one finds after application of the initial conditions on x_1 and x_2 and some manipulation that

$$
\alpha^2 x_1 = \int_{\theta_i^*}^{\theta^*} \frac{A}{\cos^2 \theta^*} \ln \left[\frac{\tan \theta^* + \sec \theta^*}{\tan \theta_i^* + \sec \theta_i^*} \right] d\theta^*
\tag{26a}
$$

$$
2\alpha^2 x_2 = g_m[2 \tan \theta_i^*(\tan \theta^* - \tan \theta_i^*) + \tan^2 \theta_i^* - \tan^2 \theta^*] \\
+ A[\tan \theta^* \sec \theta^* + \sec \theta_i^*(\tan \theta_i^* - 2 \tan \theta^*)] + \alpha x_3
\tag{26b}
$$

The terminal conditions remain to be satisfied. Since $x_2 = h$, $x_3 = U$, and $x_4 = 0$ at $\theta^* = \theta_f^*$, Eqs. (25a, b) and (26b) give, respectively,

$$
\frac{A}{g_m} = \frac{\tan \theta_f^* - \tan \theta_i^*}{\sec \theta_f^* - \sec \theta_i^*} = \frac{\cos(\theta_f^* - \theta_i^*)/2}{\sin(\theta_f^* - \theta_i^*)/2}
\tag{27a}
$$

$$
\alpha U = A \ln \left[\frac{\tan \theta_f^* + \sec \theta_f^*}{\tan \theta_i^* + \sec \theta_i^*} \right]
\tag{27b}
$$

$$
2\alpha^2 h = g_m[2 \tan \theta_i^*(\tan \theta_f^* - \tan \theta_i^*) + \tan^2 \theta_i^* - \tan \theta_f^*] \\
+ A[\tan \theta_f^* \sec \theta_f^* + \sec \theta_i^*(\tan \theta_i^* - 2 \tan \theta_f^*)] + \alpha U
\tag{27c}
$$

Equations (27a, b, c) are three equations for the determination of the three unknowns θ_i^*, θ_f^*, and α. Once they have been found the constant β is determined by setting $t = 0$ in Eq. (24). Thus

$$
\beta = \tan \theta_i^*
\tag{28}
$$

Since α, β, and θ_f^* are known, the final time t_f is obtained by evaluating Eq. (24) at $t = t_f$, yielding

$$t_f = \frac{\tan \theta_f^*}{\alpha} - \frac{\beta}{\alpha} \tag{29}$$

Determination of t_f completes the solution to the problem.

Results are presented in Fig. 3-3 and Fig. 3-4 for the particular set of terminal conditions $h = 50{,}000$ ft and $U = 5444$ ft/sec. The velocity U chosen is close to the velocity required to sustain the vehicle in a circular orbit at the altitude h. The lunar gravity was taken to be 5.32 ft/sec^2 and the vehicle was assumed to produce the constant thrust acceleration $A = 20.821$ ft/sec^2. The vehicle's thrust to lunar weight ratio, A/g_m, is thus slightly less than four.

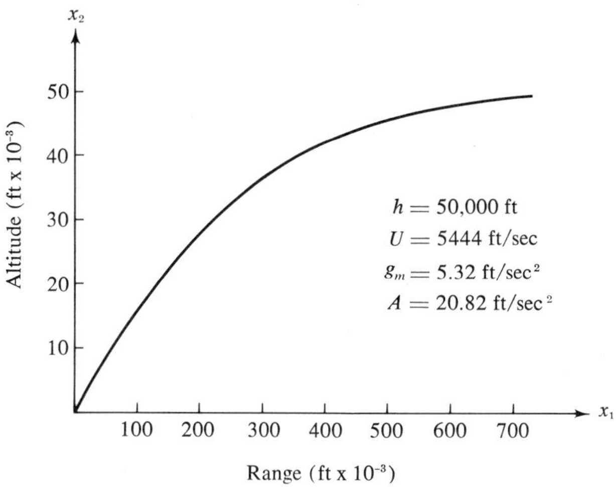

FIGURE 3-3   Optimal trajectory profile

For these values the solution of Eq. (27) yielded $\theta_i^* = 26°$, $\theta_f^* = 3°$, and $\alpha = -1.598 \times 10^{-3}$ sec^{-1}. Using these results in Eqs. (28) and (29) one finds $\beta = 0.4877$ and $t_f = 272.4$ sec. In Fig. 3-3 altitude as a function of down-range distance is given, and Fig. 3-4 shows the variation of the thrust angle with time.

Before concluding the example, let us review what has been done. With α and β determined, Eq. (24) provides the thrust angle, given as a function of time, required to transfer the spacecraft from the given initial state on the surface to the desired end point in orbit. In terms of the discussion of Chapter 1, the control function for the problem has been found.

What has not been done is to determine the control behavior which would be required to go in an optimal manner from any arbitrary initial state to the desired end state. This would be the control law for the problem.

Why is the latter desired? For the following reason. In the absence of system disturbances the spacecraft would follow the extremal trajectory with the thrust direction at any instant being given by the control function of Eq. (24). Suppose at some time t along the extremal trajectory one asked how to thrust from then on so as to reach the desired end state in minimum time. The answer for arbitrary t would be to continue following the control function originally found. No additional gain is obtained or expected.

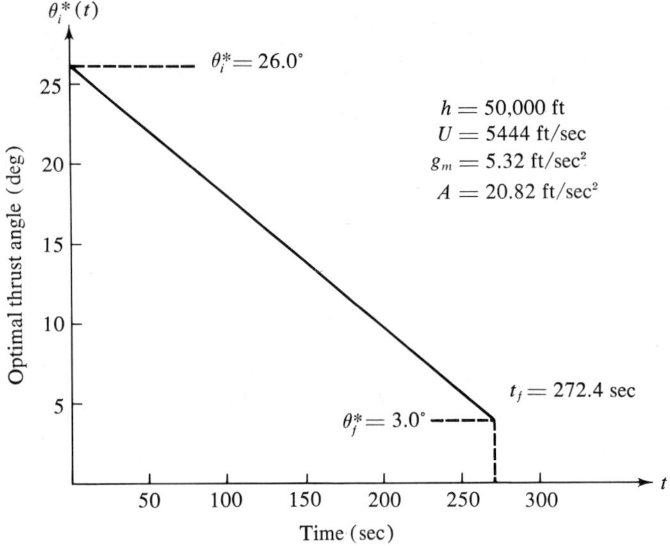

FIGURE 3-4   Optimal thrust-angle program

On the other hand, suppose more realistically that some errors or disturbances are present in the system. In this case the spacecraft will not remain on the extremal trajectory. If one continues to follow the control function it cannot be expected that the desired end conditions will be met. What would be most useful is knowledge of how to apply the control so as to go from the existing state off the extremal trajectory to the desired end point. This information is contained in the control law for the problem.

In Chapter 8 a means of finding the control law is developed for states which are in the neighborhood of the extremal trajectory as found here. This problem is termed the guidance problem.

REFERENCES

1.   G. A. Bliss, *Lectures on the Calculus of Variations*, Univ. Chicago Press, Chicago, 1946, Chap. 7.
2.   L. S. Pontryagin et al., *The Mathematical Theory of Optimal Processes*, Wiley-Interscience, New York, 1962.
3.   L. I. Rozonoer, "L. S. Pontryagin's Maximum Principle in the Theory of Optimum Systems—Parts I, II, and III" (Engl. Trans.), *Automat. Remote Control*, Vol. 20, Nos. 10, 11, and 12, Oct., Nov., and Dec., 1959, pp. 1288–1302, 1405–1421, 1517–1532.

4

THE EXTENSION TO
CONTROL-VARIABLE
INEQUALITY CONSTRAINTS

Let us proceed by first considering several examples that further illustrate the theory already developed.

4-1 A SIMPLE LINEAR PLANT

Consider the system governed by the plant equation

$$\frac{d^2x}{dt^2} + \frac{dx}{dt} = u \tag{1}$$

The system objective is taken as that of transferring the initial state point of the plant to a prescribed terminal state in such a manner as to minimize the performance index

$$\mathrm{IP} = \int_0^{t_f} \tfrac{1}{2}[x^2 + \dot{x}^2 + \mu^2 u^2]\, dt \tag{2}$$

The system can be shown to be totally controllable (*1*) so that an arbitrary initial point in state space may be chosen.

To put the governing second-order equations of (1) in standard first-order form, let $x_1 = x$ and $x_2 = \dot{x}$. Equation (1) is then equivalent to the set of first-order equations

$$\dot{x}_1 = x_2 = f_1$$

$$\dot{x}_2 = -x_2 + u = f_2 \tag{3}$$

If the quantity x is taken to be a system error, then x_1 and x_2 are the error and error rate, respectively. For the end conditions

$$t = t_i = 0 \qquad t = t_f \text{ (specified)}$$

$$x_1(t_i) = x_{1_0} \qquad x_1(t_f) = 0 \tag{4}$$

$$x_2(t_i) = x_{2_0} \qquad x_2(t_f) = 0$$

the problem may be interpreted as that of determining the control required to drive the error and error rate to zero while minimizing the IP of Eq. (2), which is seen to depend on the amount of control used.

Defining the new variable $x_3(t)$ by

$$\dot{x}_3 = \tfrac{1}{2}[x_1^2 + x_2^2 + \mu^2 u^2] = f_3$$

$$x_3(t_i) = 0 \tag{5}$$

Equation (2) becomes

$$\text{IP} = x_3(t_f) \tag{6}$$

which is the desired form for application of the theory.

Introducing $\lambda_1(t)$, $\lambda_2(t)$, and $\lambda_3(t)$ the adjoint equations of 3-8(5) give

$$\dot{\lambda}_1 = -\lambda_3 x_1 \tag{7a}$$

$$\dot{\lambda}_2 = -\lambda_1 + \lambda_2 - \lambda_3 x_2 \tag{7b}$$

$$\dot{\lambda}_3 = 0 \tag{7c}$$

and the control equation, 3-8(6), becomes

$$-\lambda_2 - \lambda_3 \mu^2 u = 0 \tag{8}$$

In addition to the end conditions contained in Eqs. (4) and (5), the transversality condition of Eq. 3-8(11) requires

$$0 = dx_3(t_f) + \lambda_3 \, dx_3 \big|_{t_f} \tag{9}$$

which yields

$$\lambda_3(t_f) = -1 \tag{10}$$

Equation (10), with Eq. (7c), is seen to give $\lambda_3(t) = -1$. Thus from Eq. (8) the control $u(t)$ is found to be

$$u(t) = \frac{\lambda_2(t)}{\mu^2} \tag{11}$$

Collecting the equations that remain to be solved we have

$$\dot{x}_1 = x_2$$

$$\dot{x}_2 = -x_2 + \frac{\lambda_2}{\mu^2}$$

$$\dot{\lambda}_1 = x_1 \tag{12}$$

$$\dot{\lambda}_2 = -\lambda_1 + \lambda_2 + x_2$$

or in matrix notation using $z_1 = x_1$, $z_2 = x_2$, $z_3 = \lambda_1$, $z_4 = \lambda_2$,

$$\dot{z} = Mz \tag{13}$$

where the matrix M is

$$M = \begin{bmatrix} 0 & 1 & 0 & 0 \\ 0 & -1 & 0 & \frac{1}{\mu^2} \\ 1 & 0 & 0 & 0 \\ 0 & 1 & -1 & 1 \end{bmatrix} \tag{14}$$

The solution to Eq. (13) is found† through use of the matrix characteristic equation given by

$$\det |M - rI| = 0$$

which in this case yields

$$r^4 - r^2\left(1 + \frac{1}{\mu^2}\right) + \frac{1}{\mu^2} = 0 \tag{15}$$

Therefore the eigenvalues are

$$r_1 = 1 \qquad r_2 = -1 \qquad r_3 = \frac{1}{\mu} \qquad r_4 = -\frac{1}{\mu} \tag{16}$$

and provided‡ $\mu^2 \neq 1$, the solution for z_i is a linear combination of the eigenfunctions

$$\varphi_1 = e^t \qquad \varphi_2 = e^{-t} \qquad \varphi_3 = e^{t/\mu} \qquad \varphi_4 = e^{-t/\mu} \tag{17}$$

---

† Alternatively, a solution to Eq. (13) may be found by first eliminating variables until a single equation on one of the variables is obtained. Differentiating the $\dot{x}_1$ equation three times one finds on using the remaining relations that

$$\ddddot{x}_1 - (1 + 1/\mu^2)\ddot{x}_1 + \frac{1}{\mu^2}x_1 = 0$$

The variable $x_2(t)$ is then obtained by differentiating the solution for $x_1(t)$. With $x_2(t)$ known, the equation for $\dot{x}_2$ provides $\lambda_2(t)$. The relation for $\dot{\lambda}_2$ may then be solved for $\lambda_1(t)$.

‡ If $\mu^2 = 1$, the z_i are linear combinations of the functions e^t, e^{-t}, te^t, and te^{-t}.

Writing

$$z_j(t) = \sum_{i=1}^{4} C_j^{(i)}\varphi_i(t) \qquad j = 1, \cdots, 4 \tag{18}$$

the constants $C_j^{(i)}$, which are the elements of the column vector $C^{(i)}$, must satisfy the matrix equation

$$r_i C^{(i)} = M C^{(i)} \tag{19}$$

obtained by using Eq. (18) in Eq. (13). For example, for $r_i = r_1 = 1$, Eq. (19) becomes

$$C_1^{(1)} = C_2^{(1)}$$

$$C_2^{(1)} = -C_2^{(1)} + \frac{C_4^{(1)}}{\mu^2} \tag{20}$$

$$C_3^{(1)} = C_1^{(1)}$$

$$C_4^{(1)} = C_2^{(1)} - C_3^{(1)} + C_4^{(1)}$$

which yields

$$C^{(1)} = \begin{bmatrix} C_1^{(1)} \\ C_1^{(1)} \\ C_1^{(1)} \\ 2\mu^2 C_1^{(1)} \end{bmatrix} \tag{21}$$

In a similar manner we find

$$C^{(2)} = \begin{bmatrix} C_1^{(2)} \\ -C_1^{(2)} \\ -C_1^{(2)} \\ 0 \end{bmatrix} \quad C^{(3)} = \begin{bmatrix} C_1^{(3)} \\ \frac{1}{\mu} C_1^{(3)} \\ \mu C_1^{(3)} \\ (1 + \mu) C_1^{(3)} \end{bmatrix} \quad C^{(4)} = \begin{bmatrix} C_1^{(4)} \\ -\frac{1}{\mu} C_1^{(4)} \\ -\mu C_1^{(4)} \\ (1 - \mu) C_1^{(4)} \end{bmatrix} \tag{22}$$

Thus the solution to Eq. (13) is

$$x_1(t) = C_1^{(1)} e^t + C_1^{(2)} e^{-t} + C_1^{(3)} e^{t/\mu} + C_1^{(4)} e^{-t/\mu}$$

$$x_2(t) = C_1^{(1)} e^t - C_1^{(2)} e^{-t} + \frac{1}{\mu} C_1^{(3)} e^{t/\mu} - \frac{1}{\mu} C_1^{(4)} e^{-t/\mu} \tag{23}$$

$$\lambda_1(t) = C_1^{(1)} e^t - C_1^{(2)} e^{-t} + \mu C_1^{(3)} e^{t/\mu} - \mu C_1^{(4)} e^{-t/\mu}$$

$$\lambda_2(t) = 2\mu^2 C_1^{(1)} e^t + 0 + (1 + \mu) C_1^{(3)} e^{t/\mu} + (1 - \mu) C_1^{(4)} e^{-t/\mu}$$

where the four constants $C_1^{(1)}$, $C_1^{(2)}$, $C_1^{(3)}$, and $C_1^{(4)}$ are to be determined so as to satisfy the four conditions of Eq. (4). Once this has been done the open-loop optimal control is obtained from Eq. (11) on using $\lambda_2(t)$ from Eq. (23).

4-2 CONTROL OF A CHEMICAL REACTOR

Consider the problem of maximizing the final amount of the product y during a two-stage chemical reaction in which $x \to y \to z$. Assuming first-order kinetics for the x and y rates, the reaction rates are taken to be

$$\dot{x} = -ax \qquad = f_1 \tag{1a}$$

$$\dot{y} = ax - by = f_2 \tag{1b}$$

$$\dot{z} = g(x, y, z) = f_3 \tag{1c}$$

where the rate coefficients a, b are related[†] so that

$$b = \rho a^k \tag{2}$$

with ρ and k positive constants.

The amount of the waste product z formed is seen not to influence the x and y reactions. As the magnitude of z is not of interest, Eq. (1c) may be dropped from further consideration.

Given that $y(t_f)$ is to be maximized by proper choice of the control $a(t)$, the IP to be minimized becomes

$$\text{IP} = -y(t_f) \tag{3}$$

where the end conditions for the problem are

$$\text{at } t = t_i = 0 \qquad t = t_f \text{ (specified)}$$

$$x(t_i) = x_0 \tag{4}$$

$$y(t_i) = y_0$$

Introducing the multipliers $\lambda_1(t)$ and $\lambda_2(t)$, the corresponding adjoint equations are

$$\dot{\lambda}_1 = a(\lambda_1 - \lambda_2)$$

$$\dot{\lambda}_2 = b\lambda_2 \tag{5}$$

---

† A common form for the rate coefficients (Arrhenius relationship) is $a = k_1 \exp(-E_1/RT)$ and $b = k_2 \exp(-E_2/RT)$, where E_1 and E_2 are activation energies, T the absolute reaction temperature, and R the universal gas constant. With these assumed forms for a and b, relation (2) follows.

and the control equation for the determination of the control $a(t)$ is

$$-\lambda_1 x + \lambda_2 x - \lambda_2 y \frac{db}{da} = 0 \tag{6}$$

Using Eq. (2) to find db/da, Eq. (6) yields the control function

$$a = \left[\frac{1}{\rho k}\frac{x}{y}\left(\frac{\lambda_2 - \lambda_1}{\lambda_2}\right)\right]^{1/(k-1)} \tag{7}$$

In addition to the end conditions of Eq. (4), two additional end conditions must be given by the transversality condition. The four conditions then specify the four integration constants of Eqs. (1a, b) and (5). For this problem the transversality condition requires

$$0 = -dy(t_f) + \lambda_1(t_f)\,dx(t_f) + \lambda_2(t_f)\,dy(t_f) \tag{8}$$

Since $dx(t_f)$ and $dy(t_f)$ are independent differentials, Eq. (8) yields

$$\lambda_1(t_f) = 0 \tag{9a}$$

$$\lambda_2(t_f) = 1 \tag{9b}$$

which are the additional conditions needed to find the solution.

For the control found from Eq. (7) to be the one which minimizes the IP of Eq. (3) the maximum principle (Weierstrass condition) must be satisfied. Forming H,

$$H = \lambda_1(-ax) + \lambda_2(ax - by) \tag{10}$$

at an interior maximum of $H(a)$ the conditions $\partial H/\partial a = 0$ and $\partial^2 H/\partial a^2 \leqslant 0$ must be met. Setting $\partial H/\partial a = 0$ yields the extremal control given in Eq. (7), while $\partial^2 H/\partial a^2$ as obtained from Eq. (10) is

$$\frac{\partial^2 H}{\partial a^2} = -\frac{d^2 b}{da^2}\lambda_2 y = -\rho k(k-1)a^{k-2}\lambda_2 y \tag{11}$$

The quantity y in Eq. (11) is positive by definition. Solving for $\lambda_2(t)$, using Eqs. (5b) and (9b), yields $\lambda_2(t) = \exp[-\int_t^{t_f} b\,dt]$, which is also a positive quantity. Assume for the present that $\lambda_2(t) > \lambda_1(t)$, so that the extremal control of Eq. (7) exists. Since ρ and k are positive, the condition that $\partial^2 H/\partial a^2 < 0$ at the maximum is seen, by Eq. (11), to be satisfied provided $k > 1$. If $k < 1$ the control of Eq. (7) does not maximize H as desired but actually minimizes it. The behavior of $H(a)$ is illustrated in Fig. 4-1 for both cases.

Consider the system when $k < 1$. Reference to Fig. 4-1 shows that the optimal control, that is, the control which maximizes $H(a)$, is infinite. Since the control available in any engineering problem is limited, this is obviously not a physically realizable solution. The solution when $k < 1$ is thus deferred until Section 4-7 while the theory necessary to treat problems with control limitations is developed.

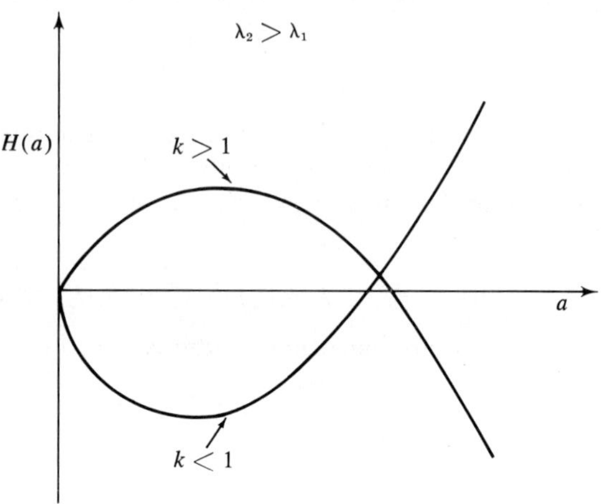

FIGURE 4-1   Behavior of $H(a)$ $(\lambda_2 > \lambda_1)$

To proceed with the solution when $k > 1$, let us summarize what remains to be done. The set of first-order, nonlinear differential equations

$$\dot{x} = -ax$$
$$\dot{y} = ax - \rho a^k y$$
$$\dot{\lambda}_1 = a(\lambda_1 - \lambda_2)$$
$$\dot{\lambda}_2 = \rho a^k \lambda_2$$

(12)

where

$$a = \left[\frac{1}{\rho k} \frac{x}{y} \frac{\lambda_2 - \lambda_1}{\lambda_2} \right]^{1/(k-1)}$$

(13)

are to be solved subject to the end conditions

$$\begin{array}{ll}
\text{at } t = t_i = 0 & \text{at } t = t_f \text{ (specified)} \\
x(t_i) = x_0 & \lambda_1(t_f) = 0 \\
y(t_i) = y_0 & \lambda_2(t_f) = 1
\end{array}$$

(14)

4-3   A TWO-POINT BOUNDARY-VALUE PROBLEM

If Eq. 4-2(12) could be integrated analytically, four integration constants would be generated to satisfy the four end conditions of Eq. 4-2(14). However, because of the nonlinear nature of the equations, it is unlikely that Eq. 4-2(12) can be integrated analytically. Resort must thus be made to numerical methods of integration to obtain the solution.

To show what is involved, imagine that a first-order integration scheme is chosen with an integration step size of Δt. Then knowing x_0, y_0, $\lambda_1(0) = \lambda_{1_0}$, and $\lambda_2(0) = \lambda_{2_0}$, the variables at time Δt are given by

$$x(\Delta t) = x_0 + \dot{x}(0)\,\Delta t$$
$$y(\Delta t) = y_0 + \dot{y}(0)\,\Delta t$$
$$\lambda_1(\Delta t) = \lambda_{1_0} + \dot{\lambda}_1(0)\,\Delta t$$
$$\lambda_2(\Delta t) = \lambda_{2_0} + \dot{\lambda}_2(0)\,\Delta t$$

$$(1)$$

where Eqs. 4-2(12) and 4-2(13) determine the rates at $t = 0$. Using the values of $z_i(\Delta t)$, where z_i denotes one of x, y, λ_1, or λ_2, the values at $t = 2\,\Delta t$ are obtained in a similar manner from $z_i(2\,\Delta t) = z_i(\Delta t) + \dot{z}_i(\Delta t)\,\Delta t$ or in general

$$z_i[(n + 1)\,\Delta t)] = z_i(n\,\Delta t) + \dot{z}_i(n\,\Delta t)\,\Delta t \qquad (2)$$

To complete the solution the procedure is continued until the specified end time t_f.

All would thus seem to be well. However, examination of the end conditions of Eq. 4-2(14) discloses that initial values of only two of the variables are known rather than all four, as required by the solution method described. The remaining two conditions are specified at the terminal time t_f.

To proceed, since x_0 and y_0 are given, an educated guess is made for λ_{1_0} and λ_{2_0} and, using Eq. (2), Eqs. 4-2(12) are integrated numerically until $t = t_f$. The values of $\lambda_1(t_f)$ and $\lambda_2(t_f)$ computed are then compared with the desired end values as given by Eq. 4-2(14). If the values computed agree with the values desired, the solution has been found.

However, as the starting values for λ_1 and λ_2 were chosen in a fairly arbitrary manner, the likelihood that this will occur is small. If the desired and computed values do not agree, one must return to $t = 0$; choose new values for λ_{1_0} and λ_{2_0}; integrate Eq. 4-2(12) until $t = t_f$; and then once again see if the desired end conditions have been met. The reader can readily see that if the initial values are chosen by only educated guesswork, there may be a long wait until the end conditions on λ_1 and λ_2 are met. A systematic method of converging on the required values is needed.

Such a method is given in what follows.† It is certainly not the most elegant approach to solving these problems, termed *two-point boundary-value problems*, but it has certain advantages. It illustrates what is required; it is easily understandable and, finally, it works for this problem; that is, it provides convergence to the desired values.

Since the values of $\lambda_1(t_f)$ and $\lambda_2(t_f)$ obtained by integration change with the choice of the initial values, one may write

$$\lambda_1(t_f) = \lambda_{1_f}[\lambda_{1_0}, \lambda_{2_0}]$$

$$\lambda_2(t_f) = \lambda_{2_f}[\lambda_{1_0}, \lambda_{2_0}]$$

(3)

Forming the differential of Eq. (3) then yields

$$d\lambda_{1_f} = \frac{\partial \lambda_{1_f}}{\partial \lambda_{1_0}} d\lambda_{1_0} + \frac{\partial \lambda_{1_f}}{\partial \lambda_{2_0}} d\lambda_{2_0} \tag{4a}$$

$$d\lambda_{2_f} = \frac{\partial \lambda_{2_f}}{\partial \lambda_{1_0}} d\lambda_{1_0} + \frac{\partial \lambda_{2_f}}{\partial \lambda_{2_0}} d\lambda_{2_0} \tag{4b}$$

Assume that an initial guess for λ_{1_0} and λ_{2_0}, denoted by $\lambda_{1_0}^{(0)}$ and $\lambda_{2_0}^{(0)}$, has been made. The resulting terminal values of λ_1 and λ_2, denoted by $\lambda_{1_f}^{(0)}$ and $\lambda_{2_f}^{(0)}$, are found by integration of Eq. 4-2(12). In general, $\lambda_{1_f}^{(0)} \neq 0$ and $\lambda_{2_f}^{(0)} \neq 1$, which are the desired values of Eq. 4-2(14).

Now consider Eq. (4), assuming for the moment that the partial derivatives $\partial \lambda_{1_f}/\partial \lambda_{1_0}$, $\partial \lambda_{1_f}/\partial \lambda_{2_0}$, $\partial \lambda_{2_f}/\partial \lambda_{1_0}$, and $\partial \lambda_{2_f}/\partial \lambda_{2_0}$ are known. Equations (4a, b) are then two simultaneous equations for the determination of $d\lambda_{1_0}^{(0)}$ and $d\lambda_{2_0}^{(0)}$, the desired change in the initial values. Since the required terminal values are $\lambda_{1_f} = 0$ and $\lambda_{2_f} = 1$, one takes for use in Eq. (4),

$$d\lambda_{1_f}^{(0)} = 0 - \lambda_{1_f}^{(0)}$$

$$d\lambda_{2_f}^{(0)} = 1 - \lambda_{2_f}^{(0)}$$

(5)

Using Eq. (5), Eq. (4) may be solved for $d\lambda_{1_0}^{(0)}$ and $d\lambda_{2_0}^{(0)}$.

The values of $d\lambda_{1_0}^{(0)}$ and $d\lambda_{2_0}^{(0)}$, when added to the original guesses, $\lambda_{1_0}^{(0)}$ and $\lambda_{2_0}^{(0)}$, provide new estimates:

$$\lambda_{1_0}^{(1)} = \lambda_{1_0}^{(0)} + d\lambda_{1_0}^{(0)}$$

$$\lambda_{2_0}^{(1)} = \lambda_{2_0}^{(0)} + d\lambda_{2_0}^{(0)}$$

(6)

of the initial values required for λ_1 and λ_2 so as to meet the specified end values. In this manner one returns to the beginning point of the discussion following

---

† An alternative method is outlined as an example in Section 6-7 following the introduction of the concept of adjoint differential forms.

Eq. (4). If after integration of Eq. 4-2(12) the new starting values $\lambda_{1_0}^{(1)}$ and $\lambda_{2_0}^{(1)}$ provide values $\lambda_{1_f}^{(1)}$ and $\lambda_{2_f}^{(1)}$ which meet the end conditions of Eq. 4-2(14) with sufficient accuracy the problem is solved.† If not, the procedure must be repeated until the nth approximation $\lambda_{1_f}^{(n)}$ and $\lambda_{2_f}^{(n)}$ does meet the end conditions.

The four first partial derivatives in Eq. (4), which were assumed known, must now be evaluated to complete the discussion of the first approximation. This is done in a straightforward manner.

Recall that $\lambda_{1_0}^{(0)}$ and $\lambda_{2_0}^{(0)}$ were the original estimates (guesses) for the starting values and that the terminal values $\lambda_{1_f}^{(0)}$ and $\lambda_{2_f}^{(0)}$ resulted from using them. Now let us integrate Eq. 4-2(12) using as initial conditions the value $\lambda_{1_0}^{(0)} + \Delta\lambda_{1_0}^{(0)}$ and $\lambda_{2_0}^{(0)}$, where $\Delta\lambda_{1_0}^{(0)}/\lambda_{1_0}^{(0)} \ll 1$.

From this integration one finds at t_f the values $\bar{\lambda}_{1_f}$ and $\bar{\lambda}_{2_f}$. Thus the change $\Delta\lambda_{1_0}^{(0)}$ in the initial value of λ_1 has caused a change in the terminal values of λ_1 and λ_2 given by

$$d\lambda_{1_f} = \bar{\lambda}_{1_f} - \lambda_{1_f}^{(0)}$$

$$d\lambda_{2_f} = \bar{\lambda}_{2_f} - \lambda_{2_f}^{(0)}$$

(8)

Then from Eq. (4) one sees, since $d\lambda_{2_0} = 0$, that

$$\frac{\partial\lambda_{1_f}}{\partial\lambda_{1_0}^{(0)}} = \frac{\bar{\lambda}_{1_f} - \lambda_{1_f}^{(0)}}{\Delta\lambda_{1_0}^{(0)}} \qquad \frac{\partial\lambda_{2_f}}{\partial\lambda_{1_0}^{(0)}} = \frac{\bar{\lambda}_{2_f} - \lambda_{2_f}^{(0)}}{\Delta\lambda_{1_0}^{(0)}}$$

(9)

In a similar manner, using as initial conditions on λ_1 and λ_2 the values $\lambda_{1_0}^{(0)}$ and $\lambda_{2_0}^{(0)} + \Delta\lambda_{2_0}^{(0)}$, where $\Delta\lambda_{2_0}^{(0)}/\lambda_{2_0}^{(0)} \ll 1$, one finds after integration of Eq. 4-2(12), the terminal values $\bar{\bar{\lambda}}_{1_f}$ and $\bar{\bar{\lambda}}_{2_f}$. Then from Eq. (4) one has, since now $d\lambda_1(0) = 0$, that

$$\frac{\partial\lambda_{1_f}}{\partial\lambda_{2_0}^{(0)}} = \frac{\bar{\bar{\lambda}}_{1_f} - \lambda_{1_f}^{(0)}}{\Delta\lambda_{2_0}^{(0)}} \qquad \frac{\partial\lambda_{2_f}}{\partial\lambda_{2_0}^{(0)}} = \frac{\bar{\bar{\lambda}}_{2_f} - \lambda_{2_f}^{(0)}}{\Delta\lambda_{2_0}^{(0)}}$$

(10)

This method of finding the first partial derivatives in Eq. (4) completes the discussion of the technique presented for the solution of two-point boundary-value problems.

---

† If the relations of Eq. (5) were linear, the first iteration for $\lambda_{1_0}^{(1)}$ and $\lambda_{2_0}^{(1)}$ would exactly meet the end conditions $\lambda_{1_f} = 0$ and $\lambda_{2_f} = 1$. However, because of the nonlinear nature of Eqs. 4-2(12) and 4-2(13), Eqs. (3) are definitely not linear in form. Hence the reason multiple iterations are required. For the same reason, one does not usually try to achieve the entire change in $\lambda_{1_f}^{(0)}$ and $\lambda_{2_f}^{(0)}$ in one step, as in Eq. (5). To avoid possible oscillations in convergence, one takes only a fraction of the required change at a time until the total change needed is achieved. Thus instead of Eq. (5) one would use

$$d\lambda_{1_f}^{(0)} = \alpha(0 - \lambda_{1_f}^{(0)}) \qquad 0 < \alpha \leq 1$$

$$d\lambda_{2_f}^{(0)} = \beta(1 - \lambda_{2_f}^{(0)}) \qquad 0 < \beta \leq 1$$

(7)

As a numerical example to illustrate the solution method, the case in which $x_0 = 1.0, y_0 = 0.01$, and $t_f = 2.0$ is presented first. To aid in our choice of $\lambda_{1_0}^{(0)}$ and $\lambda_{2_0}^{(0)}$ let us consider the behavior of $\lambda_1(t)$ and $\lambda_2(t)$. By Eqs. 4-2(12) and 4-2(14), $\lambda_2(t)$ is seen to be a positive, monotonically increasing function of time which takes on the terminal value $\lambda_{2_f} = 1$. Thus $0 < \lambda_{2_0}^{(0)} < 1$. Further, since a is real, by Eq. 4-2(13), $\lambda_2(t) > \lambda_1(t)$. With this knowledge, Eq. 4-2(12) shows that $\lambda_1(t)$ decreases monotonically in time to the terminal value $\lambda_{1_f} = 0$. Thus $1 \geqslant \lambda_2(t) > \lambda_1(t) \geqslant 0$.

On the basis of the information deduced, the values $\lambda_{1_0}^{(0)} = 0.05$ and $\lambda_{2_0}^{(0)} = 0.10$ were chosen. Summarized in Table 4-3(1) are the results of apply-

TABLE 4-3(1)   *Convergence to the Starting Conditions for the Chemical-Reactor Problem*

$$x(0) = 1.0 \qquad \alpha = \beta = 1 \text{ in Eq. (7)}$$
$$y(0) = 0.01 \qquad \Delta\lambda_{1_0} = 10^{-4}$$
$$k = 1.5, \rho = 2.5 \qquad \Delta\lambda_{2_0} = 10^{-4}$$
$$t_f = 2.0$$

run no.	λ_{1_0}	λ_{2_0}	λ_{1_f}	λ_{2_f}
1	0.0500	0.1000	-63.0940	76.7757
2	0.14927	0.22675	-41.6292	51.3940
3	0.16433	0.21579	-11.6812	15.1078
4	0.18972	0.22852	-3.6393	5.3648
5	0.22739	0.25948	-1.1765	2.3937
6	0.27037	0.29836	-0.33520	1.39394
7	0.29973	0.32529	-0.05483	1.06519
8	0.30707	0.33195	0.00019	1.00024
9	0.30709	0.33197	0.000009	0.999989

ing the convergence procedure until the desired end values are met. Figure 4-2 gives the optimal control behavior for this case as well as the behavior when $t_f = 4.0$ and $t_f = 8.0$. Figure 4-3 shows the yield y as a function of time for these cases.

4-4   CONTROL-VARIABLE INEQUALITY CONSTRAINTS

In the example just solved, Fig. 4-2 shows that as the duration of the process is shortened the peak value of the control required becomes larger. Further, as discussed toward the end of Section 4-2, when $k < 1$ the optimal control

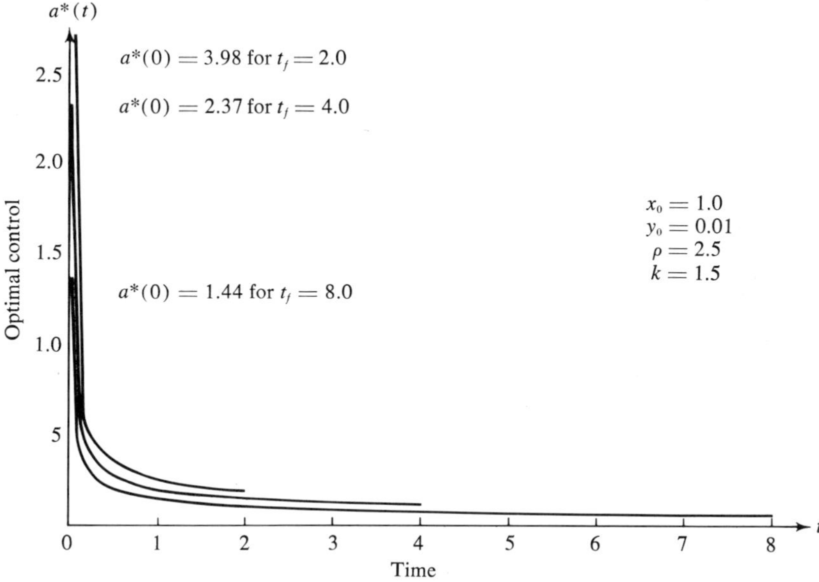

FIGURE 4-2   Control for the chemical-reactor problem

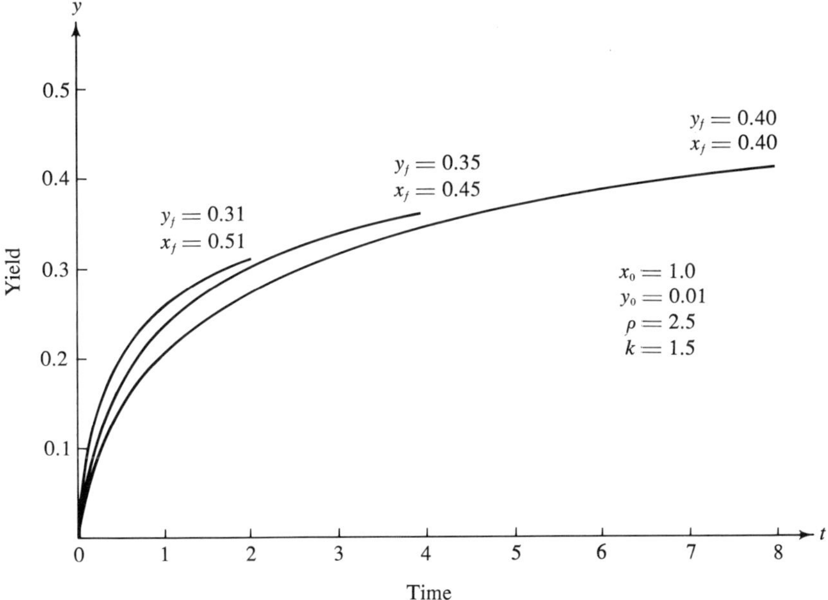

FIGURE 4-3   Yield for the chemical-reactor problem

appears to be unbounded. Although these values are called for by the solution, because of physical limitations one must expect that the control magnitudes desired cannot always be provided by the system.

One intuitively feels that a restriction on the largest value the control can assume is justifiable. For some problems, however, a limit on the smallest allowable value is needed. An example of the need for such a control restriction is a low-thrust orbit transfer in which the spacecraft engine is diffcult to restart once stopped. Thus in determining the optimal thrust behavior, the thrust might be required to be not only below some upper limit but also above a lower limit.

Actually, these control limitations can be considered to be a subclass of the more general case in which some relationship between the control and state variables of the system must be bounded. This condition may be expressed analytically by the requirement† that

$$C[\mathbf{x}(t), \mathbf{u}(t), t] \leqslant 0 \tag{1}$$

When $C(\mathbf{x}, \mathbf{u}, t) = 0$, one speaks of being on the control boundary, whereas, if $C(\mathbf{x}, \mathbf{u}, t) < 0$, the control is off the boundary.

To fix our ideas,‡ let us treat a system defined by n state variables $x_1, \cdots, x_n$ which are subject to the system equations

$$\dot{x}_i = f_i(\mathbf{x}, u, t) \qquad i = 1, \cdots, n \tag{2}$$

Only a single control, u, is taken to exist and it is assumed to be subject to the single control-variable inequality constraint§

$$C(x_1, \cdots, x_n, u, t) \leqslant 0 \tag{3}$$

The system IP to be minimized is given in standard form by

$$\mathrm{IP} = g[t_i, \mathbf{x}(t_i), t_f, \mathbf{x}(t_f)] \tag{4}$$

where the end values are required to satisfy the p end conditions

$$\psi_j[t_i, \mathbf{x}(t_i), t_f, \mathbf{x}(t_f)] = 0 \qquad j = 1, \cdots, p \leqslant 2n + 1 \tag{5}$$

Thus the problem as stated is contained within the one treated in Section 3-8,

† The notation of Eq. (1) is chosen to provide a standard for discussion and is not a restriction on the form of the inequality constraint. For example, the requirement that $u_l \leq u \leq u_u$ is, in the form of Eq. (1), given by $C = (u - u_u)(u - u_l) \leq 0$.

‡ In the main, the discussion follows that presented by Bryson, Denham, and Dreyfus, in "Optimal Programming Problems with Inequality Constraints, I: Necessary Conditions for Extremal Solutions," *AIAA J.*, Nov. 1963.

§ The limitation to a single control and a single control-variable inequality constraint is not inherent in the method to be presented. It is done only so as to not obscure the basic ideas. The reader should refer to the work, "Steepest Ascent Solution of Optimal Programming Problems," *Raytheon G. Report BR-2393*, April 1963, by W. F. Denham for the more general case.

except for the presence here of the control variable inequality constraint of Eq. (3).

To find the conditions to be satisfied by the extremal functions, as usual, one forms the augmented $\mathscr{IP}$,

$$\mathscr{IP} = g[t_i, \mathbf{x}(t_i), t_f, \mathbf{x}(t_f)] + \int_{t_i}^{t_f} F \, dt \tag{6}$$

where, as in Section 3-8,

$$F = \sum_{i=1}^{n} \lambda_i(t)[\dot{x}_i - f_i(t, \mathbf{x}, u)] \tag{7}$$

Since a control boundary exists, let us include the possibility that for at least a portion of the extremal path the control lies on the boundary. To do this let t_1 be the time the control goes on the boundary and t_2 the time it leaves the boundary.

Then Eq. (6) may be rewritten to give explicit recognition of these times as

$$\mathscr{IP} = g[t_i, \mathbf{x}(t_i), t_f, \mathbf{x}(t_f)] + \int_{t_i}^{t_1} \sum_{i=1}^{n} \lambda_i[\dot{x}_i - f_i(t, \mathbf{x}, u)] \, dt$$

$$+ \int_{t_1}^{t_2} \sum_{i=1}^{n} \lambda_i[\dot{x}_i - f_i(t, \mathbf{x}, u)] \, dt$$

$$+ \int_{t_2}^{t_f} \sum_{i=1}^{n} \lambda_i[\dot{x}_i - f_i(t, \mathbf{x}, u)] \, dt \tag{8}$$

The condition that for the extremal functions the first variation of the $\mathscr{IP}$ vanish, $\delta\mathscr{IP} = 0$, remains to be satisfied. Forming $\delta\mathscr{IP}$ from Eq. (8), yields†

$$\delta\mathscr{IP} = 0 = dg[t_i, \mathbf{x}(t_i), t_f, \mathbf{x}(t_f)]$$

$$+ \int_{t_i}^{t_1} \sum_{i=1}^{n} \lambda_i \left[\delta\dot{x}_i - \sum_{j=1}^{n} \frac{\partial f_i}{\partial x_j} \delta x_j - \frac{\partial f_i}{\partial u} \delta u \right] dt$$

$$+ \int_{t_1}^{t_2} \sum_{i=1}^{n} \lambda_i \left[\delta\dot{x}_i - \sum_{j=1}^{n} \frac{\partial f_i}{\partial x_j} \delta x_j - \frac{\partial f_i}{\partial u} \delta u \right] dt$$

$$+ \int_{t_2}^{t_f} \sum_{i=1}^{n} \lambda_i \left[\delta\dot{x}_i - \sum_{j=1}^{n} \frac{\partial f_i}{\partial x_j} \delta x_j - \frac{\partial f_i}{\partial u} \delta u \right] dt \tag{9}$$

† If doubt exists as to why dg is used in Eq. (9) and not δg, the reader should refer to the discussion following Eq. 3-2(11). The contribution to $\delta\mathscr{IP}$ caused by varying the end points of the integrals vanishes because the integrand F is identically zero.

In the $\delta \dot{x}_i$ terms the δ and derivative operations are now interchanged and the terms integrated once, by parts, to give

$$\delta \mathscr{IP} = 0 = dg[t_i, \mathbf{x}(t_i), t_f, \mathbf{x}(t_f)]$$

$$+ \sum_{i=1}^{n} \lambda_i \, \delta x_i \bigg|_{t_i}^{t_1} - \int_{t_i}^{t_1} \sum_{i=1}^{n} \left[\dot{\lambda}_i + \sum_{j=1}^{n} \lambda_j \frac{\partial f_j}{\partial x_i} \right] \delta x_i \, dt - \int_{t_i}^{t_1} \sum_{i=1}^{n} \lambda_i \frac{\partial f_i}{\partial u} \, \delta u \, dt$$

$$+ \sum_{i=1}^{n} \lambda_i \, \delta x_i \bigg|_{t_1}^{t_2} - \int_{t_1}^{t_2} \left\{ \sum_{i=1}^{n} \left[\dot{\lambda}_i + \sum_{j=1}^{n} \lambda_j \frac{\partial f_j}{\partial x_i} \right] \delta x_i + \sum_{i=1}^{n} \lambda_i \frac{\partial f_i}{\partial u} \, \delta u \right\} dt$$

$$+ \sum_{i=1}^{n} \lambda_i \, \delta x_i \bigg|_{t_2}^{t_f} - \int_{t_2}^{t_f} \sum_{i=1}^{n} \left[\dot{\lambda}_i + \sum_{j=1}^{n} \lambda_j \frac{\partial f_j}{\partial x_i} \right] \delta x_i \, dt - \int_{t_2}^{t_f} \sum_{i=1}^{n} \lambda_i \frac{\partial f_i}{\partial u} \, \delta u \, dt$$

$$\tag{10}$$

Correspondence has been maintained between the lines in Eq. (9) and those in Eq. (10).

To continue, imagine for the moment that the problem has been solved and the optimum times t_i, t_1, t_2, and t_f as well as the optimum values of the x_j at these times determined. Now consider as a new problem one in which these values are specified. Then

$$dt_i = dx_j(t_i) = 0 \qquad dt_2 = dx_j(t_2) = 0$$
$$dt_1 = dx_j(t_1) = 0 \qquad dt_f = dx_j(t_f) = 0$$

$$\tag{11}$$

Since†

$$dx_j(t_i) = \dot{x}_j(t_i) \, dt_i + \delta x_j(t_i)$$
$$dx_j(t_1) = \dot{x}_j(t_1) \, dt_1 + \delta x_j(t_1)$$
$$dx_j(t_2) = \dot{x}_j(t_2) \, dt_2 + \delta x_j(t_2)$$
$$dx_j(t_f) = \dot{x}_j(t_f) \, dt_f + \delta x_j(t_f)$$

$$\tag{12}$$

then with Eq. (11) one has

$$\delta x_j(t_i) = \delta x_j(t_1) = \delta x_j(t_2) = \delta x_j(t_f) = 0 \tag{13}$$

Now by hypothesis the control is unconstrained in the interval $t_i \leqslant t < t_1$ before going on the control boundary, and in $t_2 < t \leqslant t_f$ after coming off the boundary. Thus in these intervals the adjoint equations of Eq. 3-8(5) and the

---

† The relations of Eq. (12) are obtained by the same argument as that given following Eq. 3-2(8). The second and third relations in Eq. (12) reflect the fact that the times at which the control goes on and off the boundary as well as the values of the x_j at these times are to be varied in looking for the extremal solution.

control equation of 3-8(6) must be satisfied, by the extremal functions; that is,

$$
\text{Off control} \atop \text{boundary}
\begin{cases}
\dot{\lambda}_i = -\sum_{j=1}^{n} \lambda_j \frac{\partial f_j}{\partial x_i} & i = 1, \cdots, n \\[2em]
\sum_{j=1}^{n} \lambda_j \frac{\partial f_j}{\partial u} = 0 &
\end{cases}
\tag{14}
$$

Using Eqs. (11), (13), and (14) the relation for $\delta \mathscr{IP}$ given in Eq. (10) reduces to an integral over the time the control is on the boundary.

$$
\delta \mathscr{IP} = 0 = \int_{t_1}^{t_2} \left\{ \sum_{i=1}^{n} \left[\dot{\lambda}_i + \sum_{j=1}^{n} \lambda_j \frac{\partial f_j}{\partial x_i} \right] \delta x_i + \sum_{i=1}^{n} \lambda_j \frac{\partial f_j}{\partial u} \delta u \right\} dt
\tag{15}
$$

Although in preceding discussions the δx_j and δu were independent variations, leading to the conclusion that each of their coefficients must vanish, this is not true in Eq. (15). For, on the control boundary the condition $C(\mathbf{x}, u, t) = 0$ provides a relation between the x_i and u. Taking the first variation of this condition yields

$$
\sum_{i=1}^{n} \frac{\partial C}{\partial x_i} \delta x_i + \frac{\partial C}{\partial u} \delta u = 0
$$

from which it is seen that on the control boundary

$$
\delta u = \frac{-\sum_{i=1}^{n} (\partial C / \partial x_i) \, \delta x_i}{\partial C / \partial u}
\tag{16}
$$

Using Eq. (16) to eliminate δu from Eq. (15), the resulting relation is

$$
\delta \mathscr{IP} = 0 = \int_{t_1}^{t_2} \sum_{i=1}^{n} \left[\dot{\lambda}_i + \sum_{j=1}^{n} \lambda_j \left\{ \frac{\partial f_j}{\partial x_i} - \frac{\partial f_j / \partial u \; \partial C}{\partial C / \partial u \; \partial x_i} \right\} \right] \delta x_i \, dt
\tag{17}
$$

where the δx_i may now all be considered as independent variations. Since this is the case, Eq. (17) requires that the coefficient of each δx_i vanish. Thus, on the control boundary the adjoint equations and the control condition are given by

$$
\text{On control} \atop \text{boundary}
\begin{cases}
\dot{\lambda}_i = -\sum_{j=1}^{n} \lambda_j \left[\frac{\partial f_j}{\partial x_i} - \frac{\partial f_j / \partial u \; \partial C}{\partial C / \partial u \; \partial x_i} \right] \\[1.5em]
C(\mathbf{x}, u, t) = 0
\end{cases}
\tag{18}
$$

instead of the relations of Eq. (14) which hold off the boundary.

The requirement that the extremals satisfy Eqs. (14) and (18) has been established here for the case where t_i, t_1, t_2, t_f, and the x_i at these times are specified. However, since the fixed values chosen were assumed to be the optimum values for the problem, the extremals must certainly satisfy these equations when the optimum quantities are unknown and must be found if the two solutions are to coincide.

To continue our original problem, for the first variation of our $\mathscr{IP}$ to vanish when not all of t_i, t_1, t_2, t_f, and the x_i at these times are fixed, requires, by Eq. (10), in addition to satisfying Eqs. (14) and (12) that

$$0 = dg + \sum_{i=1}^{n} \lambda_i \, \delta x_i \bigg|_{t_1}^{t_i} + \sum_{i=1}^{n} \lambda_i \, \delta x_i \bigg|_{t_1}^{t_2} + \sum_{i=1}^{n} \lambda_i \, \delta x_i \bigg|_{t_2}^{t_f} \tag{19}$$

As Eq. (19) contains both dx_i and δx_i, which are related by Eq. (12), let us use Eq. (12) to eliminate δx_i at t_i, t_1, t_2, and t_f. Further, at t_1 and t_2, Eq. (19) contains two types of terms; those resulting from approaching t_1 and t_2 from below and those resulting from approaching these times from above. Terms of the former type will bear the superscript $^{(-)}$; those obtained by approaching t_1 or t_2 from above will bear the superscript $^{(+)}$. Equation (19) may now be written

$$0 = dg - \sum_{i=1}^{n} \lambda_i f_i \, dt \bigg|_{t_i}^{t_f} + \sum_{i=1}^{n} \lambda_i \, dx_i \bigg|_{t_i}^{t_f}$$

$$+ \sum_{i=1}^{n} \lambda_i f_i \bigg|_{t_1^{(-)}}^{t_1^{(+)}} dt_1 - \sum_{i=1}^{n} \lambda_i \bigg|_{t_1^{(-)}}^{t_1^{(+)}} dx_i(t_1)$$

$$+ \sum_{i=1}^{n} \lambda_i f_i \bigg|_{t_2^{(-)}}^{t_2^{(+)}} dt_2 - \sum_{i=1}^{n} \lambda_i \bigg|_{t_2^{(-)}}^{t_2^{(+)}} dx_i(t_2) \tag{20}$$

where use has been made of the fact that each of the x_i and t are to be continuous. Equation (20), which is the transversality condition for the problem, must be satisfied for all differentials consistent with the p conditions of Eq. (5). Introducing the p constant multipliers v_i, $i = 1, \cdots, p$, and forming

$$G = g + \sum_{i=1}^{p} v_i \psi_i \tag{21}$$

Eq. (20) may be written with dG replacing dg, as in Section 3-4 or 3-5.

Introduction of the undetermined multipliers, as usual, enables us to treat the differentials as though they were independent in Eq. (20). This requires that the coefficient of each of the differentials vanish. With G replacing g, setting the coefficients of dt_i, $dx_i(t_i)$, dt_f, and $dx_i(t_f)$ to zero in Eq. (20) yields

$$\sum_{i=1}^{n} \lambda_i f_i \bigg|_{t_i} + \frac{\partial G}{\partial t_i} = 0 \tag{22a}$$

$$\sum_{i=1}^{n} \lambda_i f_i \bigg|_{t_f} - \frac{\partial G}{\partial t_f} = 0 \tag{22b}$$

$$\lambda_j(t_i) = \frac{\partial G}{\partial x_j(t_i)} \qquad j = 1, \cdots, n \tag{22c}$$

$$\lambda_j(t_f) = -\frac{\partial G}{\partial x_j(t_f)} \qquad j = 1, \cdots, n \tag{22d}$$

Equation (22) is seen to be the same as the relations obtained in Eq. 3-8(10) for the problem when no control boundary is present.

Noting that none of the constraints of Eq. (5) involve t_1, t_2, $x_i(t_1)$, or $x_i(t_2)$, the condition that the coefficient of dt_1, dt_2, $dx_i(t_1)$, and $dx_i(t_2)$ in Eq. (20) vanish yields

$$\sum_{i=1}^{n} \lambda_i f_i \bigg|_{t_1^{(+)}} = \sum_{i=1}^{n} \lambda_i f_i \bigg|_{t_1^{(-)}} \qquad \sum_{i=1}^{n} \lambda_i f_i \bigg|_{t_2^{(+)}} = \sum_{i=1}^{n} \lambda_i f_i \bigg|_{t_2^{(-)}}$$

$$\lambda_i \bigg|_{t_1^{(+)}} = \lambda_i \bigg|_{t_1^{(-)}} \tag{23}$$

$$i = 1, \cdots, n$$

$$\lambda_i \bigg|_{t_2^{(+)}} = \lambda_i \bigg|_{t_2^{(-)}}$$

The relations of Eq. (23) provide the juncture conditions required to join the solutions of Eqs. (2) and (18) on the control boundary to the solution of Eqs. (2) and (14) in the unconstrained region on either side. Note that the conditions contained in Eq. (23) are just those obtained by satisfying the Weierstrass–Erdmann corner conditions of Eq. 3-8(8).

Thus, although the adjoint equations to be satisfied on and off the control boundary are different, no new conditions must be satisfied at the juncture of the regions or at the end points.

4-5 AN ALTERNATIVE FORMULATION

In Section 4-4 care had to be exercised when the control is on the boundary since the state variables x_i and the control variable u are then not independent but related through the condition $C(\mathbf{x}, u, t) = 0$. As has been done in each of the other cases when constraints between the variables occurred, let us introduce an undetermined multiplier so that the variables may be treated as though they are independent. Although no new results are obtained by formulation of the problem in this manner, it allows us to see how problems with multiple inequality constraints might be treated. In addition, since the problem will then be within the framework of our earlier theory, the presentation to be made in Section 4-6, extending the Weierstrass condition to problems with control inequality constraints, is simplified.

Thus, for the problem as formulated in Section 4-4, on introducing the multiplier μ the control inequality constraint $C(\mathbf{x}, u, t) \leqslant 0$ is replaced by the condition

$$\mu C(\mathbf{x}, u, t) = 0 \tag{1}$$

When the control is off the boundary, $\mu = 0$; for the control on the boundary, $C = 0$. The augmented integrand is now given by

$$F = \sum_{i=1}^{n} \lambda_i(t)[\dot{x}_i - f_i(t, \mathbf{x}, u)] + \mu(t)C(\mathbf{x}, u, t) \tag{2}$$

Each of the N system variables, where N is made up of the n state variables $x_i(t)$ and the single control $u(t)$, $N = n + 1$, must satisfy the characteristic equations of 3-5(5), which are

$$\frac{d}{dt}\frac{\partial F}{\partial \dot{x}_i} - \frac{\partial F}{\partial x_i} = 0 \qquad i = 1, \cdots, N = n + 1 \tag{3}$$

Equations (3), expressed in terms of the problem variables through use of Eq. (2), then give

$$\dot{\lambda}_i = -\sum_{j=1}^{n} \lambda_j \frac{\partial f_j}{\partial x_i} + \mu \frac{\partial C}{\partial x_i} \qquad i = 1, \cdots, n \tag{4a}$$

$$\sum_{j=1}^{n} \lambda_j \frac{\partial f_j}{\partial u} - \mu \frac{\partial C}{\partial u} = 0 \tag{4b}$$

which, along with

$$C(\mathbf{x}, u, t) = 0 \tag{5}$$

must be satisfied everywhere in $t_i \leqslant t \leqslant t_f$.

With $\mu \neq 0$, that is, for u on the control boundary, elimination of μ between Eqs. (4a) and (4b) yields the characteristic equations for the problem as given by Eqs. 4-4(18). For $\mu = 0$, that is, for u off the control boundary Eqs. (4a) and (4b) are seen to be the same as the characteristic equations of 4-4(14).

For the problem formulated here the transversality condition and the corner conditions have already been derived and are given by Eqs. 4-4(22) and 4-4(23), respectively. In the previous section it has already been noted that these are the same conditions as those derived in Eq. 3-8(10 or 11) and Eq. 3-8(8).

4-6   THE WEIERSTRASS CONDITION WITH CONTROL-VARIABLE INEQUALITY CONSTRAINTS

Obtained in Sections 4-4 and 4-5 by two different methods were the characteristic equations to be used in determining extremal solutions to problems with control-variable inequality constraints. Remember though, that in finding these solutions, no criterion beyond the necessary condition that the first variation of the IP vanish has been applied.

Let us review our previous work of Section 3-6 in the development of the Weierstrass condition for problems without inequality constraints. The analogous condition is to be obtained here for problems with inequality constraints. In particular, the problem as already formulated with a single control inequality constraint and a single control variable is discussed, although the results are not limited to that case.

In Section 3-6 the Weierstrass condition was developed under the assumption that, except for differential constraints, the N system variables are unconstrained in $t_i \leqslant t \leqslant t_f$. Using the condition that the N extremal functions satisfy the characteristic equations

$$\frac{d}{dt}\frac{\partial F}{\partial \dot{x}_i} - \frac{\partial F}{\partial x_i} = 0 \qquad i = 1, \cdots, N \tag{1}$$

Eq. 3-6(11) is developed. It shows that minimization of the IP by the extremal functions, x_i, requires

$$E = F(t, \mathbf{x}, \dot{\mathbf{X}}) - F(t, \mathbf{x}, \dot{\mathbf{x}}) - \sum_{i=1}^{N} \frac{\partial F}{\partial \dot{x}_i}(t, \mathbf{x}, \dot{\mathbf{x}})(\dot{X}_i - \dot{x}_i) \geqslant 0 \tag{2}$$

In the development of Eq. (2) in Section 3-6 each of the N variables x_i is treated as if it could be varied independently of the others. But this is just what was achieved in Section 4-5 by introduction of the multiplier μ and formation of the augmented integrand as given by Eq. 4-5(2). Thus the Weierstrass condition of Eq. (2) may be applied directly to the problem with inequality constraints as formulated in Section 4-5.

Substituting from Eq. 4-5(2) into Eq. (2), remembering that $F \equiv 0$, then yields†

$$E = -\sum_{j=1}^{n} \lambda_j [\dot{X}_j - \dot{x}_j] - \left[-\sum_{i=1}^{n} \lambda_i \frac{\partial f_i}{\partial u} + \mu \frac{\partial C}{\partial u} \right] (U - u) \geqslant 0 \qquad (3)$$

However, since by Eq. 4-5(4b) the last term on the right vanishes, Eq. (3) reduces to

$$E = -\sum_{j=1}^{n} \lambda_j [\dot{X}_j - \dot{x}_j] \geqslant 0 \qquad (4)$$

Using Eq. 4-4(2),

$$\dot{X}_j(t) = f_j(t, \mathbf{x}, U)$$

and $\qquad (5)$

$$\dot{x}_j(t) = f_j(t, \mathbf{x}, u)$$

where t, $\mathbf{x}$, and u are the extremal values while U is any allowable control.

On employing Eq. (5) in Eq. (4) there results

$$\sum_{i=1}^{n} \lambda_i f_i(t, \mathbf{x}, u) \geqslant \sum_{i=1}^{n} \lambda_i f_i(t, \mathbf{x}, U) \qquad (6)$$

from which one concludes that the extremal control u must maximize the left side of Eq. (6) if Eq. (6) is to be valid for all allowable U. Defining

$$H = \sum_{i=1}^{n} \lambda_i(t) f_i(t, \mathbf{x}, u) \qquad (7)$$

as in the problem with no inequality constraints, one concludes that the extremal control must maximize H for the extremal solutions to minimize the IP.

Although H must be maximized whether the control is on or off the boundary, the conditions that must be satisfied to assure this are not the same in both cases. If the control is off the boundary, that is, $C(\mathbf{x}, u, t) < 0$, then in accord with Chapter 2 one requires

$$\text{Off control boundary} \qquad \begin{cases} \dfrac{\partial H}{\partial u} = \displaystyle\sum_{j=1}^{n} \lambda_j \dfrac{\partial f_j}{\partial u} = 0 \\[2em] \dfrac{\partial^2 H}{\partial u^2} < 0 \end{cases} \qquad (8)$$

to maximize H.

† Recall also (see Sections 3-5 and 3-6) that since only u appears in F and not $\dot{u}$, the variable u itself is to be used in E.

On the other hand, if the control is on the boundary, that is, $C(\mathbf{x}, u, t) = 0$, then Eq. (8) need not be satisfied to maximize H. Rather, one must require that if the boundary constraint is violated by the control a larger value of H would result than that on the boundary. To see what this implies let us expand H and C about their values when the control is on the boundary, that is, about $H = H_b$, $C = C_b = 0$. Then, to first order,

$$H = H_b + \left.\frac{\partial H}{\partial u}\right|_b \delta u$$

$$(9)$$

$$C = C_b + \left.\frac{\partial C}{\partial u}\right|_b \delta u = \left.\frac{\partial C}{\partial u}\right|_b \delta u$$

Taking δu as positive in Eq. (9), for the control boundary to be exceeded, requires $\partial C/\partial u|_b > 0$. Then for H to be larger than its value on the boundary requires $\partial H/\partial u|_b > 0$. Taking δu as negative would simply reverse the sign of the inequalities. The resulting condition, that on the control boundary $\partial C/\partial u$ and $\partial H/\partial u$ have the same sign, is expressed by requiring

$$\text{On control boundary}\quad \frac{\partial H/\partial u|_b}{\partial C/\partial u|_b} = \frac{\sum_{j=1}^{n} \lambda_j (\partial f_j/\partial u)|_b}{\partial C/\partial u|_b} > 0 \qquad (10)$$

Solving for μ from Eq. 4-5(4b) and employing Eq. (10) one finds that on the control boundary

$$\mu = \frac{\sum_{j=1}^{n} \lambda_j (\partial f_j/\partial u)|_b}{\partial C/\partial u|_b} > 0$$

4-7 THE CHEMICAL REACTOR—BOUNDED CONTROL

We return to the problem first formulated in Section 4-2. To review, the reactor is defined by the equations

$$\dot{x} = -ax$$

$$(1a)$$

$$\dot{y} = ax - by$$

with

$$b = \rho a^k; \qquad \rho > 0, \qquad k > 0 \qquad (1b)$$

Let us now also assume that the allowable control $a(t)$ must lie in the range

$$a_l \leqslant a(t) \leqslant a_u \qquad (2)$$

In the notation of Section 4-4, the constraint of Eq. (2) is equivalent to requiring that the control satisfy the condition

$$C = (a - a_u)(a - a_l) \leqslant 0 \tag{3}$$

The system is assumed to start from the state x_0, y_0 at $t_i = 0$ and have a fixed time of operation t_f. The task is then to determine $a(t)$, subject to the limitation of Eq. (3), so as to maximize the final value of the product y. Thus

$$\text{IP} = -y(t_f) \tag{4}$$

Introducing the multipliers $\lambda_1(t)$ and $\lambda_2(t)$ to the $\dot{x}$ and $\dot{y}$ constraints of Eq. (1a), respectively, the solution of our problem requires that the characteristic equations be satisfied. When the control is within the admissible region ($C < 0$) these are given by Eqs. 4-4(14); when the control is on the boundary of the region ($C = 0$), Eqs. 4-4(18) must be met.

Forming the characteristic equations, one finds that they are the same both on and off the control boundary. This follows from Eq. 4-4(18), using the control constraint of Eq. (3), since $\partial C/\partial x = \partial C/\partial y \equiv 0$. Thus as in Section 4-2 one obtains for all t, $0 \leqslant t \leqslant t_f$, that

$$\dot{\lambda}_1 = a(\lambda_1 - \lambda_2)$$

$$\dot{\lambda}_2 = b\lambda_2 \tag{5}$$

To solve Eqs. (1) and (5) the end conditions

$$t_i = 0 \qquad t_f \text{ (specified)}$$

$$x(t_i) = x_0$$

$$y(t_i) = y_0$$

have been given. Two additional conditions at t_f are found from the transversality condition. As in Eq. 4-2(9) one obtains

$$\lambda_1(t_f) = 0$$

$$\lambda_2(t_f) = 1 \tag{6}$$

At points where the control goes on and off the control boundary during the process, the juncture conditions of Eq. 4-4(23) must be met. They yield that

$$\lambda_1|_{t^{(+)}} = \lambda_1|_{t^{(-)}} \qquad \lambda_2|_{t^{(+)}} = \lambda_2|_{t^{(-)}}$$

$$[\lambda_1(-ax) + \lambda_2(ax - by)]|_{t^{(+)}} = [\lambda_1(-ax) + \lambda_2(ax - by)]|_{t^{(-)}} \tag{7}$$

at points of entrance or departure from the control boundary. As noted at the end of Section 4-4, the relations of Eq. (7) are nothing more than the regular continuity conditions of Eq. 3-8(8) to be satisfied at *all* points along the extremal arc.

The control itself during the process must be chosen so as to maximize H,

$$H = \lambda_1(-ax) + \lambda_2(ax - by) \tag{8}$$

To see how this can be done let us examine the behavior of $H(a)$. Taking dH/da one has

$$\frac{dH}{da} = x(\lambda_2 - \lambda_1) - \rho k \lambda_2 y a^{k-1} \tag{9}$$

The solution for $\lambda_2(t)$ from Eqs. (5), (6), and (7) is given by $\lambda_2(t) = \exp[-\int_t^{t_f} b \, dt]$ which is always positive. Thus $\lambda_2(t)$, ρ, k, $a(t)$, and $y(t)$ can all be treated as positive quantities.

Then from Eq. (9), if $\lambda_2 > \lambda_1$, there is a zero of dH/da where

$$a(t) = \left[\frac{1}{\rho k} \frac{x}{y} \frac{\lambda_2 - \lambda_1}{\lambda_2}\right]^{1/(k-1)} \tag{10}$$

Continuing to assume $\lambda_2 > \lambda_1$, examination of Eq. (9) shows that

$$\lambda_2(t) > \lambda_1(t)$$

$$a(t) = 0 \qquad\qquad a(t) = \infty$$

$$k > 1: \frac{dH}{da} > 0 \qquad\qquad k > 1: \frac{dH}{da} = -\infty$$

$$k < 1: \frac{dH}{da} = -\infty \qquad\qquad k < 1: \frac{dH}{da} > 0$$

Using this information one can determine the behavior of $H(a)$. By Eq. (8) the curve will start from $H(0) = 0$ and pass through the point $dH/da = 0$ given by Eq. (10).

The result is shown in Figs. 4-4 and Fig. 4-5. Also shown there is $H(a)$ when $\lambda_2 < \lambda_1$. For this case Eq. (10) shows there is no point at which $dH/da = 0$. This knowledge plus that of dH/da given by

$$\lambda_2(t) < \lambda_1(t)$$

$$a(t) = 0 \qquad\qquad a(t) = \infty$$

$$k > 1: \frac{dH}{da} < 0 \qquad\qquad k > 1: \frac{dH}{da} = -\infty$$

$$k < 1: \frac{dH}{da} = -\infty \qquad\qquad k < 1: \frac{dH}{da} < 0$$

is sufficient to construct the form of $H(a)$.

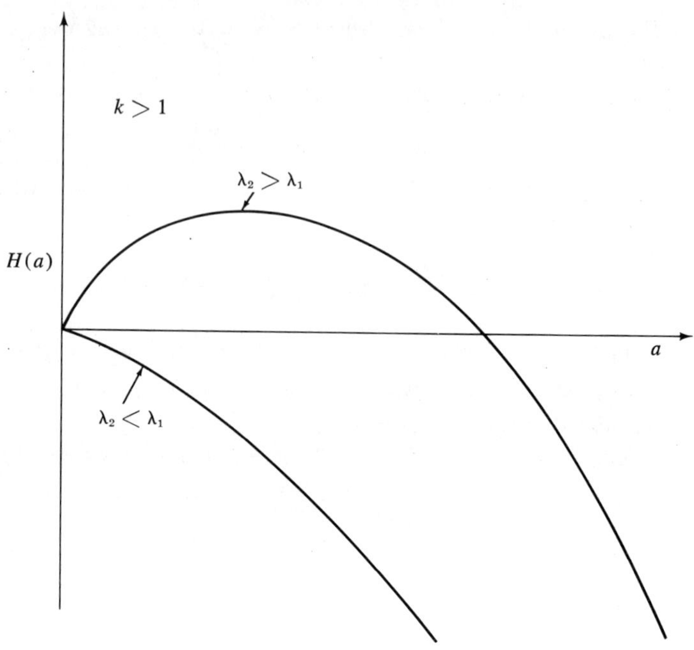

FIGURE 4-4   Behavior of $H(a)$ $(k > 1)$

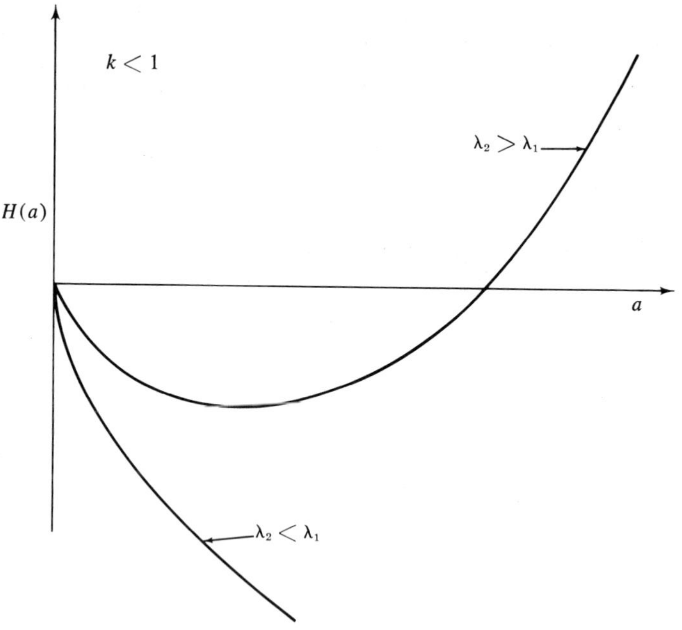

FIGURE 4-5   Behavior of $H(a)$ $(k < 1)$

Knowing the behavior of $H(a)$ let us examine the possibilities for the extremal control. Denoting this control by a^*, it is determined from the maximum principle by the condition that it maximize $H(a)$ within the range $a_l \leqslant a(t) \leqslant a_u$.

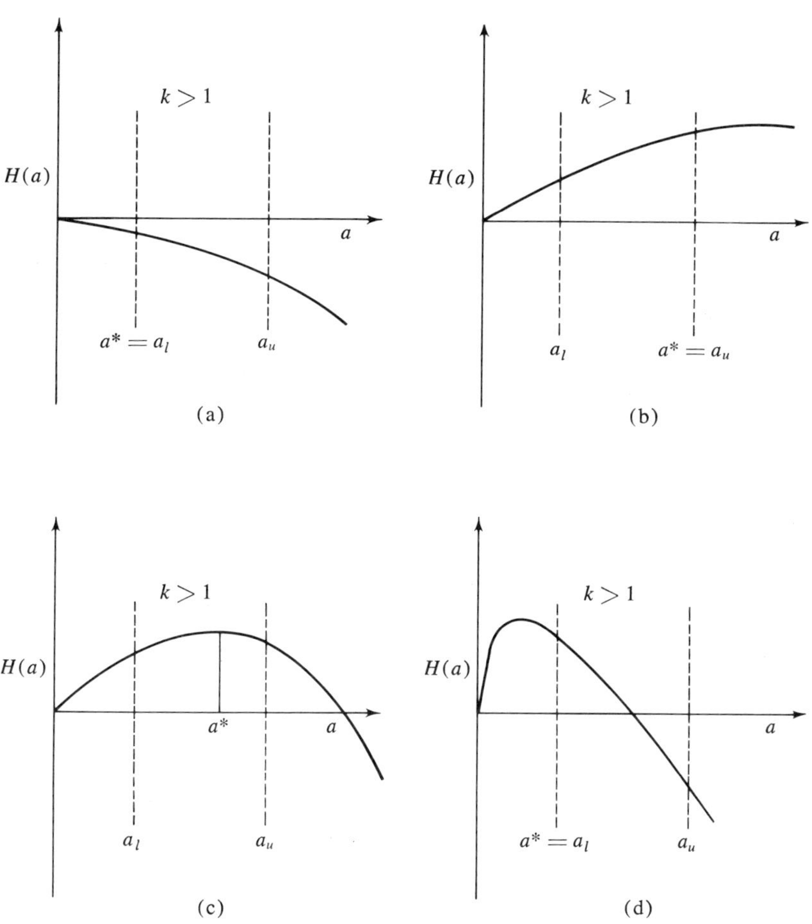

FIGURE 4-6   Maximum of $H(a)$ for $k > 1$, $a_l \leqslant a \leqslant a_u$

The case $k > 1$ is presented first in Fig. 4-6. It is seen that either $a^* = a_u$, $a^* = a_l$, or a^* is given by Eq. (10) as in Fig. 4-6c. Alternatively, if $k < 1$, Fig. 4-7 shows that the only possibilities are $a^* = a_u$ or $a^* = a_l$.

As a particular example, the problem first solved in Section 4-3 with $x_0 = 1.0$, $y_0 = 0.01$, $\rho = 2.5$, and $k = 1.5$ is redone with the constraint $0.1 \leqslant a(t) \leqslant 0.5$. Using as initial guesses for $\lambda_1(0)$ and $\lambda_2(0)$ the values found for the unconstrained problem, the technique of Section 4-3 is used to obtain

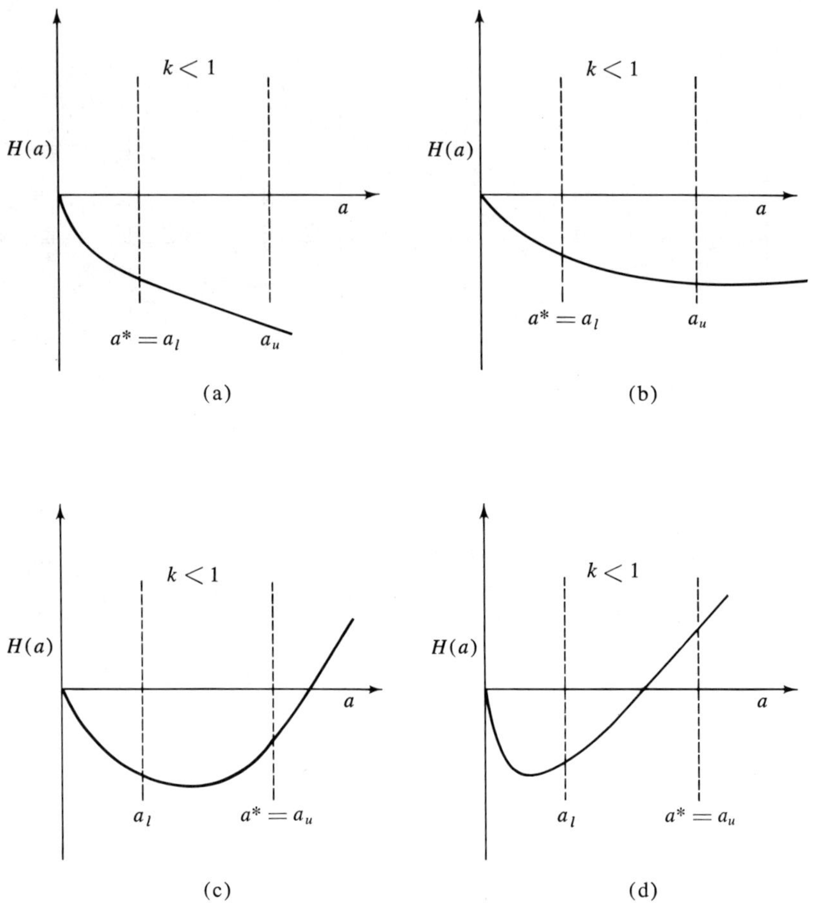

FIGURE 4-7   Maximum of $H(a)$ for $k < 1$, $a_l \leqslant a \leqslant a_u$

convergence to the values which yield the required end conditions $\lambda_1(t_f) = 0$, $\lambda_2(t_f) = 1.0$. At each integration point $H(a)$ is computed using $a_l = 0.1$, $a_u = 0.5$, and the control of Eq. (10), provided that in the latter case the value is in the admissible range $0.1 \leqslant a \leqslant 0.5$. Of these three possible candidates,

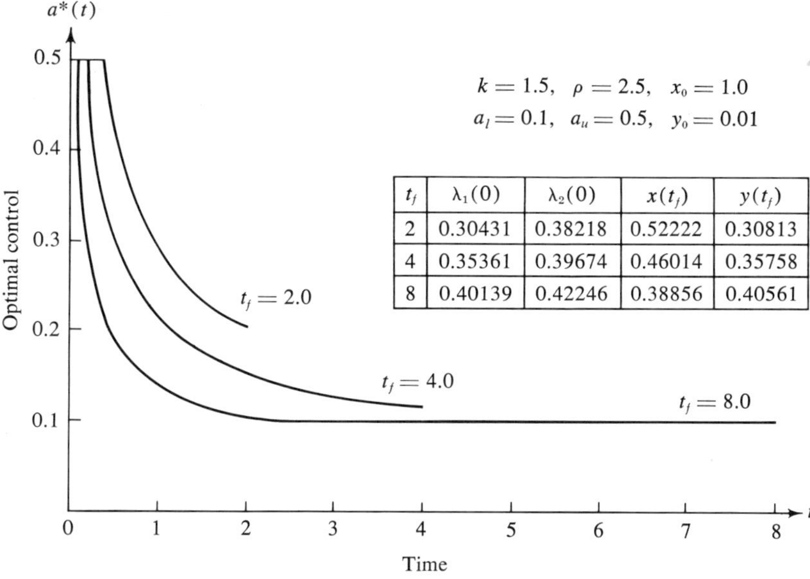

t_f	$\lambda_1(0)$	$\lambda_2(0)$	$x(t_f)$	$y(t_f)$
2	0.30431	0.38218	0.52222	0.30813
4	0.35361	0.39674	0.46014	0.35758
8	0.40139	0.42246	0.38856	0.40561

FIGURE 4-8   Chemical-reactor problem ($k = 1.5$)

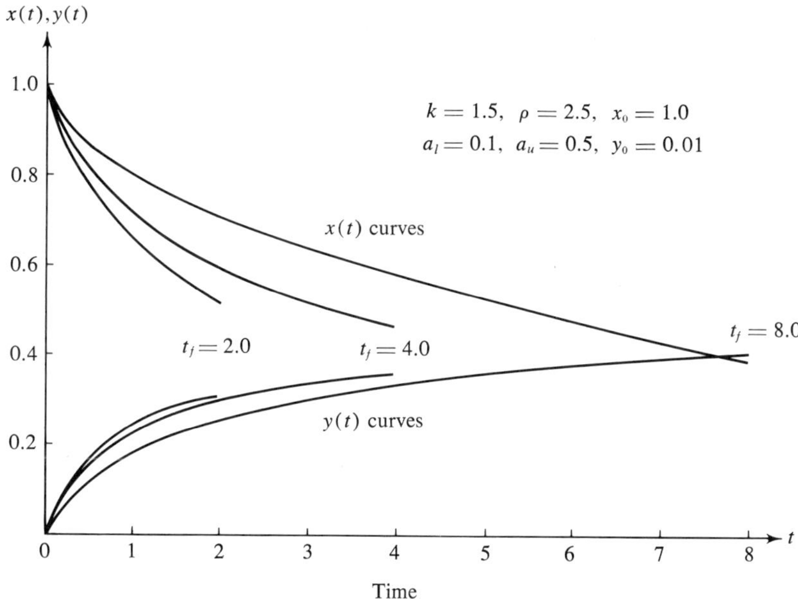

FIGURE 4-9   Chemical-reactor problem ($k = 1.5$)

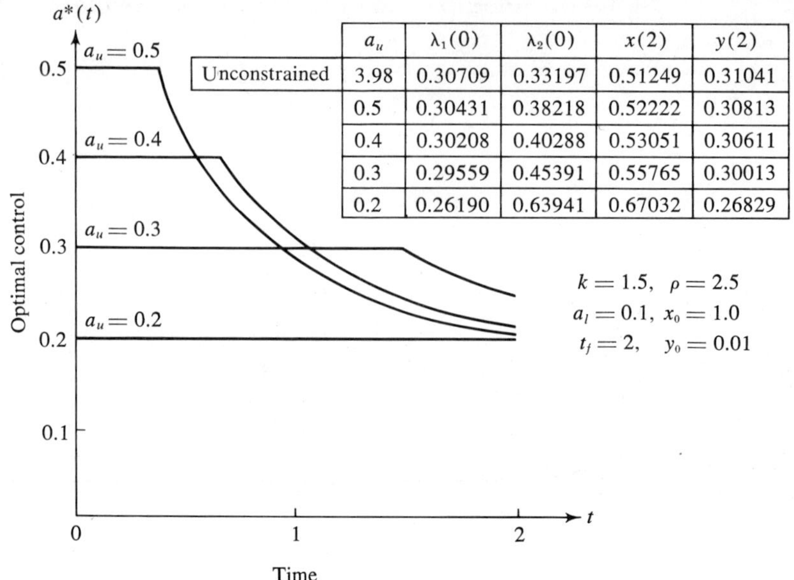

	a_u	$\lambda_1(0)$	$\lambda_2(0)$	$x(2)$	$y(2)$
Unconstrained	3.98	0.30709	0.33197	0.51249	0.31041
	0.5	0.30431	0.38218	0.52222	0.30813
	0.4	0.30208	0.40288	0.53051	0.30611
	0.3	0.29559	0.45391	0.55765	0.30013
	0.2	0.26190	0.63941	0.67032	0.26829

FIGURE 4-10   Chemical-reactor problem ($k = 1.5$)

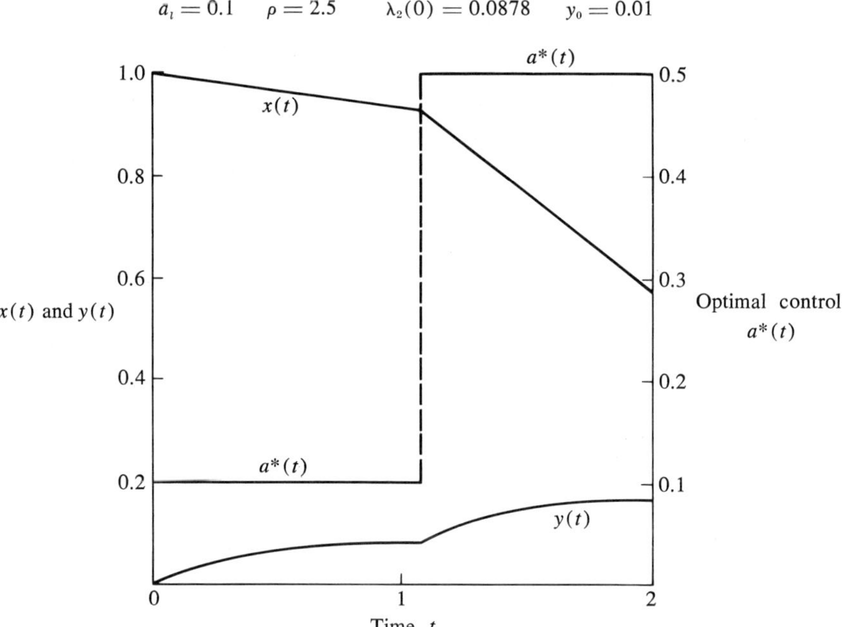

FIGURE 4-11   Chemical-reactor problem ($k = 0.5$)

the one which gives the largest value of H at each time is used as the extremal control in the integration of Eqs. (1) and (5).

The solution for the control is shown in Fig. 4-8 for $t_f = 2.0$, 4.0, and 8.0. The corresponding behavior of $x(t)$ and $y(t)$ is given in Fig. 4-9. Shown in Fig. 4-10 are solutions for the case $t_f = 2.0$ for several different values of a_u ranging from that required by the unconstrained solution of 3.98 down to $a_u = 0.2$. Note that the final yield $y(2)$ for the case $a_u = 0.5$ only differs in the third decimal place from the unconstrained solution where the control reaches 3.98. Although imposition of the constraint has resulted in a suboptimal solution relative to the unconstrained problem, the penalty is not severe.

The final case treated is shown in Fig. 4-11, where the solution to a problem when $k < 1$ ($k = 0.5$) is presented. As noted before, the control is either on the upper or lower boundary when $k < 1$. This is borne out in Fig. 4-11, where the control starts at $a_l = 0.1$ and then switches to the upper boundary $a_u = 0.5$ for the remainder of the process.

4-8   THE PURE INERTIA PLANT

The optimum control of a pure inertia plant† has received much attention in the literature (2, 3). Such a plant is defined by the system equation

$$\ddot{x} = u \tag{1}$$

The early attention paid this system is due, of course, to the simple nature of the plant. The current interest in the control of spacecraft outside the earth's atmosphere now gives new and practical meaning to this particular problem. Considering both these reasons, the problem of the pure inertia plant will be treated here as an example involving control variable inequality constraints.

In the derivation that follows, certain aspects of the problem have been worked in great detail. The extra detail is provided first; to permit comparison of the solution given here with the considerable quantity of literature on this particular problem, and, second, to show that although the maximum principle does indeed quickly yield the "form" of the solution as is often stated, the complete solution is somewhat more evasive.

The system equation of (1) is written in standard form by defining $x_1 = x$, $x_2 = \dot{x}$. Then Eq. (1) is equivalent to the set of equations

$$\dot{x}_1 = x_2 = f_1$$
$$\dot{x}_2 = u = f_2 \tag{2}$$

† The discussion here follows, in general, that given by C. D. Leedham, "Pontriagin's Maximum Principle and the Design of Optimum Systems," Ph.D. thesis, Purdue Univ., Lafayette, Ind., 1963.

The available control u will be limited by upper and lower bounds so that $|u| \leqslant U$. In terms of our notation for control variable inequality constraints, the condition may be written

$$C(x, u, t) = (u - U)(u + U) \leqslant 0 \tag{3}$$

The problem is taken to be that of minimizing the time required to transfer the system between initial and final states, where

$$
\begin{array}{cc}
\text{at } t = t_i = 0 & \text{at } t = t_f \text{ (unspecified)} \\
x_1(0) = x_{1_0} & x_1(t_f) = 0 \\
x_2(0) = x_{2_0} & x_2(t_f) = 0
\end{array}
\tag{4}
$$

The system performance index is then

$$\text{IP} = t_f \tag{5}$$

To obtain the solution the adjoint equations for the problem must be found, both for the case when the control is on the boundary and the separate case when the control is off the boundary. However, for this problem, with C given by Eq. (3), it is seen that $\partial C/\partial x_i = 0$, $i = 1, 2$. Thus, referring to Eqs. 4-4(14) and 4-4(18), the characteristic equations for the problem are seen to

$$
\begin{aligned}
\dot{\lambda}_1 &= 0 \\
\dot{\lambda}_2 &= -\lambda_1
\end{aligned}
\tag{6}
$$

be both on and off the control boundary.

To supplement the end conditions of Eq. (4) the transversality condition for the problem as given by Eq. 3-8(11) must be satisfied. Using Eq. (4), Eq. 3-8(11) reduces to

$$0 = dt_f - \sum_{i=1}^{2} \lambda_i f_i \bigg|_{t_f} dt_f \tag{7}$$

from which one concludes

$$\sum_{i=1}^{2} \lambda_i f_i \bigg|_{t_f} = 1 \tag{8}$$

Once the control is specified, the four conditions of Eq. (4) and the single condition of Eq. (8) provide the number of conditions required to determine t_f and the solution to the four first-order equations of Eqs. (2) and (6).

To find the optimum control, let us apply the maximum principle (Weierstrass condition). Forming the H function

$$H = \sum_{i=1}^{2} \lambda_i f_i = \lambda_1(t)x_2(t) + \lambda_2(t)u(t) \tag{9}$$

the control $u(t)$ is to be chosen to maximize H. As the first term of H does not involve $u(t)$, the problem becomes that of maximizing the second term. It is readily seen that this is done, within the limitations of Eq. (3) by choosing $u(t) = U$ when $\lambda_2(t) > 0$ and $u(t) = -U$ when $\lambda_2(t) < 0$. Thus the control function is given by

$$u(t) = U \operatorname{sign} \lambda_2(t) \tag{10}$$

Equation (10) shows that, provided $\lambda_2(t) \not\equiv 0$, the control takes on its maximum magnitude at all times; that is, the control is "bang-bang." Either maximum positive control or maximum negative control is to be employed. In addition, it is seen that the control will switch from the upper value to the lower value at the zero crossings of the multiplier $\lambda_2(t)$. In this light, $\lambda_2(t)$ is termed the *switching function* for the control.

The system of first-order equations to be solved can now be written

$$\begin{aligned}
\dot{x}_1 &= x_2 \\
\dot{x}_2 &= U \operatorname{sign} \lambda_2(t) \\
\dot{\lambda}_1 &= 0 \\
\dot{\lambda}_2 &= -\lambda_1
\end{aligned} \tag{11}$$

Difficulty arises in the solution of Eq. (11) due to the quantity "sign $\lambda_2(t)$," which destroys the linearity of the equations. The form of $\lambda_2(t)$ can, however, be determined directly from the last two equations of (11). Integrating the third relation one has

$$\lambda_1(t) = \bar{\lambda}_1 \tag{12}$$

where $\bar{\lambda}_1$ is a constant to be determined.

Using this result one finds from Eq. (11) that

$$\lambda_2(t) = -\bar{\lambda}_1 t + \bar{\lambda}_2 \tag{13}$$

where $\bar{\lambda}_2$ is another constant to be determined. Finally, from Eq. (10),

$$u(t) = U \operatorname{sign}(-\bar{\lambda}_1 t + \bar{\lambda}_2) \tag{14}$$

Equation (13) is the equation of a straight line and, therefore, has at most one zero crossing. Figure 4-12 shows the two possible ways the line can be drawn if the control is to switch at all. It is seen that if the sign of $\bar{\lambda}_2$ is positive, the constant $\bar{\lambda}_1$ must be positive and vice versa; that is, for switching,

$$\text{sign } \bar{\lambda}_2 = \text{sign } \bar{\lambda}_1 \tag{15}$$

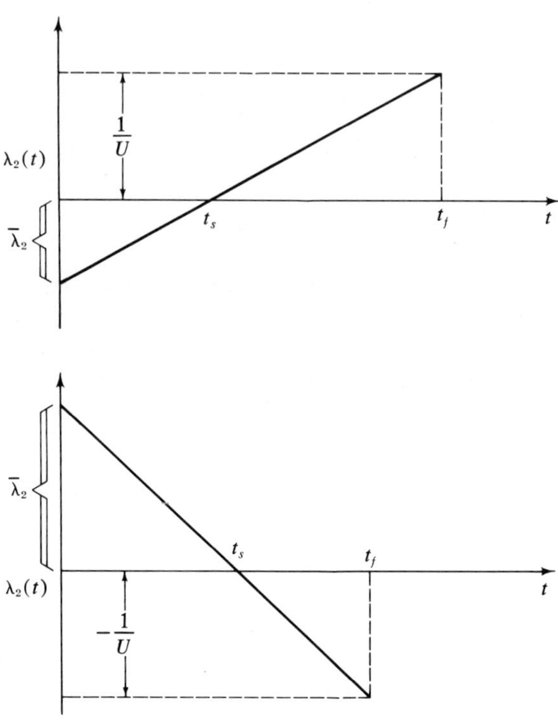

FIGURE 4-12   Termination of the control problem

To continue the solution let us make use of a result obtained in Section 3-8 but not as yet employed in any of our examples. The result referred to is that of Eq. 3-8(7), expressing the fact that if for a system with n state variables $\partial f_i/\partial t \equiv 0, i = 1, \cdots, n$, then a first integral of the adjoint equations exists and is given by

$$H(t) = \sum_{i=1}^{n} \lambda_i f_i = \text{const.} \tag{16}$$

Since for this problem Eq. (2) show that neither f_1 nor f_2 are explicit functions of time, a first integral exists. From Eq. (16) one then has

$$H(t) = \lambda_1 x_2 + \lambda_2 u = \text{const.} \tag{17}$$

Now, applying the transversality condition of Eq. (8), the constant in Eq. (17) is seen to be unity, giving

$$H(t) = 1 \tag{18}$$

Since Eq. (18) is valid for all t, it is valid, in particular, for $t = 0$ and t_f; thus

$$H(0) = \lambda_1(0)x_2(0) + \lambda_2(0)u(0) = 1$$
$$H(t_f) = \lambda_1(t_f)x_2(t_f) + \lambda_2(t_f)u(t_f) = 1 \tag{19}$$

Substituting the known boundary conditions and the control signal from Eqs. (4) and (14), respectively, Eq. (19) become

$$\bar{\lambda}_1 x_{2_0} + \bar{\lambda}_2 U \operatorname{sign} \bar{\lambda}_2 = 1$$
$$\lambda_2(t_f) U \operatorname{sign} \lambda_2(t_f) = 1 \tag{20}$$

Note that the second equation in (20) is equivalent to the condition

$$|\lambda_2(t_f)| = |-\bar{\lambda}_1 t_f + \bar{\lambda}_2| = \frac{1}{U} \tag{21}$$

Let us assume that the solution will require the control to switch† between

---

† If there is to be no switching, then the second relation of Eqs. (11) may be integrated directly to yield

$$x_2(t) = x_{2_0} \pm Ut \tag{23}$$

Applying the condition $x_2(t_f) = 0$, Eq. (23) yields

$$t_f = \frac{-x_{2_0}}{\pm U} = \frac{x_{2_0}}{U \operatorname{sign} x_{2_0}} \tag{24}$$

where since $t_f > 0$, the control u has been taken in Eq. (24) to be the opposite sign of x_{2_0}; that is, $u = -U \operatorname{sign} x_{2_0}$. Now let us integrate the first relation of Eq. (11) using Eq. (23) and evaluate the result at t_f. Eliminating t_f through use of Eq. (24) one finds, since $x_1(t_f) = 0$, that

$$x_{1_0} = \frac{-x_{2_0}^2}{2U} \operatorname{sign} x_{2_0} \tag{25}$$

Examination of Eq. (25) shows that only a single value of x_{1_0} satisfies the relation, once x_{2_0} is given. Thus it can be concluded that the initial conditions must be quite special if there is to be no switching; that is, they must lie on the curve defined by Eq. (25).

Any initial conditions x_{1_0} and x_{2_0} off the curve defined by Eq. (25) require that the control switch before reaching the origin. Since only a single switch is allowed, the solution proceeds with one value of u, say U, until the relation of Eq. (25) is satisfied at which point the control switches to $u = -U$. The solution then approaches the origin with u equal to $-U$ along the curve defined by Eq. (25). This is illustrated in Fig. 4-13. As the switch in the control occurs when the relation of Eq. (25) is satisfied, the curve obtained from Eq. (25) is termed the switching boundary.

its upper and lower boundary values. Drawing a horizontal line at plus and minus $1/U$ on the two plots of Fig. 4-12 determines the termination of the control problem as the intersection of this horizontal line with the straight line plotted. It can be concluded by examination of Fig. 4-12 that if $\bar{\lambda}_2$ is positive, then $\lambda_2(t_f) = -1/U$, whereas if $\bar{\lambda}_2$ is negative, then $\lambda_2(t_f) = 1/U$; that is,

$$\lambda_2(t_f) = -\frac{1}{U}\operatorname{sign}\bar{\lambda}_2 \tag{22}$$

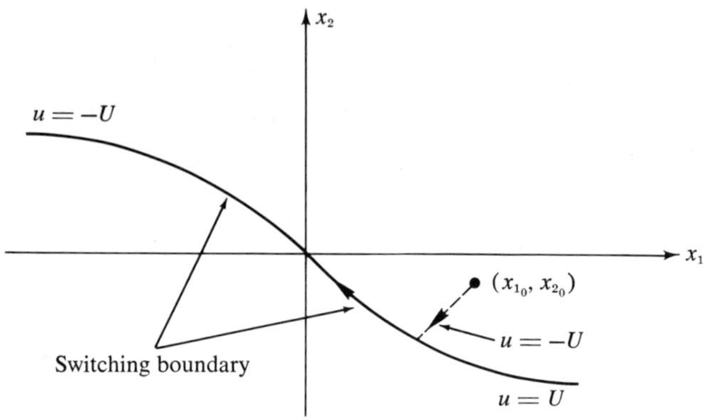

FIGURE 4-13   Switching boundary

Using this result in place of the second relation in Eq. (20), the equations to be solved become

$$\bar{\lambda}_1 x_{2_0} + \bar{\lambda}_2 U \operatorname{sign}\bar{\lambda}_2 = 1$$

$$-\bar{\lambda}_1 t_f + \bar{\lambda}_2 = -\frac{1}{U}\operatorname{sign}\bar{\lambda}_2 \tag{26}$$

Simultaneous solution of Eq. (26) for $\bar{\lambda}_1$ and $\bar{\lambda}_2$ then yields

$$\bar{\lambda}_1 = \frac{2}{x_{2_0} + t_f U \operatorname{sign}\bar{\lambda}_2}$$

$$\bar{\lambda}_2 = \frac{t_f U - x_{2_0}\operatorname{sign}\bar{\lambda}_2}{U(x_{2_0} + t_f U \operatorname{sign}\bar{\lambda}_2)} \tag{27}$$

The right side of Eq. (27) is seen to involve t_f, the duration of the process. To find this quantity the system equations of (2) must now be integrated. Recall that only one control switch is allowable. Let us assume that it occurs at t_s.

Then, using Eq. (10), the control function to be used for integration of Eq. (2) may be taken as

$$u(t) = U \operatorname{sign} \bar{\lambda}_2 \qquad 0 \leqslant t < t_s$$
$$u(t) = -U \operatorname{sign} \bar{\lambda}_2 \qquad t_s < t \leqslant t_f \tag{28}$$

Integration of the second relation of Eq. (2) now yields

$$x_2(t) = x_{2_0} + U \operatorname{sign} \bar{\lambda}_2 t \qquad 0 \leqslant t \leqslant t_s$$
$$x_2(t) = x_{2_0} + 2U \operatorname{sign} \bar{\lambda}_2 t_s - U \operatorname{sign} \bar{\lambda}_2 t \qquad t_s \leqslant t \leqslant t_f \tag{29}$$

Employing Eq. (29) to integrate the first relation of Eq. (2), one obtains

$$x_1(t) = x_{1_0} + x_{2_0} t + U \operatorname{sign} \bar{\lambda}_2 \frac{t^2}{2} \qquad 0 \leqslant t \leqslant t_s$$

$$x_1(t) = x_{1_0} - U \operatorname{sign} \bar{\lambda}_2 t_s^2 + [x_{2_0} + 2U \operatorname{sign} \bar{\lambda}_2 t_s]t \tag{30}$$

$$- U \operatorname{sign} \bar{\lambda}_2 \frac{t^2}{2} \qquad t_s \leqslant t \leqslant t_f$$

The expressions given for x_1 and x_2 in Eqs. (29) and (30) must, in addition, satisfy the terminal conditions from Eq. (4). Therefore,

$$0 = x_{2_0} + 2U \operatorname{sign} \bar{\lambda}_2 t_s - U \operatorname{sign} \bar{\lambda}_2 t_f$$

$$0 = x_{1_0} - U \operatorname{sign} \bar{\lambda}_2 t_s^2 + [x_{2_0} + 2U \operatorname{sign} \bar{\lambda}_2 t_s]t_f - U \operatorname{sign} \bar{\lambda}_2 \frac{t_f^2}{2} \tag{31}$$

The first equation in Eq. (31) may be solved for the switching time, giving

$$t_s = \frac{-1}{2U \operatorname{sign} \bar{\lambda}_2} [x_{2_0} - U \operatorname{sign} \bar{\lambda}_2 t_f] \tag{32}$$

Employing Eq. (32) to eliminate t_s from the second relation of Eq. (31) one obtains

$$0 = \left[4x_{1_0} - \frac{x_{2_0}^2 \operatorname{sign} \bar{\lambda}_2}{U}\right] + 2x_{2_0} t_f + U \operatorname{sign} \bar{\lambda}_2 t_f^2 \tag{33}$$

from which the duration of the process may be found. Thus

$$t_f = -\frac{x_{2_0}}{U} \operatorname{sign} \bar{\lambda}_2 \pm \left[2\left(\frac{x_{2_0}}{U}\right)^2 - 4\left(\frac{x_{1_0}}{U}\right) \operatorname{sign} \bar{\lambda}_2\right]^{1/2} \tag{34}$$

Equation (34) along with the relations of Eq. (27), which are repeated here for convenience,

$$\bar{\lambda}_1 = \frac{2}{x_{2_0} + t_f U \operatorname{sign} \bar{\lambda}_2} \tag{35a}$$

$$\bar{\lambda}_2 = \frac{t_f U - x_{2_0} \operatorname{sign} \bar{\lambda}_2}{U(x_{2_0} + t_f U \operatorname{sign} \bar{\lambda}_2)} \tag{35b}$$

are to be solved for t_f, $\bar{\lambda}_1$, and $\bar{\lambda}_2$. However, as the quantity "sign $\bar{\lambda}_2$" appears on the right side of Eqs. (34) and (35b), it is seen that t_f and $\bar{\lambda}_2$ cannot be evaluated independently.

To obtain the solution one proceeds by choosing a sign for $\bar{\lambda}_2$. Using this in Eq. (34), two values of t_f are calculated. Negative values may be discarded and any positive value is used with Eq. (35b) to calculate $\bar{\lambda}_2$. If the sign of the constant so determined does not check with that assumed at the start, the corresponding value of t_f is eliminated. Then choosing the alternate sign for $\bar{\lambda}_2$ the procedure is repeated. When this is done the smallest value of t_f from among the acceptable values for both choices of "sign $\bar{\lambda}_2$" provides the desired solution for t_f and the quantity "sign $\bar{\lambda}_2$." Using Eq. (32) the switching time for the process is then obtained, which concludes the solution.

REFERENCES

1. E. Kriendler and P. E. Sarachik, "On the Concepts of Controllability and Observability of Linear Systems," *IEEE Trans. Automat. Control*, April 1964, pp. 124–136.
2. A. M. Hopkin, "A Phase Plane Approach to the Compensation of Saturating Servomechanisms," *Trans. AIEE*, Part 2, Vol. 70, 1951.
3. I. Bogner and L. Kazda, "An Investigation of the Switching Criteria for Higher Order Contacter Servomechanisms," *Trans. AIEE*, Part 2, July 1954.

5

STATE-VARIABLE
INEQUALITY CONSTRAINTS

5-1 PROBLEMS WITH INTERMEDIATE CONDITIONS

Because of system requirements in some problems, specified relations between the state variables must be satisfied at an intermediate time in the process. The present section will find the modifications in our theory needed to treat these problems. The development, in addition to being important in its own right, is a necessary preliminary for the work to follow in Section 5-2 on state-variable inequality constraints.

As an example of a problem with intermediate conditions consider a space rescue mission. The rescue craft is required to transfer in minimum time from the space station to the distressed craft and then back to the space station. It is reasonable to imagine that at the time of rescue the velocity and coordinates of the two crafts must be the same. This requirement is, in effect, an intermediate condition on the state variables of the rescue ship. The word intermediate is used because the condition is applied between the end times of the mission, rather than, as previously discussed, at the terminal times.

With this illustration as motivation for what follows let us proceed with the task. As usual the system is defined by n state variables $x_1, \cdots, x_n$ which are subject to the n differential constraints

$$\dot{x}_i = f_i(\mathbf{x}, \mathbf{u}, t) \qquad i = 1, \cdots, n \tag{1}$$

There are m control variables $u_1, \cdots, u_m$ which along with the state variables are required to satisfy the r control inequality constraints

$$C_i(\mathbf{x}, \mathbf{u}, t) \leqslant 0 \qquad i = 1, \cdots, r \tag{2}$$

139

The system IP to be minimized is given in standard form by

$$\text{IP} = g[t_i, \mathbf{x}(t_i), t_f, \mathbf{x}(t_f)] \tag{3}$$

where the end values are required to satisfy the p end conditions

$$\psi_j[t_i, \mathbf{x}(t_i), t_f, \mathbf{x}(t_f)] = 0 \qquad j = 1, \cdots, p \leqslant 2n + 1 \tag{4}$$

In addition to the terminal conditions of Eq. (4), at some intermediate time t_1, $t_i < t_1 < t_f$, the state variables will be required to satisfy the conditions

$$\bar{\psi}_i[t_1, \mathbf{x}(t_1)] = 0 \qquad i = 1, \cdots, q \leqslant n + 1 \tag{5}$$

To find the extremal solution the augmented $\mathscr{IP}$ is formed

$$\mathscr{IP} = g + \int_{t_i}^{t_1} F \, dt + \int_{t_1}^{t_f} F \, dt \tag{6}$$

where the augmented integrand is given by

$$F = \sum_{i=1}^{n} \lambda_i[\dot{x}_i - f_i(t, \mathbf{x}, \mathbf{u})] + \sum_{i=1}^{r} \mu_i C_i(t, \mathbf{x}, \mathbf{u}) \tag{7}$$

Introduction of the multipliers λ_i and μ_i, as before, allows us to treat δx_i and δu_i as though they are independent during the process. The inequality constraint of Eq. (2) has been replaced by the condition

$$\mu_i C_i(t, \mathbf{x}, \mathbf{u}) = 0 \qquad \text{for } i = 1, \cdots, r \tag{8}$$

where μ_i equals zero when $C_i(t, \mathbf{x}, \mathbf{u})$ is less than zero.

Now, taking the first variation of Eq. (6) so as to apply the condition that the first variation of the $\mathscr{IP}$ vanish for the extremal functions, one has

$$\delta\mathscr{IP} = 0 = dg + \int_{t_i}^{t_1} \left\{ \sum_{i=1}^{n} \left[\frac{\partial F}{\partial x_i} \delta x_i + \frac{\partial F}{\partial \dot{x}_i} \delta \dot{x}_i \right] + \sum_{i=1}^{m} \frac{\partial F}{\partial u_i} \delta u_i \right\} dt$$

$$+ \int_{t_1}^{t_f} \left\{ \sum_{i=1}^{n} \left[\frac{\partial F}{\partial x_i} \delta x_i + \frac{\partial F}{\partial \dot{x}_i} \delta \dot{x}_i \right] + \sum_{i=1}^{m} \frac{\partial F}{\partial u_i} \delta u_i \right\} dt \tag{9}$$

Then after the usual integration by parts of the $\delta \dot{x}_i$ term, Eq. (9) becomes

$$\delta\mathscr{IP} = 0 = dg + \sum_{i=1}^{n} \frac{\partial F}{\partial \dot{x}_i} \delta x_i \Big|_{t_i}^{t_1} + \sum_{i=1}^{n} \frac{\partial F}{\partial \dot{x}_i} \delta x_i \Big|_{t_1}^{t_f}$$

$$+ \int_{t_i}^{t_1} \left\{ \sum_{i=1}^{n} \left[\frac{\partial F}{\partial x_i} - \frac{d}{dt}\frac{\partial F}{\partial \dot{x}_i} \right] \delta x_i + \sum_{i=1}^{m} \frac{\partial F}{\partial u_i} \delta u_i \right\} dt$$

$$+ \int_{t_1}^{t_f} \left\{ \sum_{i=1}^{n} \left[\frac{\partial F}{\partial x_i} - \frac{d}{dt}\frac{\partial F}{\partial \dot{x}_i} \right] \delta x_i + \sum_{i=1}^{m} \frac{\partial F}{\partial u_i} \delta u_i \right\} dt \tag{10}$$

In Eq. (10), repeating a line of reasoning already followed several times before, the integrals must vanish by themselves. Since the δx_i and δu_i may be treated as independent variations, using Eq. (7) this gives the standard set of characteristic equations

$$\frac{d}{dt}\frac{\partial F}{\partial \dot{x}_i} - \frac{\partial F}{\partial x_i} = 0 \quad \text{yielding } \dot{\lambda}_i = -\sum_{j=1}^{n} \lambda_j \frac{\partial f_j}{\partial x_i} + \sum_{j=1}^{r} \mu_j \frac{\partial C_j}{\partial x_i} \quad i = 1, \cdots, n$$

$$(11)$$

$$\frac{\partial F}{\partial u_i} = 0 \quad \text{yielding } -\sum_{j=1}^{n} \lambda_j \frac{\partial f_j}{\partial u_i} + \sum_{j=1}^{r} \mu_j \frac{\partial C_j}{\partial u_i} = 0 \quad i = 1, \cdots, m$$

Among the terms remaining in Eq. (10) are those which result from approaching t_1 from above and those which result from approaching t_1 from below. The former are designated with the superscript $^{(+)}$, the latter with the superscript $^{(-)}$. Then using the relations

$$dx_i(t_i) = \dot{x}_i(t_i)\, dt_i + \delta x_i(t_i)$$

$$dx_i(t_1) = \dot{x}_i(t_1)\, dt_1 + \delta x_i(t_1) \qquad (12)$$

$$dx_i(t_f) = \dot{x}_i(t_f)\, dt_f + \delta x_i(t_f)$$

to eliminate δx_i at t_i, t_1, and t_f from Eq. (10) one has

$$0 = dg - \sum_{i=1}^{n} \lambda_i f_i\, dt \Big|_{t_i}^{t_f} + \sum_{i=1}^{n} \lambda_i\, dx_i \Big|_{t_i}^{t_f}$$

$$+ \sum_{i=1}^{n} \lambda_i f_i \Big|_{t_1^{(-)}}^{t_1^{(+)}} dt_1 - \sum_{i=1}^{n} \lambda_i \Big|_{t_1^{(-)}}^{t_1^{(+)}} dx_i(t_1) \quad (13)$$

where continuity of t and the x_i themselves has been assumed.

Because of the conditions given in Eqs. (4) and (5) the differentials appearing in Eq. (13) are not all independent. Rather they are related by the $p + q$ conditions obtained from Eqs. (4) and (5) by forming the differential relations

$$d\psi_j[t_i, \mathbf{x}(t_i), t_f, \mathbf{x}(t_f)] = 0 \qquad j = 1, \cdots, p$$

$$(14)$$

$$d\overline{\psi}_j[t_1, \mathbf{x}(t_1)] = 0 \qquad j = 1, \cdots, q$$

To enable us to treat the differentials in Eq. (13) as though they were independent, the product is formed of each relation in Eq. (14) with an undetermined multiplier. Using ν_i and $\bar{\nu}_i$ for the multipliers the sum of these products, $\sum_{i=1}^{p} \nu_i\, d\psi_i + \sum_{i=1}^{q} \bar{\nu}_i\, d\overline{\psi}_i$, is added to Eq. (13).

Defining

$$G = g + \sum_{i=1}^{p} v_i \psi_i + \sum_{i=1}^{q} \bar{v}_i \bar{\psi}_i \tag{15}$$

Equation (13) may then be rewritten

$$0 = dG - \sum_{i=1}^{n} \lambda_i f_i\, dt \bigg|_{t_i}^{t_f} + \sum_{i=1}^{n} \lambda_i\, dx_i \bigg|_{t_i}^{t_f}$$

$$+ \sum_{i=1}^{n} \lambda_i f_i \bigg|_{t_1^{(-)}}^{t_1^{(+)}}\, dt_1 - \sum_{i=1}^{n} \lambda_i \bigg|_{t_1^{(-)}}^{t_1^{(+)}}\, dx_i(t_1) \tag{16}$$

Introduction of the $p + q$ multipliers v_i and $\bar{v}_i$ enables us to treat each of the differentials in Eq. (16) as though it is independent of the others. Equation (16) then requires the coefficient of each differential to vanish, yielding

$$\sum_{i=1}^{n} \lambda_i f_i \bigg|_{t_i} + \frac{\partial G}{\partial t_i} = 0 \tag{17a}$$

$$\sum_{i=1}^{n} \lambda_i f_i \bigg|_{t_f} - \frac{\partial G}{\partial t_f} = 0 \tag{17b}$$

$$\lambda_j(t_i) = \frac{\partial G}{\partial x_j(t_i)} \qquad j = 1, \cdots, n \tag{17c}$$

$$\lambda_j(t_f) = -\frac{\partial G}{\partial x_j(t_f)} \qquad j = 1, \cdots, n \tag{17d}$$

and

$$\sum_{i=1}^{n} \lambda_i f_i \bigg|_{t_1^{(+)}} = \sum_{i=1}^{n} \lambda_i f_i \bigg|_{t_1^{(-)}} - \frac{\partial G}{\partial t_1} = \sum_{i=1}^{n} \lambda_i f_i \bigg|_{t_1^{(-)}} - \sum_{k=1}^{q} \bar{v}_k \frac{\partial \bar{\psi}_k}{\partial t_1}$$

$$\lambda_i \bigg|_{t_1^{(+)}} = \lambda_i \bigg|_{t_1^{(-)}} + \frac{\partial G}{\partial x_i(t_1)} = \lambda_i \bigg|_{t_1^{(-)}} + \sum_{k=1}^{q} \bar{v}_k \frac{\partial \bar{\psi}_k}{\partial x_i} \tag{18}$$

Equation (17) is the usual transversality condition at t_i and t_f. On the other hand, Eq. (18), which is the extension of the Weierstrass–Erdmann corner conditions, provides a new result. Rather than requiring continuity, as in the problem with no intermediate conditions [see Eq. 4-4(23)], the juncture conditions of Eq. (18) show that jumps may occur at times at which there are intermediate conditions to be satisfied.

5-2 STATE-VARIABLE INEQUALITY CONSTRAINTS

We are now ready to treat problems in which system requirements are such that not all parts of state space can be reached in finding the optimum solution. Thus imagine the acceptable state-variable values are surrounded by a boundary surface that must not be penetrated by the solution variables. Mathematically this is expressed by requiring that the state variables satisfy the condition

$$S[\mathbf{x}(t), t] \leqslant 0 \tag{1}$$

The constraint of Eq. (1) is termed a state-variable inequality constraint. It differs from the control-variable inequality constraint $C[\mathbf{x}(t), u(t), t] \leqslant 0$ in that the control $u(t)$ does not appear explicitly in S, whereas it does in C.

The problem to be considered here is that formulated in Section 4-4 with the single exception that the state-variable inequality constraint of Eq. (1) replaces the control-variable inequality constraint of Eq. 4-4(1). Reviewing the solution procedure of Section 4-4 it is apparent that difficulty exists if one attempts to follow the same line of attack on the present problem. Recall that to stay on the control boundary $C[\mathbf{x}(t), u(t), t] = 0$ the control $u(t)$ is determined by the condition it maintain $C = 0$. In the present problem the state-variable boundary condition $S[\mathbf{x}(t), t] = 0$ provides no means of determining the control required to stay on the state boundary. The analogous condition to Eq. 4-4(16), expressing the relation required between changes in the state and control variables to stay on the boundary does not follow, because $\partial S / \partial u \equiv 0$. Yet it is intuitively obvious that if $S = 0$ at time t not all controls $u(t)$ will maintain $S = 0$ at time $t + \Delta t$.

To overcome this difficulty let us consider the technique presented by Dreyfus (1). Since on the state-variable boundary

$$S[\mathbf{x}(t), t] = 0 \tag{2}$$

it follows that on the boundary

$$\frac{d^r S}{dt^r} = 0 \qquad r = 1, 2, \cdots$$

In particular, consider the first derivative

$$\frac{dS}{dt} = \frac{\partial S}{\partial t} + \sum_{i=1}^{n} \frac{\partial S}{\partial x_i} \dot{x}_i \tag{3}$$

Substituting for $\dot{x}_i(t)$ from the motion equations of 4-4(2), Eq. (3) may be written

$$\frac{dS}{\partial t} = \frac{\partial S}{dt} + \sum_{i=1}^{n} \frac{\partial S}{\partial x_i} f_i(t, \mathbf{x}, u) \tag{4}$$

Now dS/dt as given by Eq. (4) may explicitly involve u or the terms involving u may cancel, leaving dS/dt an explicit function of only the x_i and t. If the latter is the case, the quantity d^2S/dt^2 is formed and its dependence on the control u determined. If d^2S/dt^2 is not an explicit function of u, the procedure is continued until the lowest-order time derivative of S is obtained which explicitly involves the control. Let us say that it is the qth one. Equation (1) is then termed a qth-order state-variable inequality constraint. The condition

$$\frac{d^q S}{dt^q} [\mathbf{x}(t), u(t), t] = 0 \tag{5}$$

will now be used in a role analogous to the condition $C[\mathbf{x}(t), u(t), t] = 0$ to determine the control $u(t)$ required to maintain the variables on the boundary.

Let t_1 be the time the state variables go on the boundary $S = 0$ and consider what is required to maintain $S = 0$ for $t > t_1$. Equation (5) provides a qth-order ordinary differential equation for the determination of S for $t > t_1$. With the control $u(t)$ determined so as to satisfy Eq. (5), the solution of this qth-order equation will only yield $S = 0$ for $t > t_1$ provided the q initial conditions on S are zero. That is, in addition to having the control satisfy Eq. (5), to stay on the state boundary after t_1 one must require that at t_1 the initial conditions

$$S^{(0)} = S[\mathbf{x}(t_1), t_1] = 0$$

$$S^{(1)} = \frac{dS}{dt} [\mathbf{x}(t_1), t_1] = 0 \tag{6}$$

$$\vdots$$

$$S^{(q-1)} = \frac{d^{q-1}S}{dt^{q-1}} [\mathbf{x}(t_1), t_1] = 0$$

are satisfied.

Using these results, the problem can be fitted into the theory as previously developed. Let t_1 be the time the solution goes on the state variable boundary and t_2 the time it goes off the boundary. (The times t_1 and t_2 are to be determined during the course of the solution.) Then in the interval $t_1 \leqslant t < t_2$ the control $u(t)$ is determined from the relation

$$\frac{d^q S}{dt^q} [\mathbf{x}, u, t] = 0 \qquad t_1 \leqslant t < t_2 \tag{7}$$

The condition of Eq. (7) on the state-variable boundary replaces the condition $C(\mathbf{x}, u, t) = 0$ on the control-variable boundary for the determination of the control.

In like manner the quantity $d^q S(\mathbf{x}, u, t)/dt^q$ replaces the quantity $C(\mathbf{x}, u, t)$ in the adjoint equations of Eqs. 4-4(18). Thus, writing $S^{(q)} = d^q S/dt^q$, on the state-variable boundary the adjoint equations and the control condition become

$$\text{On state boundary} \quad \begin{cases} \dot{\lambda}_i = -\sum_{j=1}^{n} \lambda_j \left[\dfrac{\partial f_j}{\partial x_i} - \dfrac{\partial f_j/\partial u}{\partial S^{(q)}/\partial u} \dfrac{\partial S^{(q)}}{\partial x_i} \right] \quad i = 1, \cdots, n \\[2em] S^{(q)}[\mathbf{x}(t), u(t), t] = 0 \end{cases} \tag{8}$$

and off the boundary, one has from Eq. 4-4(14) that

$$\text{Off state boundary} \quad \begin{cases} \dot{\lambda}_i = -\sum_{j=1}^{n} \lambda_j \dfrac{\partial f_j}{\partial x_i} \quad i = 1, \cdots, n \\[2em] \sum_{j=1}^{n} \lambda_j \dfrac{\partial f_j}{\partial u} = 0 \end{cases} \tag{9}$$

At time t_1, when going on the boundary the conditions of Eq. (6) must be satisfied. These are seen to be q conditions on the state variables at t_1; hence the juncture conditions of Eq. 5-1(18) must be satisfied at t_1. For convenience G is written using the intermediate conditions of Eq. (6) as

$$G = g + \sum_{i=1}^{p} \nu_i \psi_i + \sum_{i=0}^{q-1} \bar{\nu}_i S^{(i)} \tag{10}$$

Equation 5-1(18) then requires at t_1 that

$$\sum_{i=1}^{n} \lambda_i f_i \bigg|_{t_1^{(+)}} = \sum_{i=1}^{n} \lambda_i f_i \bigg|_{t_1^{(-)}} - \left[\bar{\nu}_0 \frac{\partial S^{(0)}}{\partial t_1} + \bar{\nu}_1 \frac{\partial S^{(1)}}{\partial t_1} + \cdots + \bar{\nu}_{q-1} \frac{\partial S^{(q-1)}}{\partial t_1} \right]$$

$$\lambda_i \bigg|_{t_1^{(+)}} = \lambda_i \bigg|_{t_1^{(-)}} + \left[\bar{\nu}_0 \frac{\partial S^{(0)}}{\partial x_i(t_1)} + \bar{\nu}_1 \frac{\partial S^{(1)}}{\partial x_i(t_1)} + \cdots + \bar{\nu}_{q-1} \frac{\partial S^{(q-1)}}{\partial x_i(t_1)} \right] \tag{11}$$

Since no additional conditions are specified when coming off the boundary at t_2, G does not contain either t_2 or $x_i(t_2)$. It follows that $\partial G/\partial t_2 = \partial G/\partial x_i(t_2) = 0$. Thus at t_2 Eq. 5-1(18) reduces to the regular corner conditions, yielding

$$\sum_{i=1}^{n} \lambda_i f_i \bigg|_{t_2^{(+)}} = \sum_{i=1}^{n} \lambda_i f_i \bigg|_{t_2^{(-)}}$$

$$\lambda_i \bigg|_{t_2^{(+)}} = \lambda_i \bigg|_{t_2^{(-)}} \tag{12}$$

Finally, at the initial and terminal times t_i and t_f, the usual transversality condition of Eq. 5-1(17) must be satisfied.

The conditions of Eqs. (11) and (12) serve to join the solution to the set of equations 5-1(1) and (8), on the state boundary, to the solution to the set of equations 5-1(1) and (9), off the boundary.

5-3 CONDITIONS FOR PROBLEMS WITH DISCONTINUOUS CONTROLS

The technique developed in Section 5-1 is now used to show that jumps in the multipliers can occur at a time t_1 when the rates $\dot{x}_i = f_i(\mathbf{x}, u, t)$ have finite discontinuities due to discontinuous control changes; that is, $\dot{x}_i(t_1^{(+)}) \neq \dot{x}_i(t_1^{(-)})$, provided that there is defined a surface $\bar{\psi}(\mathbf{x}, t) = 0$ on which such control jumps can occur. At t_1 when this surface is reached the requirement that

$$\bar{\psi}[\mathbf{x}(t_1), t_1] = 0 \tag{1}$$

can be treated as an intermediate condition for the problem.

Inserting Eq. (1) into Eq. 5-1(18), the continuity conditions at t_1 are

$$\sum_{i=1}^{n} \lambda_i \dot{x}_i \bigg|_{t_1^{(+)}} = \sum_{i=1}^{n} \lambda_i \dot{x}_i \bigg|_{t_1^{(-)}} - \bar{\nu} \frac{\partial \bar{\psi}}{\partial t} \tag{2a}$$

$$\lambda_i \bigg|_{t_1^{(+)}} = \lambda_i \bigg|_{t_1^{(-)}} + \bar{\nu} \frac{\partial \bar{\psi}}{\partial x_i(t_1)} \tag{2b}$$

Then using Eq. (2b) to eliminate $\lambda_i(t_1^{(+)})$ from Eq. (2a) one finds

$$\sum_{i=1}^{n} \lambda_i(t_1^{(-)})[\dot{x}_i(t_1^{(+)}) - \dot{x}_i(t_1^{(-)})] = -\bar{\nu} \frac{d\bar{\psi}}{dt}(t_1^{(+)}) \tag{3}$$

where the fact that

$$\frac{d\bar{\psi}}{dt}(t_1^{(+)}) = \frac{\partial \bar{\psi}}{\partial t_1} + \sum_{i=1}^{n} \frac{\partial \bar{\psi}}{\partial x_i(t_1)} \dot{x}_i(t_1^{(+)})$$

has been employed. Since in general $d\bar{\psi}(t_1^{(+)})/dt$ will not be zero, if there is a discontinuous change in $\dot{x}_i$ at t_1 one finds from Eq. (3) that $\bar{\nu}$ takes on a nonzero value.† With this result, Eq. (2) then shows that jumps in the multipliers can occur at points at which the control is discontinuous.

On the other hand, if $\dot{x}_i$ is continuous at t_1 the left side of Eq. (3) vanishes, requiring that $\bar{\nu} = 0$. From the conditions of Eq. (2) the multipliers are then seen to be continuous and in this case the time t_1 is of no special interest.

† This is true, provided $\lambda_i(t_1^{(-)})$ is not zero for each nonzero term $[\dot{x}_i(t_1^{(-)}) - \dot{x}_i(t_1^{(+)})]$.

5-4 AN EXERCISE IN ROAD BUILDING

Imagine that a road is to be built over uneven terrain. To ease the traveler's lot the contractor will be required to construct the road in such a manner that the slope at no point exceeds a maximum value, θ_m, in magnitude.

Within this system limitation, it is assumed to be to the contractor's advantage to minimize the cost of building the road. To provide a specific example, the cost of construction will be assumed proportional to the total amount of dirt removed in building the road without any distinction as to the method of movement.

Let $y(x)$ be the equation describing the road to be built and $t(x)$ the equation of the terrain. Forming $[t(x) - y(x)]^2$, the contractor's cost is taken as the integral of this quantity over the distance L in which the road will be built. Thus the contractor will attempt to design the road to minimize the cost,

$$\text{IP} = \int_0^L [t(x) - y(x)]^2 \, dx \tag{1}$$

which by its form is seen to be greater than or equal to zero.

A cursory examination of the IP shows that if the roadway can follow the existing terrain the cost to the contractor is minimized. For, if $y(x) \equiv t(x)$, then the IP takes on the value zero, its lower limit. However, although this would certainly be desired by the contractor, the system constraint

$$\left| \frac{dy(x)}{dx} \right| \leqslant \theta_m \tag{2}$$

limits the types of terrain that can be followed exactly to those for which $|dt(x)/dx| \leqslant \theta_m$. The problem of this section is to determine how to minimize the IP of Eq. (1) when the terrain does not meet this condition.

As a simple example let the terrain be given by

$$
\begin{aligned}
t(x) &= 0 & 0 &\leqslant x \leqslant x_1 \\
t(x) &= \theta_t(x - x_1) & x_1 &\leqslant x \leqslant x_2 \\
t(x) &= h = \theta_t(x_2 - x_1) & x_2 &\leqslant x \leqslant L
\end{aligned}
\tag{3}
$$

as shown in Fig. 5-1 where $\theta_t > \theta_m$.

At its ends the construction must join with roadways already built, requiring that the end values of $y(x)$ be specified. In this problem let us take $y(0) = 0, y(L) = h$.

For construction of the road, determination must be made of $y(x)$ within the system limitation of Eq. (2). Proceeding so as to obtain the equivalent of the system equations of motion, the constraint given by Eq. (2) is written

$$\frac{dy(x)}{dx} = \theta(x) \tag{4a}$$

$$|\theta(x)| \leqslant \theta_m \tag{4b}$$

Equation (4a) provides the governing (motion) equation. The variable $\theta(x)$ introduced in Eq. (4) is identified as the control variable of the problem while the independent variable x in Eq. (4) replaces the variable t in our preceding work. The control $\theta(x)$ is to be determined within the control variable inequality constraint of Eq. (4b) so as to minimize the IP.

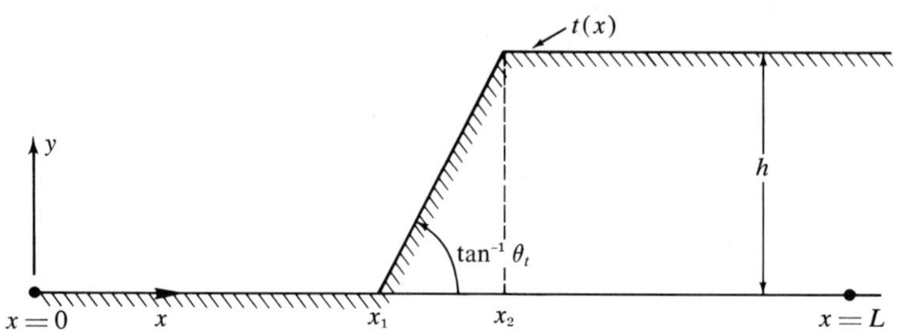

FIGURE 5-1   Terrain function

Now, putting the performance index in standard form by introducing the variable $z(x)$, minimizing the IP of Eq. (1) is seen to be the same as minimizing

$$IP = z_f = z(L)$$

where

$$\frac{dz}{dx} = [t(x) - y(x)]^2$$

$$z(0) = 0$$

Thus the problem to be solved becomes that of minimizing

$$IP = z_f \tag{5}$$

subject to the differential constraints,

$$\frac{dy}{dx} = \theta(x) = f_y \tag{6a}$$

$$\frac{dz}{dx} = [t(x) - y(x)]^2 = f_z \tag{6b}$$

the end conditions,

$$\text{at } x = x_i = 0 \qquad \text{at } x = x_f = L$$

$$y(0) = 0 \qquad\qquad y(L) = h \tag{7}$$

$$z(0) = 0$$

and the control-variable inequality constraint,

$$C(y, z, \theta, x) = (\theta - \theta_m)(\theta + \theta_m) \leqslant 0 \tag{8}$$

with the terrain function $t(x)$ given by Eq. (3).

Let $\lambda_y(x)$ and $\lambda_z(x)$ be the multipliers associated with Eqs. (6a) and (6b), respectively. The adjoint equations for the problem are obtained from Eq. 4-4(14) when the control is off the boundary $(C < 0)$ and from Eq. 4-4(18) when the control is on the boundary $(C = 0)$. However, for the C of Eq. (8), $\partial C/\partial y = \partial C/\partial z \equiv 0$. Therefore, the adjoint equations on the control boundary reduce to those to be satisfied off the boundary. Thus both on and off the control boundary the adjoint equations are given by

$$\frac{d\lambda_y}{dx} = 2\lambda_z[t(x) - y(x)] \tag{9a}$$

$$\frac{d\lambda_z}{dx} = 0 \tag{9b}$$

In addition to the adjoint equations of (9) the optimum solution must satisfy the transversality condition for the problem. As the end points and all the values of the state variables at these points are specified except z_f, the condition becomes

$$0 = dz_f + \lambda_z(L)\, dz_f$$

yielding

$$\lambda_z(L) = -1 \tag{10}$$

From Eqs. (9b) and (10) one then finds

$$\lambda_z(x) \equiv -1 \tag{11}$$

The control $\theta(x)$ is to be determined at each x so as to satisfy the maximum principle. Forming

$$H = \lambda_y f_y + \lambda_z f_z = \lambda_y \theta - [t(x) - y(x)]^2 \tag{12}$$

θ must be chosen so as to maximize H. Since only the first term in H involves the control θ, the condition requires that θ maximize

$$\tilde{H} = \theta \lambda_y$$

Thus for λ_y greater than zero θ should be chosen equal to θ_m, while for λ_y less than zero the choice should be θ equal $-\theta_m$. Therefore, provided λ_y is not zero over some x interval, the control is "bang-bang." This is expressed by requiring

$$\theta(x) = \theta_m \operatorname{sign} \lambda_y(x) \qquad \text{for } \lambda_y(x) \neq 0 \tag{13}$$

The quantity $\lambda_y(x)$ is termed the switching function for the control; that is, switches from the upper to the lower control boundary occur at the zero crossings of $\lambda_y(x)$.

The possibility that the switching function be identically zero over some finite x interval,

$$\lambda_y(x) \equiv 0 \qquad \text{in interval } I \tag{14}$$

remains to be investigated. If this is the case, termed *singular control*, it immediately follows that over this interval the rate of change of the switching function with x would also have to vanish:

$$\frac{d\lambda_y}{dx} \equiv 0 \qquad \text{in interval } I \tag{15}$$

Now applying Eq. (9a) one has that for Eq. (15) to hold requires

$$y(x) \equiv t(x) \qquad \text{in interval } I \tag{16}$$

But if Eq. (16) is valid, then $dy(x)/dx = dt(x)/dx$ in I and, on applying Eq. (6a) one has

$$\theta(x) = \frac{dt(x)}{dx} \qquad \text{in interval } I \tag{17}$$

Summarizing, it is seen that there are two control possibilities. For $\lambda_y(x) \neq 0$, $\theta(x) = \theta_m \operatorname{sign} \lambda_y(x)$, while for $\lambda_y(x) \equiv 0$ over some interval (singular control), then $\theta(x) = dt(x)/dx$ in that interval. To rephrase the results in a more physically descriptive manner, the analysis shows that either the road takes on its maximum positive or negative slope or the ground terrain is identically followed.

To determine the solution consider Eq. (9a) governing $\lambda_y(x)$. Using Eq. (11) it becomes

$$\frac{d\lambda_y}{dx} = 2[y(x) - t(x)] \tag{18}$$

From the end conditions of Eq. (7) and the definition of $t(x)$ given by Eq. (3) it is seen that at x equal to zero, $d\lambda_y/dx\big|_{x=0} = 0$. Assume first that at x equal to zero $\lambda_y(0) < 0$. Then from Eq. (13) the slope of the road at x equal to zero is seen to be negative. Since at x equal to zero, $y(0)$ equals zero, then at a distance

Δx along the roadway, $y(\Delta x) < 0$. Further, as $t(x)$ is always greater than or equal to zero, Eq. (18) shows that whenever $y(x)$ is negative, then $d\lambda_y/dx$ is negative. Therefore, if λ_y starts with a negative value at x equal to zero, its value becomes progressively more negative with increasing x. But if λ_y is alway negative, then the slope of the roadway will also always be negative, leaving no means of satisfying the terminal condition $y(L) = h$. This contradiction eliminates the possibility that $\lambda_y(0) < 0$.

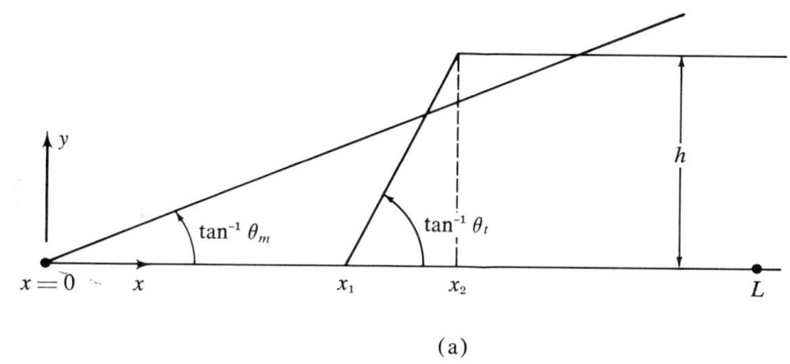

(a)

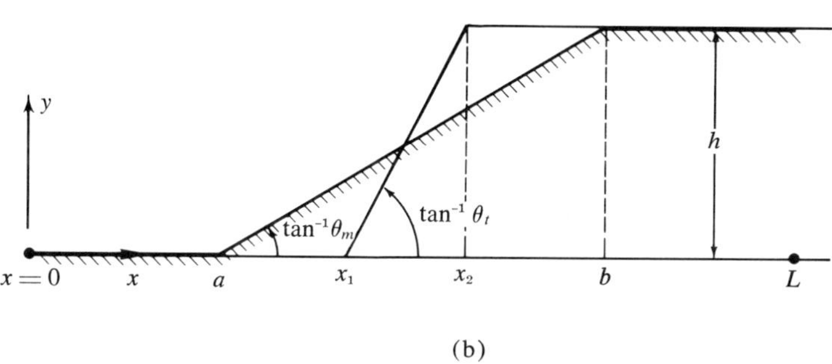

(b)

FIGURE 5-2   (a) Case $\lambda_y(0) > 0$; (b) solution $\lambda_y(0) = 0$

Assume now that $\lambda_y(0) > 0$. Then using Eq. (13) one finds that $y(\Delta x) > 0$ and from Eq. (18) one has that $d\lambda_y/dx > 0$. Thus λ_y is becoming more positive, causing the slope of y at Δx to remain at θ_m. The general result is illustrated in Fig. 5-2a. Now, by Eq. (18)

$$\lambda_y(x) = \lambda_y(0) + 2 \int_0^x [y(x) - t(x)] \, dx \qquad (19)$$

where the integral gives the area between the curves $y(x)$ and $t(x)$ from zero to x. With the values of x_1, x_2, θ_t, L, and θ_m indicated in Fig. 5-2a it is apparent that the contribution to $\lambda_y(x)$ of the negative area $[y(x) < t(x)]$ is not sufficient at any x to drive $\lambda_y(x)$ to zero.

In particular, as $\lambda_y(x)$ is not zero when the road crosses the terrain at elevation h it will never go to zero as the contribution to $\lambda_y(x)$ of the area between $y(x)$ and $t(x)$ for increasing x will always be positive. But if $\lambda_y(x)$ never goes through zero, the analysis has shown that the solution cannot switch from $\theta = \theta_m$ to either $\theta = -\theta_m$ or to following the terrain. Thus when the road reaches the elevation h, λ_y must be zero for the roadway to meet the terminal condition $y(L) = h$.

To achieve this let $\lambda_y(0) = 0$, the only remaining possibility. The road then follows the terrain a distance $x = a$, at which point it starts climbing at maximum allowable slope θ_m. The point $x = a$ is chosen so that at $x = b$, where $y(b) = h$, one also has $\lambda_y(b) = 0$. By Eq. (19), as $\lambda_y(a) = 0$, the result $\lambda_y(b) = 0$ is seen to be accomplished by making the terrain removed equal to the fill required. The solution is shown in Fig. 5-2b. Analytically the result is

$$y(x) = \begin{cases} t(x) = 0 & 0 \leqslant x \leqslant a \\ \theta_m(x - a) & a \leqslant x \leqslant b \\ t(x) = h & b \leqslant x \leqslant L \end{cases} \tag{20a}$$

where

$$a = x_1 - \frac{h}{2}\frac{\theta_t - \theta_m}{\theta_t\theta_m}$$

$$\tag{20b}$$

$$b = x_1 + \frac{h}{2}\frac{\theta_t + \theta_m}{\theta_t\theta_m}$$

5-5   TOWARD SMOOTHER ROADS

The contractor in Section 5-4 has minimized his cost while satisfying the only specified system constraint on the road; that is, $|\theta| \leqslant \theta_m$. After submitting the design solution of Fig. 5-2b for approval, he might reasonably expect to relax. Such optimism on his part would be unwarranted.

Scale-model testing of the proposed design by the approving agency would undoubtedly show the ride at points $x = a$ and $x = b$ to be unacceptable. This difficulty implies that besides requiring a design limitation on the slope of the road, a restriction should also be imposed on the magnitude of the rate of change of the road's slope. Thus the design must be redone so as to now meet the new system constraints on the road,

$$\left|\frac{dy}{dx}\right| \leqslant \theta_m \tag{1a}$$

$$\left|\frac{d^2y}{dx^2}\right| \leqslant U \tag{1b}$$

To obtain the governing equations, Eq. (1b) may be rewritten

$$\frac{d^2y}{dx^2} = u(x) \tag{2a}$$

$$|u(x)| \leqslant U \tag{2b}$$

Then defining $\theta(x) \equiv dy/dx$ as in Section 5-4, Eq. (2a) is equivalent to the set of two first-order equations

$$\frac{dy}{dx} = \theta(x)$$

$$\tag{3}$$

$$\frac{d\theta}{dx} = u(x)$$

Referring to Eq. (1), it is seen that Eq. (3) is to be solved for $y(x)$ subject to the two inequality constraints

$$-\theta_m \leqslant \theta(x) \leqslant \theta_m \tag{4a}$$

$$-U \leqslant u(x) \leqslant U \tag{4b}$$

In Section 5-4 the variable $\theta(x)$ served as the control variable for the problem. In the present section its role has been changed. A comparison of Eq. (3) with the standard form for the governing first-order equations, as in Eq. 5-1(1), reveals that $y(x)$ and now also $\theta(x)$ are state variables while $u(x)$ is the control variable. Thus in this problem the solution for $y(x)$ must be found subject to the state-variable inequality constraint of Eq. (4a) and the control-variable inequality constraint of Eq. (4b).

Placing the performance index of Eq. 5-4(1) in standard form, the problem becomes that of minimizing

$$\text{IP} = z_f \tag{5}$$

subject to the differential constraints,

$$\frac{dy}{dx} = \theta(x) \tag{6a}$$

$$\frac{d\theta}{dx} = u(x) \tag{6b}$$

$$\frac{dz}{dx} = [t(x) - y(x)]^2 \tag{6c}$$

the end conditions,

$$\text{at } x = x_i = 0 \qquad \text{at } x = x_f = L$$

$$y(0) = 0 \qquad\qquad y(L) = h \tag{7}$$

$$z(0) = 0$$

and the inequality constraints,

$$S = (\theta - \theta_m)(\theta + \theta_m) \leqslant 0 \tag{8a}$$

$$C = (u - U)(u + U) \leqslant 0 \tag{8b}$$

The modification of the solution to the problem originally posed in Section 5-4 so as to meet the dual inequality constraints of Eq. (8) is intuitively clear. Refer to Fig. 5-2b, which shows the solution to the original problem. The solution to the modified problem removes the jump in slope at points a and b, replacing the jumps by arcs on which $|d^2y/dx^2| = U$. On the remainder of the trajectory the road either follows the terrain or is on the state boundary and climbs with constant slope θ_m.

Verification of this result is left to the reader. In doing so, use must be made of the concepts of Section 5-3.

5-6 A MINIMUM-WEIGHT BEAM

Consider the problem of designing a minimum-weight elastic beam under a given loading. The particular system to be treated is shown in Fig. 5-3. The

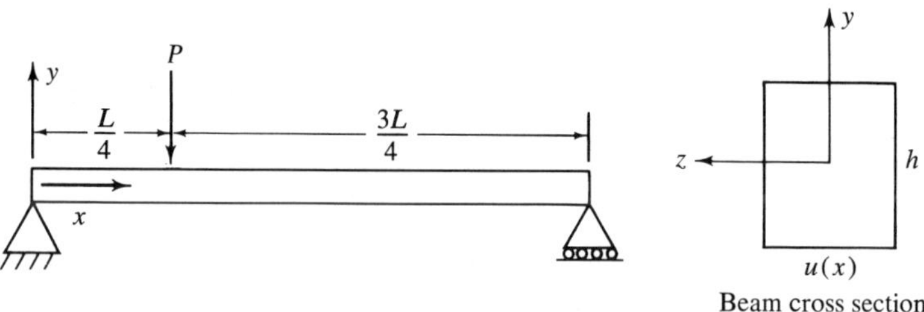

FIGURE 5-3 System configuration

beam, which is of length L, is taken to be simply supported and hence can withstand no bending moment at its ends. Further, as shown in the figure, the deflection at both ends is required to be zero:

$$y(0) = y(L) = 0 \tag{1}$$

From strength of materials theory (2) for the bending of elastic beams it is known that for small deflections and slopes the relation between the curvature of the beam and the moment at any section is given by

$$EI(x) \frac{d^2y}{dx^2} = M(x) \tag{2}$$

where
$$E = \text{modulus of elasticity}$$
$$I(x) = I_{zz}, \text{ moment of inertia about } z \text{ axis}$$
$$M(x) = \text{moment at any cross section } x \text{ about the } z \text{ axis}$$

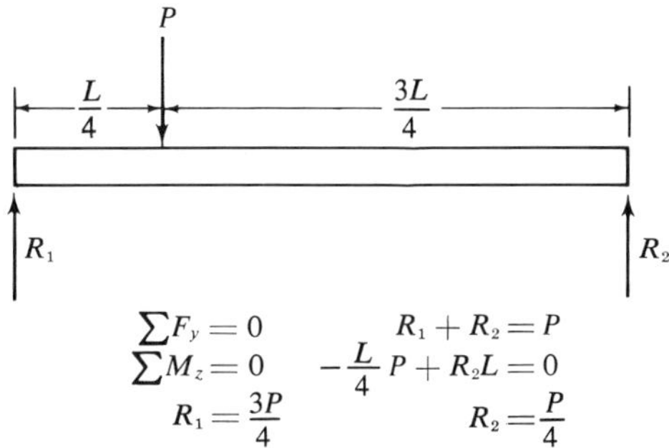

FIGURE 5-4 System equilibrium

Using the free-body diagram of Fig. 5-4 one finds that the moment at any section x is given by

$$M(x) = \begin{cases} \dfrac{3Px}{4} & \text{for } x \leqslant L/4 \\[2ex] \dfrac{P(L-x)}{4} & \text{for } x \geqslant L/4 \end{cases} \tag{3}$$

We consider a rectangular cross section of constant height h and variable width $u(x)$ as shown in Fig. 5-3. For this cross section one finds

$$I(x) = \tfrac{1}{12}u(x)h^3 \tag{4}$$

The width $u(x)$, which is the control variable for the problem, is to be determined so as to minimize the beam weight within system constraints. Thus

$$\text{IP} = \int_0^L \rho h u \, dx \tag{5}$$

where ρ = weight density of the beam.

The system constraints to be imposed are two. The maximum deflection of the beam is to be restricted by the requirement that

$$|y(x)| \leqslant Y \qquad \text{where } Y > 0 \tag{6}$$

In addition, a minimum allowable beam width is specified:

$$|u(x)| \geqslant U \qquad \text{where } U > 0 \tag{7}$$

The latter constraint, by fixing the minimum cross-sectional area, limits the stresses that need be supported by the beam.

Before continuing, Eq. (2) is put in standard form by defining

$$y_1(x) = y(x)$$

$$y_2(x) = y'(x)$$

where the prime indicates d/dx. Equation (2) may then be replaced by the set of equations

$$y_1' = y_2$$

$$y_2' = \frac{M(x)}{EI(x)} \tag{8}$$

In addition, on introducing $y_3(x)$ by the conditions

$$y_3'(x) = \rho h u(x)$$

$$y_3'(0) = 0 \tag{9}$$

the IP of Eq. (5) becomes

$$\text{IP} = y_3(L) \tag{10}$$

To summarize the problem conditions, the IP of Eq. (10) is to be minimized by $u(x)$ subject to the differential constraints,

$$y_1' = y_2$$

$$y_2' = \frac{\alpha M(x)}{u} \qquad \text{where } \alpha = \frac{12}{Eh^3} \tag{11}$$

$$y_3' = \rho h u$$

the end conditions,

$$y_1(0) = 0$$

$$y_1(L) = 0 \tag{12}$$

$$y_3(0) = 0$$

and the control and state-variable inequality constraints,

$$C = U - u(x) \leqslant 0 \tag{13a}$$

$$S = [y_1(x) - Y][y_1(x) + Y] \leqslant 0 \tag{13b}$$

A CONSTANT-WIDTH SOLUTION: Let us first determine the solution to the problem posed under the additional condition that

$$u(x) = \bar{u}, \text{ a constant}$$

$$\bar{u} \geqslant U$$

In this case the system equations of Eq. (11) may be integrated directly. Because of the form of $M(x)$ as given by Eq. (3), the integrations over $x \leqslant L/4$ and $x \geqslant L/4$ are carried out separately. Using Eq. (3) and the condition at $x = 0$ one finds for $x \leqslant L/4$ that

$$y_2(x) = \frac{\alpha}{\bar{u}} \frac{3Px^2}{8} + C_1$$

$$x \leqslant \frac{L}{4} \tag{14}$$

$$y_1(x) = \frac{\alpha}{\bar{u}} \frac{Px^3}{8} + C_1 x$$

Then for $x \geqslant L/4$, integrating Eq. (11) using the condition at $x = L$ and $M(x)$ from Eq. (3) one finds

$$y_2(x) = -\frac{\alpha}{\bar{u}} \frac{P(x-L)^2}{8} + C_2$$

$$x \geqslant \frac{L}{4} \tag{15}$$

$$y_1(x) = -\frac{\alpha}{\bar{u}} \frac{P(x-L)^3}{24} + C_2(x-L)$$

The state variables $y_1(x)$ and $y_2(x)$ must be continuous everywhere. Evaluating the solutions of Eqs. (14) and (15) at $x = L/4$ where they overlap, the constants C_1 and C_2 are chosen so that $y_1(L/4)$ and $y_2(L/4)$ meet the continuity condition. This requires

$$\frac{\alpha}{\bar{u}}\frac{3P}{8}\frac{L^2}{16} + C_1 = -\frac{\alpha}{\bar{u}}\frac{9P}{8}\frac{L^2}{16} + C_2$$

$$\frac{\alpha}{\bar{u}}\frac{P}{8}\frac{L^3}{64} + C_1\frac{L}{4} = \frac{\alpha}{\bar{u}}\frac{9P}{8}\frac{L^3}{64} - \frac{3C_2 L}{4} \tag{16}$$

from which one finds

$$C_1 = \frac{-7}{128}\frac{\alpha P L^2}{\bar{u}}$$

$$C_2 = \frac{5}{128}\frac{\alpha P L^2}{\bar{u}} \tag{17}$$

Thus Eqs. (14) and (15) yield

$$y(x) = \frac{\alpha}{\bar{u}}\frac{Px}{8}\left(x^2 - \tfrac{7}{16}L^2\right) \qquad\qquad x \leqslant \frac{L}{4} \tag{18a}$$

$$y(x) = \frac{\alpha}{\bar{u}}\frac{P(L-x)}{24}\left[(L-x)^2 - \tfrac{15}{16}L^2\right] \qquad x \geqslant \frac{L}{4} \tag{18b}$$

The state-variable inequality constraint of Eq. (13b) has yet to be satisfied. This must be done in such a manner as to minimize our performance index, the beam weight. Since $u(x) = \bar{u}$, a constant, using Eq. (11) the IP of Eq. (10) is found to be

$$\text{IP} = \rho h L \bar{u}$$

Hence the smaller $\bar{u}$, the lower the IP. But from Eq. (18) it is seen that with decreasing $\bar{u}$, the $|y(x)|$ grows larger, until eventually the state constraint will be exceeded.

To balance the two conflicting requirements, the minimizing beam width $\bar{u}^*$ is determined as the value for which the magnitude of the maximum deflection equals Y, the allowable limit. From Eq. (18) the point of maximum deflection magnitude is determined by setting $dy/dx = 0$. One finds†

$$(L - x)|_{|y|\max} = \sqrt{\tfrac{5}{16}}\,L \tag{19}$$

---

† From Eq. (18a) one finds on setting $dy/dx = 0$ that $x = (7/48)L$, which is greater than $L/4$. As Eq. (18a) is only valid for $x \leq L/4$, the value found is not an admissible solution for the point of maximum deflection.

With the result of Eq. (19), which gives the point of peak deflection at $x > L/4$, the magnitude of the maximum deflection is found from Eq. (18b) to be

$$\left. |y(x)| \right|_{\max} = \frac{5\sqrt{5}\alpha PL^3}{768\bar{u}} = 0.01456 \frac{\alpha PL^3}{\bar{u}}$$

Then determining $\bar{u}^*$ by the requirement that $\left. |y(x)| \right|_{\max} = Y$ one obtains

$$\bar{u}^* = 0.01456 \frac{\alpha PL^3}{Y} = 0.01456 \frac{12}{Eh^3}\frac{PL^3}{Y} \tag{20}$$

which is the desired solution.

5-7 A MINIMUM-WEIGHT BEAM—SOLUTION CONTINUED

In the present section the restriction that the beam width be constant is removed. With knowledge of the solution of Eq. 5-6(20), the control constraint of Eq. 5-6(13a) is formulated in terms of the optimal constant width solution $\bar{u}^*$. Thus we let the minimum beam width be

$$U = k\bar{u}^* \qquad 0 < k < 1 \tag{1}$$

For $k > 1$ it is apparent that the minimum-weight beam is obtained by using $u(x) = U$ for $0 \leqslant x \leqslant L$.

Before starting the solution it is desirable to put the problem in dimensionless form. We define

$$\tilde{x} = \frac{x}{L} \qquad \tilde{y}_1 = \frac{y}{Y} \qquad \tilde{y}_2 = y_2 \frac{L}{Y} \qquad \tilde{y}_3 = \frac{y_3}{h\bar{u}^*L}$$

$$\tilde{u} = \frac{u}{\bar{u}^*} \qquad \tilde{U} = \frac{U}{\bar{u}^*} \qquad \tilde{M} = \frac{M}{PL} \tag{2}$$

In terms of these variables the system equations of Eq. 5-6(11) become

$$\tilde{y}_1' = \tilde{y}_2 \tag{3a}$$

$$\tilde{y}_2' = \beta \frac{\tilde{M}(\tilde{x})}{\tilde{u}(\tilde{x})} \qquad \beta = \frac{768}{5\sqrt{5}} \tag{3b}$$

$$\tilde{y}_3' = \tilde{u} \tag{3c}$$

where the prime now indicates $d/d\tilde{x}$. From Eq. 5-6(3) one finds

$$\tilde{M}(\tilde{x}) = \begin{cases} \frac{3}{4}\tilde{x} & \text{for } \tilde{x} \leqslant \frac{1}{4} \\[2ex] \frac{1}{4}(1 - \tilde{x}) & \text{for } \tilde{x} \geqslant \frac{1}{4} \end{cases} \tag{4}$$

The system constraints of Eq. 5-6(13) become

$$C = \tilde{U} - \tilde{u}(\tilde{x}) \leqslant 0 \tag{5a}$$

$$S = [\tilde{y}_1(\tilde{x}) - 1][\tilde{y}_1(\tilde{x}) + 1] \leqslant 0 \tag{5b}$$

while the end conditions require

$$\tilde{y}_1(0) = 0$$
$$\tilde{y}_1(1) = 0 \tag{6}$$
$$\tilde{y}_3(0) = 0$$

The dimensionless index of performance is then

$$\mathbf{IP} = \tilde{y}_3(1) \tag{7}$$

To begin the solution let us determine the order of the state-variable inequality constraint of Eq. (5b). On the boundary $S \equiv 0$. Taking the derivative of S one obtains

$$S' = S^{(1)} = 2\tilde{y}_1\tilde{y}_1' = 2\tilde{y}_1\tilde{y}_2 = 0$$

$S^{(1)}$ does not involve the control $\tilde{u}$. Differentiating again one finds

$$S^{(2)} = 2\tilde{y}_1'\tilde{y}_2 + 2\tilde{y}_1\tilde{y}_2' = 0$$

or

$$S^{(2)} = 2\tilde{y}_2^2 + 2\beta\tilde{y}_1 \frac{\tilde{M}}{\tilde{u}} = 0$$

$S^{(2)}$ contains the control $\tilde{u}$. Thus the constraint is a second-order state-variable inequality constraint. Since on the state boundary $|\tilde{y}_1| = 1$, the expression for $S^{(1)}$ yields

$$\tilde{y}_2 = 0 \qquad \text{on state boundary}$$

Using this result one obtains for $S^{(2)}$,

$$S^{(2)} = \pm 2\beta \frac{\tilde{M}}{\tilde{u}} = 0 \qquad \text{on state boundary} \tag{8}$$

From Eq. (8) it is seen that to satisfy the requirement that $S^{(2)}$ equal zero over some interval will require the beam width to be infinite over that interval. Since this leads to an infinite weight, the possibility that this occur is ruled out. The alternative is for the beam deflection to only be at the allowed maximum at a single point. The solution will then not be on the state boundary over a finite interval; hence there is no necessity to require that $S^{(2)}$ equal zero over any interval.

We next form the adjoint equations to the system equations of (3) for determination of the optimal solution. From Eq. (5a) it is seen that the control inequality constraint function C does not involve the state variables. Thus by Eq. 4-4(18) the adjoint equations on the control boundary reduce to those off the boundary. These are

$$\lambda_1' = 0$$

$$\lambda_2' = -\lambda_1 \tag{9}$$

$$\lambda_3' = 0$$

The transversality condition for the problem requires that at the end points

$$\lambda_2(0) = 0$$

$$\lambda_2(1) = 0 \tag{10}$$

$$\lambda_3(1) = -1$$

From Eq. 4-4(23) one has that all the λ's and the product $\sum_{i=1}^{3} \lambda_i f_i$ are continuous going on and off the control boundary. Consider next what occurs on reaching the state boundary at say $\tilde{x}_1$. Expressing Eq. 5-2(6) in terms of this problem, the following conditions must be satisfied at $\tilde{x}_1$:

$$S^{(0)} = \tilde{y}_1^2(\tilde{x}_1) - 1 = 0 \tag{11a}$$

$$S^{(1)} = 2\tilde{y}_1(\tilde{x}_1)\tilde{y}_2(\tilde{x}_1) = 0 \tag{11b}$$

from which it is seen that

$$\tilde{y}_2(\tilde{x}_1) = 0 \tag{12}$$

Using Eq. (11), the conditions of Eq. 5-2(11) on the multipliers at $\tilde{x}_1$ become

$$\sum_{i=1}^{3} \lambda_i f_i \bigg|_{\tilde{x}_1^{(-)}} = \sum_{i=1}^{3} \lambda_i f_i \bigg|_{\tilde{x}_1^{(+)}} \tag{13a}$$

$$\lambda_1(\tilde{x}_1^{(+)}) = \lambda_1(\tilde{x}_1^{(-)}) + 2\bar{\nu}_0\tilde{y}_1(\tilde{x}_1) \tag{13b}$$

$$\lambda_2(\tilde{x}_1^{(+)}) = \lambda_2(\tilde{x}_1^{(-)}) + 2\bar{\nu}_1\tilde{y}_1(\tilde{x}_1) \tag{13c}$$

$$\lambda_3(\tilde{x}_1^{(+)}) = \lambda_3(\tilde{x}_1^{(-)}) \tag{13d}$$

where Eq. (12) has been used to eliminate $\bar{\nu}_1\tilde{y}_2(\tilde{x}_1)$ from Eq. (13b). Thus both λ_1 and λ_2 may have jumps on reaching the state boundary. Everywhere else in $0 \leqslant \tilde{x} \leqslant 1$ it can now be concluded that the λ's and the quantity $\sum_{i=1}^{3} \lambda_i f_i$ are continuous.

With these conclusions, the equations of (9) can be integrated subject to Eqs. (10) and (12), yielding

$$
\lambda_1(\tilde{x}) = \begin{cases} \bar{\lambda}_1 & 0 \leqslant \tilde{x} < \tilde{x}_1 \\ -\bar{\bar{\lambda}}_1 & \tilde{x}_1 < \tilde{x} \leqslant 1 \end{cases}
$$

$$
\lambda_2(\tilde{x}) = \begin{cases} -\bar{\lambda}_1 x & 0 \leqslant \tilde{x} < \tilde{x}_1 \\ -\bar{\bar{\lambda}}_1(1 - x) & \tilde{x}_1 < \tilde{x} \leqslant 1 \end{cases} \tag{14}
$$

$$
\lambda_3(\tilde{x}) = -1 \qquad 0 \leqslant \tilde{x} \leqslant 1
$$

where $\bar{\lambda}_1$ and $\bar{\bar{\lambda}}_1$ are to be determined.

Using Eq. (12), the continuity condition of Eq. (13a) requires

$$
\left[\frac{\lambda_2 \beta \tilde{M}}{\tilde{u}} - \tilde{u} \right] \Bigg|_{\tilde{x}_1^{(-)}} = \left[\frac{\lambda_2 \beta \tilde{M}}{\tilde{u}} - \tilde{u} \right] \Bigg|_{\tilde{x}_1^{(+)}} \tag{15}
$$

Equation (15) provides one relation for determination of the three constants $\tilde{x}_1$, $\bar{\lambda}_1$, and $\bar{\bar{\lambda}}_1$. The additional relations required come from integrating the system equations of (3) so as to satisfy the specified conditions on $\tilde{y}_1$ and $\tilde{y}_2$.

To integrate the system equations the extremal control must be found. It is determined by the condition that the extremal control maximize the H function of Eq. 4-6(7). For this problem

$$
H = \lambda_1 \tilde{y}_2 + \frac{\lambda_2 \beta \tilde{M}}{\tilde{u}} - \tilde{u} \tag{16}
$$

Determination of the dependence of H on $\tilde{u}$ is aided by examining $\partial H / \partial \tilde{u}$,

$$
\frac{\partial H}{\partial \tilde{u}} = -\frac{\lambda_2 \beta \tilde{M}}{\tilde{u}^2} - 1 \tag{17}
$$

Assume first that λ_2 is negative. By Eq. (4), $\tilde{M}(\tilde{x})$ is always positive. Thus in this case as $\tilde{u}$ approaches zero, $\partial H / \partial \tilde{u}$ approaches positive infinity. As $\tilde{u}$ increases from zero $\partial H / \partial \tilde{u}$ becomes less positive, passes through zero, and then approaches -1 as $\tilde{u}$ becomes infinite. Further by Eq. (16), H is negatively infinite for both $\tilde{u} = 0$ and $\tilde{u} = \infty$. The behavior of H in this case is shown in Fig. 5-5a. Alternatively, if λ_2 is positive, $\partial H / \partial \tilde{u}$ does not pass through zero but starting with $\tilde{u} = 0$, where $\partial H / \partial \tilde{u} = -\infty$ it increases monotonically with $\tilde{u}$ until $\partial H / \partial \tilde{u} = -1$ when $\tilde{u} = \infty$. The function H itself is seen by Eq. (16) to change from positive to negative infinity over this range of $\tilde{u}$. The behavior of H in this case is shown in Fig. 5-5b.

To start the solution let us assume that $\lambda_2(\tilde{x}) \leqslant 0$ for the entire interval of interest, $0 \leqslant \tilde{x} \leqslant 1$. Then by Fig. 5-7(1a) the maximum of H is found where $\partial H/\partial \tilde{u}$ equals zero. From Eq. (17) one finds that this occurs when

$$\tilde{u}(\tilde{x}) = [-\beta \lambda_2(\tilde{x}) M(\tilde{x})]^{\frac{1}{2}} \tag{18}$$

Using Eqs. (4) and (14), assuming for the moment that the point of maximum deflection, that is, the point $\tilde{x}_1$ at which the state boundary is reached, is greater than $\frac{1}{4}$, Eq. (18) may be written

$$\tilde{u}(\tilde{x}) = [\tfrac{3}{4}\beta \bar{\lambda}_1]^{\frac{1}{2}} \tilde{x} \qquad\qquad 0 \leqslant \tilde{x} \leqslant \tfrac{1}{4} \tag{19a}$$

$$\tilde{u}(\tilde{x}) = \left[\frac{\beta}{4} \bar{\lambda}_1 (1 - \tilde{x}) \tilde{x}\right]^{\frac{1}{2}} \qquad \tfrac{1}{4} \leqslant \tilde{x} < \tilde{x}_1 \tag{19b}$$

$$\tilde{u}(\tilde{x}) = \left[\frac{\beta}{4} \bar{\bar{\lambda}}_1\right]^{\frac{1}{2}} (1 - \tilde{x}) \qquad \tilde{x}_1 < \tilde{x} \leqslant 1 \tag{19c}$$

where $\bar{\lambda}_1$ and $\bar{\bar{\lambda}}_1$ must be positive to satisfy the assumption $\lambda_2(x) \leqslant 0$.

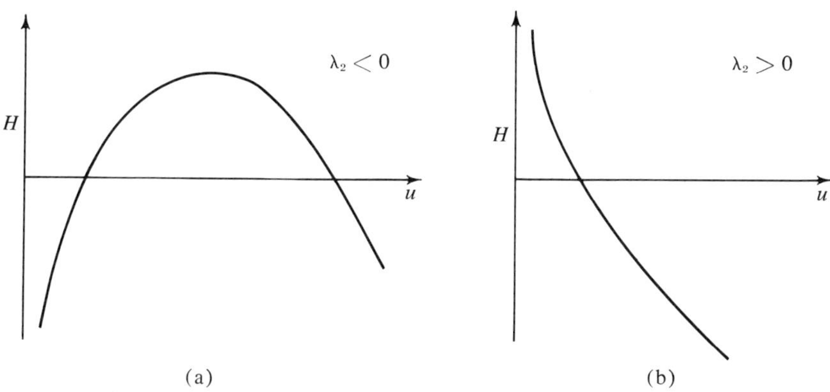

FIGURE 5-5   (a) Behavior of H ($\lambda_2 < 0$); (b) behavior of H ($\lambda_2 > 0$)

From Eq. (19a) it is seen that as $\tilde{x}$ approaches zero, $\tilde{u}(\tilde{x})$ does also, which violates the control inequality constraint $\tilde{u}(\tilde{x}) \geqslant \tilde{U}$. The same occurs as $\tilde{x}$ approaches 1. For these regions the solution of Eq. (19) is replaced by the requirement that the width $\tilde{u}$ equal the minimum allowable value $\tilde{U}$. Let $\tilde{x}_L$ and $\tilde{x}_R$ be the points at which the control boundary is left about $\tilde{x}$ equal to zero and returned to about $\tilde{x}$ equal to 1. Thus by Eqs. (19a, c)

$$\tilde{x}_L = \frac{\tilde{U}}{[\tfrac{3}{4}\beta \bar{\lambda}_1]^{\frac{1}{2}}} \qquad\qquad \text{for } \tilde{x}_L < \tfrac{1}{4}$$

$$\tilde{x}_R = 1 - \frac{\tilde{U}}{[(\beta/4)\bar{\bar{\lambda}}_1]^{\frac{1}{2}}} \qquad \text{for } \tilde{x}_R > \tilde{x}_1 \tag{20}$$

Then, assuming $\tilde{x}_L < \frac{1}{4}$ and $\tilde{x}_R > x_1$, the desired control is of the form

$$\tilde{u}(\tilde{x}) = \tilde{U} \qquad\qquad\qquad 0 \leqslant \tilde{x} \leqslant \tilde{x}_L$$

$$\tilde{u}(\tilde{x}) = [\tfrac{3}{4}\beta\bar{\lambda}_1]^{\frac{1}{2}}\tilde{x} \qquad\qquad \tilde{x}_L \leqslant \tilde{x} \leqslant \tfrac{1}{4}$$

$$\tilde{u}(\tilde{x}) = \left[\frac{\beta}{4}\bar{\lambda}_1(1 - \tilde{x})\tilde{x}\right]^{\frac{1}{2}} \qquad \tfrac{1}{4} \leqslant \tilde{x} < \tilde{x}_1 \qquad\qquad (21)$$

$$\tilde{u}(\tilde{x}) = \left[\frac{\beta}{4}\bar{\bar{\lambda}}_1\right]^{\frac{1}{2}}(1 - \tilde{x}) \qquad \tilde{x}_1 < \tilde{x} \leqslant \tilde{x}_R$$

$$\tilde{u}(\tilde{x}) = \tilde{U} \qquad\qquad\qquad \tilde{x}_R \leqslant \tilde{x} \leqslant 1$$

Let us now outline the procedure for determining the solution. In particular, let us check that the numbers of unknown constants is sufficient to satisfy the specified conditions. In the integration of the system equations of (3) the value of $\tilde{y}_2(0)$ is an unknown parameter. Thus using $\tilde{u}(\tilde{x})$ from Eq. (21), in the region $0 \leqslant \tilde{x} < \tilde{x}_1$ there are two unknown parameters in the solution, $\bar{\lambda}_1$ and $\tilde{y}_2(0)$. At $\tilde{x}_1$ there are three conditions to be met. These are the requirements that the solution be on the state boundary at $\tilde{x}_1$ with zero slope as well as the continuity condition on H as given by Eq. (15). The latter can be simplified through use of Eq. (18), yielding

$$\tilde{u}(\tilde{x}_1^{(-)}) = \tilde{u}(\tilde{x}_1^{(+)})$$

Thus at $\tilde{x}_1$ the three conditions to be satisfied are

$$\tilde{y}_1(\tilde{x}_1) = -1 \qquad\qquad (22a)$$

$$\tilde{y}_2(\tilde{x}_1) = 0 \qquad\qquad (22b)$$

$$\tilde{u}(\tilde{x}_1^{(-)}) = \tilde{u}(\tilde{x}_1^{(+)}) \qquad\qquad (22c)$$

The solution is extended beyond $\tilde{x}_1$ using the requirement for continuity of $\tilde{y}_1(\tilde{x})$ and $\tilde{y}_2(\tilde{x})$ at $\tilde{x}_1$. For $\tilde{x} > \tilde{x}_1$ the parameter $\bar{\bar{\lambda}}_1$ is introduced through the control of Eq. (21). At the end of the beam the condition

$$\tilde{y}_1(1) = 0 \qquad\qquad (23)$$

remains to be satisfied. To meet the four conditions of Eqs. (22a, b, c) and (23) there are four parameters $\tilde{y}_2(0)$, $\bar{\lambda}_1$, $\bar{\bar{\lambda}}_1$, and $\tilde{x}_1$, which are the requisite number.

For this problem the system equations of (3) can be integrated analytically using the extremal control. When this is done and the results used in Eqs. (22) and (23), a set of transcendental equations is obtained for the unknown parameters. Solution of these equations provides $\tilde{y}_2(0)$, $\bar{\lambda}_1$, $\bar{\bar{\lambda}}_1$, $\tilde{x}_1$, and with them the deflection $\tilde{y}(\tilde{x})$ and the optimal beam width $\tilde{u}(\tilde{x})$.

Although the parameters may be determined in the manner indicated, the process is involved. Because $\widetilde{M}(\tilde{x})$ and $\tilde{u}(\tilde{x})$ are discontinuous, the analytic solution for $\tilde{y}_1(\tilde{x})$ and $\tilde{y}_2(\tilde{x})$ must be found in sections and pieced together. This is a tedious task. An alternative scheme is to treat the problem as though analytic solution of the system equations were not possible. The system equations are then integrated by computer using, within the constraint of Eq. (5a), the control computed from Eq. (18) with $\widetilde{M}(\tilde{x})$ and $\lambda_2(\tilde{x})$ given by Eqs. (4) and (14), respectively.

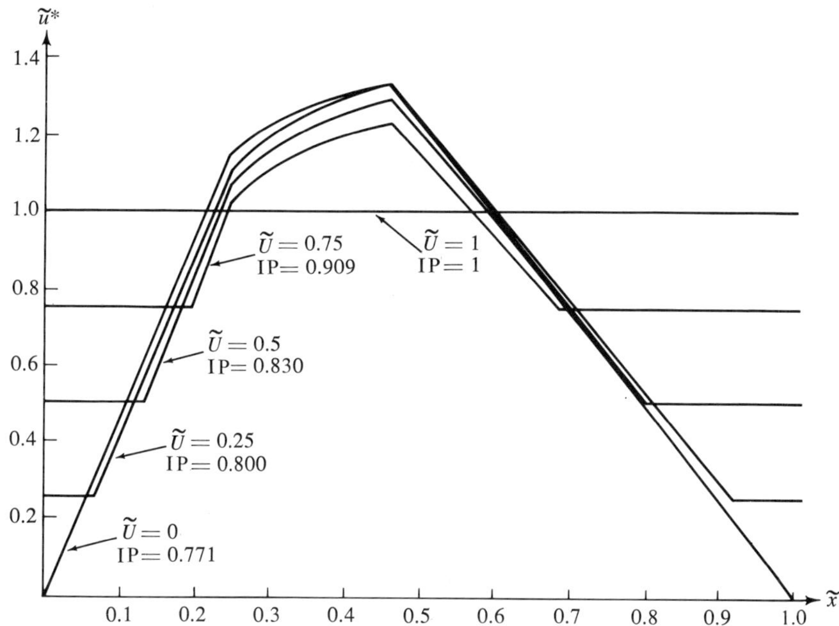

FIGURE 5-6 Optimal beam width for various levels of the control constraint

Starting with $\tilde{y}_1(0) = 0$ and guessing $\tilde{y}_2(0)$ and $\bar{\lambda}_1$, the equations for $\tilde{y}_1(\tilde{x})$ and $\tilde{y}_2(\tilde{x})$ are integrated forward until condition (22b) is met. The point at which this occurs is our first approximation to $\tilde{x}_1$. However, since $\tilde{y}_2(0)$ and $\bar{\lambda}$ were guessed, it cannot be expected that Eq. (22a) will be met; that is,

$$\tilde{y}_1(\tilde{x}_1) \neq -1 \tag{24}$$

Consider Eq. (22c), which by Eq. (18) is equivalent to requiring continuity of $\lambda_2(\tilde{x}_1)$. From Eq. (14) one then has that for $\lambda_2(\tilde{x}_1)$ to be continuous

$$\bar{\bar{\lambda}}_1 = \bar{\lambda}_1 \frac{\tilde{x}_1}{1 - \tilde{x}_1} \tag{25}$$

Using the value of $\bar{\bar{\lambda}}_1$ given by Eq. (25) and the continuity requirements on $\tilde{y}_1(\tilde{x}_1)$ and $\tilde{y}_2(\tilde{x}_1)$, the system equations are integrated until $\tilde{x} = 1$. Again, because of the initial arbitrariness in the choice of $\tilde{y}_2(0)$ and $\bar{\lambda}_1$ it cannot be expected that the terminal condition on $\tilde{y}_1(1)$ will be met. Thus in general

$$\tilde{y}_1(1) \neq 0 \tag{26}$$

Two of the problem conditions, Eqs. (22a) and (23), have not been met. To satisfy these conditions the values of the two guessed parameters $\tilde{y}_2(0)$ and $\bar{\lambda}_1$ are available and can be adjusted to achieve the desired result. Thus we consider

$$\tilde{y}_1(\tilde{x}_1) = \tilde{y}_{1_{\tilde{x}_1}}[\tilde{y}_2(0), \bar{\lambda}_1]$$

$$\tilde{y}_1(1) = \tilde{y}_{1_1}[\tilde{y}_2(0), \bar{\lambda}_1]$$

A technique of driving the parameters to the values that meet the specified conditions has been discussed in Section 4-3. Actually the present problem is a three-point boundary-value problem as it involves conditions at the points $\tilde{x}$ equal to 0, $\tilde{x}_1$, and 1. However, the technique developed in Section 4-3 is seen to be directly applicable here.

In Fig. 5-6 solutions to the problem for several values of the control constraint $\tilde{U}$ are presented and compared to the optimal constant-beam-width solution.

REFERENCES

1. S. Dreyfus, "Variational Problem with State Variable Inequality Constraints," *Rand Corp. Paper P-2605*, July 1962; also Ph.D. dissertation, Harvard Univ., Cambridge, Mass., 1962.
2. N. J. Cernica, *Strength of Materials*, Holt, Rinehart and Winston, New York, 1966, Secs. 4.1, 4.2, 3.4.

6

DIRECT METHODS—
THE TECHNIQUE OF
STEEPEST DESCENT

6-1 INTRODUCTION

In several of the examples treated thus far, the problem of finding the minimum of the system IP has required that the techniques of numerical analysis be employed. This occurred because the equations obtained from satisfying the condition that the first variation of the system IP vanish were too difficult to solve analytically. Solution of the equations arising from setting $\delta IP = 0$ thus led to consideration of numerical techniques for solving the resulting two-point boundary-value problems (see Sections 4-2 and 4-3).

Since determination of the minimizing solution by requiring that it satisfy the condition $\delta IP = 0$ necessitates numerical techniques, the question arises whether such techniques might not be employed directly in the minimization problem. That is, does a means exist of approximately determining the minimizing solution by numerical techniques without requiring in the procedure knowledge of those functions for which the first variation of the IP vanishes? Techniques of this type are termed *direct methods*, whereas those employing the requirement that $\delta IP = 0$ to find the minimizing solution are termed *indirect methods*.

The present chapter develops the direct method of steepest descent. In this technique an initial "estimate" (guess) of the minimizing solution is first made and inserted into the IP. The technique then improves this approximation by determining what changes are required to drive the system IP to a still lower value along the path of steepest descent.

167

6-2 A SIMPLE EXAMPLE FROM THE MINIMA OF FUNCTIONS

To introduce the technique let us return to the problem of minimizing a function of a single variable. Consider the system index

$$IP = IP(x) \tag{1}$$

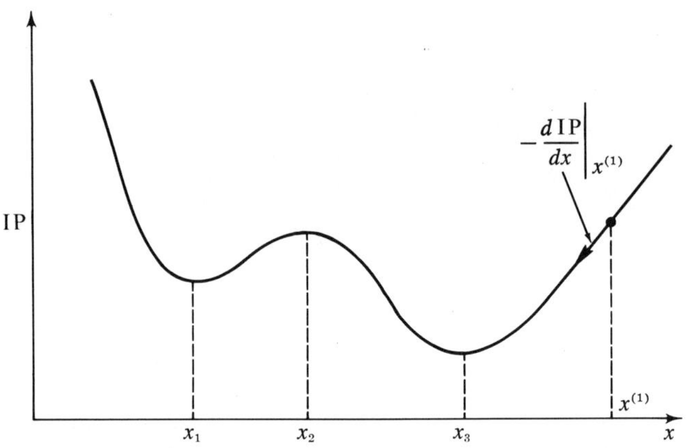

FIGURE 6-1   Application of the method to minimizing an unconstrained function

shown in Fig. 6-1, the minimum value of which is seen to occur at x_3. Let us compare how this result is obtained in both the indirect method and the direct method of steepest descent.

INDIRECT METHOD: Following the development in Chapter 2, to find the value of x which minimizes the IP, one first determines those values of x which render the IP stationary ($\delta IP = 0$). This requires that in the interior of the admissible x range the minimizing value $x = x^*$ be a solution of the equation

$$\frac{dIP}{dx}\bigg|_{x^*} = 0 \tag{2}$$

Consistent with Fig. 6-1, denote the solutions found by x_1, x_2, and x_3. These values must then be tested against the additional condition that for a minimum of the IP at $x = x^*$, $\delta^2 IP|_{x^*} \geqslant 0$, requiring

$$\frac{d^2 IP}{dx^2}\bigg|_{x^*} \geqslant 0 \tag{3}$$

In this manner the value x_2 is eliminated and local minima are found to exist at x_1 and x_3. To determine the absolute minimum the values x_1 and x_3 are inserted into Eq. (1) and the smallest value of the IP chosen.

In practice, the ease with which the minimum IP can be found by this method depends on the ease with which solutions that satisfy the condition $\delta\text{IP} = 0$ [Eq. (2)], can be found.

DIRECT METHOD: An initial estimate, $x^{(1)}$, is made of the minimizing value of x as shown in Fig. 6-1. To improve this approximation to the solution the direction of maximum rate of decrease of the IP at $x^{(1)}$ is determined. An improved estimate of the minimizing value is then found by changing $x^{(1)}$ in that direction.

In one dimension the direction of steepest descent of the system is determined by the slope of the IP. Evaluating $d\text{IP}/dx$ at $x^{(1)}$ it is seen from Fig. 6-1 to be positive; hence a lower value of the IP can be obtained by moving from $x^{(1)}$ in the direction of decreasing x. That is, one may obtain a better approximation to the minimizing value of x by using

$$x^{(2)} = x^{(1)} + \Delta x^{(1)} = x^{(1)} - k_1 \left.\frac{d\text{IP}}{dx}\right|_{x^{(1)}} \qquad k_1 > 0$$

where the constant k_1 is chosen so as to limit the magnitude of $\Delta x^{(1)}$. The procedure can then be repeated using the improved estimate $x^{(2)}$ to determine the third approximation from

$$x^{(3)} = x^{(2)} - k_2 \left.\frac{d\text{IP}}{dx}\right|_{x^{(2)}} \qquad k_2 > 0$$

One continues in this manner until $d\text{IP}/dx|_{x^{(n)}}$ falls below some acceptable lower level, indicating that the vicinity of the minimizing solution has been reached, that is, that $x^* \simeq x^{(n)}$.

Before going on to more complicated examples, consider some of the features of the method. In using steepest descent one is only assured that the solution reached is an approximation to the solution for a local minimum in the IP. For example, if the initial estimate $x^{(1)}$ was chosen to be smaller than x_2, the procedure would have converged to x_1, a local minimum in the IP but not the absolute minimum.

In addition, when choosing the allowable magnitude of the change $\Delta x^{(i)}$, care must be taken to ensure that the range in which $d\text{IP}/dx|_{x^{(i)}}$ is a reasonable approximation to the rate of change of the IP is not exceeded. If this is not done, the possibility exists that the approximate solutions may oscillate, as between the two sides of the valley about x_3 or even jump over to a neighboring valley such as about x_1 in Fig. 6-1.

Finally, consider the condition for determining when the vicinity of the minimum has been reached, that is, that $d\mathrm{IP}/dx|_{x^{(i)}}$ be below some given value. If the valley about the minimum point does not have steep sides, it is apparent that $d\mathrm{IP}/dx|_{x^{(i)}}$ may be quite small, while the solution for x^* still differs considerably from the minimizing one. When it can be done, the IP should be chosen to eliminate or at least minimize this possibility.

All these points should be kept in mind when using the method of steepest descent.

6-3   THE TREATMENT OF CONSTRAINTS

The problem of determining the state variables x_1 and x_2 which minimize the performance index

$$\mathrm{IP} = -x_1 x_2 \tag{1}$$

subject to the restriction that the variables satisfy the constraint

$$\psi(x_1, x_2) = \frac{x_1^2}{a^2} + \frac{x_2^2}{b^2} - 1 = 0 \tag{2}$$

is solved first as an illustrative example. Utilizing Eq. (2) it is possible in this simple case to eliminate one of the variables from the IP of Eq. (1). In this manner the IP could be reduced to a function of a single independent variable and the problem solved as in Section 6-2. However, since eliminating the dependent variables in the IP is usually impractical, let us not employ that technique here.

Proceeding in the direction indicated in Section 6-2, an initial estimate, denoted by

$$\mathbf{x}^{(1)} = \begin{bmatrix} x_1^{(1)} \\ x_2^{(1)} \end{bmatrix} \tag{3}$$

is first made of the solution variables. Now it is quite unlikely that $\mathbf{x}^{(1)}$ either satisfies the constraint of Eq. (2) or minimizes the IP. Thus the problem becomes that of determining the change to be made in $\mathbf{x}^{(1)}$ in forming the new approximation $\mathbf{x}^{(2)}$ so as to better satisfy the constraint while lowering the value of the IP. Let the change be denoted by $d\mathbf{x}$. Then

$$\mathbf{x}^{(2)} = \mathbf{x}^{(1)} + d\mathbf{x} = \begin{bmatrix} x_1^{(1)} + dx_1 \\ x_2^{(1)} + dx_2 \end{bmatrix} \tag{4}$$

To determine the changes in the values of $IP(\mathbf{x}^{(1)})$ and $\psi(\mathbf{x}^{(1)})$ that result from changing $\mathbf{x}^{(1)}$ by $d\mathbf{x}$ one forms

$$dIP = \left.\frac{\partial IP}{\partial x_1}\right|_{\mathbf{x}^{(1)}} dx_1 + \left.\frac{\partial IP}{\partial x_2}\right|_{\mathbf{x}^{(1)}} dx_2 \tag{5a}$$

$$d\psi = \left.\frac{\partial \psi}{\partial x_1}\right|_{\mathbf{x}^{(1)}} dx_1 + \left.\frac{\partial \psi}{\partial x_2}\right|_{\mathbf{x}^{(1)}} dx_2 \tag{5b}$$

The value of $d\psi$ is so far unspecified and thus can be selected so that the anticipated value $\psi(\mathbf{x}^{(2)})$ given by

$$\psi(\mathbf{x}^{(2)}) = \psi(\mathbf{x}^{(1)}) + d\psi \tag{6}$$

better satisfies the constraint of Eq. (2). Then since the IP is to be minimized in determining $\mathbf{x}^{(2)}$, the best choice of $d\mathbf{x}$ would appear to be that which by Eq. (5a) minimizes dIP (drives it as far negative as possible) while satisfying Eq. (5b) with the specified value of $d\psi$.

However, as pointed out in Section 6-2, if too large a change $d\mathbf{x}$ is made at one time, the representation of dIP and $d\psi$ given by Eq. (5) will be inaccurate. The possibility then exists that the desired changes in $IP(\mathbf{x}^{(1)})$ and $\psi(\mathbf{x}^{(1)})$ will not occur. To lessen this possibility the magnitude of $d\mathbf{x}$ is restricted so that

$$dS^2 = dx_1^2 + dx_2^2 \tag{7}$$

where the value of dS (the "safe" step size) is chosen on the basis of physical insight (intuition) into the problem.

The problem thus reduces to choosing $d\mathbf{x}$ so as to minimize

$$dIP = -x_2^{(1)}\, dx_1 - x_1^{(1)}\, dx_2 \tag{8}$$

while satisfying†

$$d\psi = \frac{2x_1^{(1)}}{a^2}\, dx_1 + \frac{2x_1^{(1)}}{b^2}\, dx_2 \tag{9}$$

$$dS^2 = dx_1^2 + dx_2^2 \tag{10}$$

where $d\psi$ and dS are specified values selected in the manner just described.

Forming the augmented differential IP by adding the product of the

---

† As only two variables, dx_1 and dx_2, are available, the two constraints of Eqs. (9) and (10) could be used to determine them directly. However, to provide an introduction to the more general problem that follows, this procedure is not used here, although the result is the same as in the method employed.

constraints of Eqs. (9) and (10) with the undetermined multipliers ν and μ one has

$$d\mathscr{IP} = d\text{IP} = -x_2^{(1)}\,dx_1 - x_1^{(1)}\,dx_2$$

$$+ \nu\left[d\psi - \left(\frac{2x_1^{(1)}}{a^2}\,dx_1 + \frac{2x_2^{(1)}}{b^2}\,dx_2\right)\right]$$

$$+ \mu[dS^2 - (dx_1^2 + dx_2^2)] \tag{11}$$

Then for $d\mathbf{x}$ to minimize $d\mathscr{IP}$ requires that $d^2\mathscr{IP}$ vanish:

$$d^2\mathscr{IP} = 0 = \frac{\partial d\mathscr{IP}}{\partial dx_1}\,dx_1 + \frac{\partial d\mathscr{IP}}{\partial dx_2}\,dx_2 \tag{12}$$

where since dx_1 and dx_2 may be treated as independent due to introduction of the multipliers ν and μ it follows that

$$\frac{\partial d\mathscr{IP}}{\partial dx_1} = 0 \tag{13a}$$

$$\frac{\partial d\mathscr{IP}}{\partial dx_2} = 0 \tag{13b}$$

Equations (13a, b) yield

$$dx_1 = \frac{-1}{2\mu}\left(x_2^{(1)} + \frac{2x_1^{(1)}}{a^2}\nu\right) \tag{14a}$$

$$dx_2 = \frac{-1}{2\mu}\left(x_1^{(1)} + \frac{2x_2^{(1)}}{a^2}\nu\right) \tag{14b}$$

Equations (9), (10), (14a), and (14b) are four equations for the determination of the four unknowns ν, μ, dx_1, and dx_2.

Using Eq. (14) to eliminate dx_1 and dx_2 in the constraint of Eq. (9) one obtains

$$d\psi = \frac{-1}{2\mu}\left[2x_1x_2\left(\frac{1}{a^2} + \frac{1}{b^2}\right) + 4\nu\left(\frac{x_1^2}{a^4} + \frac{x_2^2}{b^4}\right)\right]_{\mathbf{x}^{(1)}} \tag{15}$$

which may be written

$$\nu = \left[\frac{-\mu\,d\psi}{2} - \frac{x_1^{(1)}x_2^{(1)}}{2}\left(\frac{1}{a^2} + \frac{1}{b^2}\right)\right]\frac{1}{[(x_1^2/a^4) + (x_2^2/b^4)]_{\mathbf{x}^{(1)}}} \tag{16}$$

To satisfy the remaining constraint on the step size, dx_1 and dx_2 are substituted from Eq. (14) into Eq. (10). Then using Eq. (16) to eliminate ν, Eq. (10) can be solved for the only remaining unknown μ, yielding

$$2\mu = \pm \frac{\left[(x_1^2 + x_2^2) - \dfrac{x_1^2 x_2^2[(1/a^2) + (1/b^2)]^2}{(x_1^2/a^4) + (x_2^2/b^4)}\right]^{\frac{1}{2}}_{\mathbf{x}^{(1)}}}{\left[dS^2 - \dfrac{d\psi^2}{4[(x_1^2/a^4) + (x_2^2/b^4)]}\right]^{\frac{1}{2}}_{\mathbf{x}^{(1)}}} \tag{17}$$

The sign of μ remains to be determined.

With ν given by Eq. (16), the results of Eq. (14) become

$$dx_1 = \frac{1}{2\mu}\left[-x_2 + \frac{(x_1^2 x_2/a^2)[(1/a^2) + (1/b^2)]}{(x_1^2/a^4) + (x_2^2/b^4)}\right]_{\mathbf{x}^{(1)}}$$

$$+ \frac{x_1^{(1)}\, d\psi}{2a^2[(x_1^2/a^4) + (x_2^2/b^4)]_{\mathbf{x}^{(1)}}} \tag{18a}$$

$$dx_2 = \frac{1}{2\mu}\left[-x_1 + \frac{(x_1 x_2^2/b^2)[(1/a^2) + (1/b^2)]}{(x_1^2/a^4) + (x_2^2/b^4)}\right]_{\mathbf{x}^{(1)}}$$

$$+ \frac{x_2^{(1)}\, d\psi}{2b^2[(x_1^2/a^4) + (x_2^2/b^4)]_{\mathbf{x}^{(1)}}} \tag{18b}$$

Substituting dx_1 and dx_2 from Eq. (18) into Eq. (8) for $d\mathrm{IP}$, one finds that

$$d\mathrm{IP} = \pm\left[(x_1^2 + x_2^2) - \frac{x_1^2 x_2^2[(1/a^2) + (1/b^2)]^2}{(x_1^2/a^4) + (x_2^2/b^4)}\right]^{\frac{1}{2}}_{\mathbf{x}^{(1)}}$$

$$\times \left[dS^2 - \frac{d\psi^2}{4[(x_1^2/a^4) + (x_2^2/b^4)]}\right]^{\frac{1}{2}}_{\mathbf{x}^{(1)}}$$

$$- \frac{x_1^{(1)} x_2^{(1)}\, d\psi}{2[(x_1^2/a^4) + (x_2^2/b^4)]_{\mathbf{x}^{(1)}}}\left(\frac{1}{a^2} + \frac{1}{b^2}\right) \tag{19}$$

where Eq. (17) has been used to eliminate μ. The choice of sign in front of the first term in Eq. (19) corresponds to the choice of sign to be made for μ in Eq. (17). Thus, since $d\mathrm{IP}$ is to be minimized, the minus sign must be used in Eqs. (19) and (17).

With this determination only the value of $d\psi$ remains to be specified. Typically one lets

$$d\psi = -k\psi(\mathbf{x}^{(1)}) \qquad 0 < k \leqslant 1 \tag{20}$$

with the parameter k chosen so that the value of $d\psi$ used is within the range for which Eq. (5b), which represents small changes in ψ, can be expected to be valid. An upper limit is placed on the magnitude of $d\psi$ by the requirement that the denominator in Eq. (17) be real, so that $d\mathrm{IP}$ will be real.

The development of the first step in the procedure is now complete. After

TABLE 6-3(1)   *Minimization of* $IP = -X_1 X_2$ *with* $\psi = X_1^2 + \dfrac{X_2^2}{4} - 1 = 0$.

Run No.	X_1	X_2	IP	$\left\|\dfrac{dIP}{dS}\right\|_{d\psi=0}$	dS
1	1.0000	0.0000	-0.0000	1.0000	
2	1.0000	0.1000	-0.1000	0.9971	
3	0.9962	0.1999	-0.1991	0.9849	
4	0.9899	0.2997	-0.2967	0.9645	
5	0.9811	0.3993	-0.3918	0.9357	
6	0.9697	0.4986	-0.4836	0.8982	
7	0.9556	0.5976	-0.5712	0.8518	
8	0.9388	0.6962	-0.6537	0.7962	
9	0.9192	0.7943	-0.7302	0.7308	
10	0.8967	0.8917	-0.7996	0.6550	0.1
11	0.8710	0.9884	-0.8609	0.5682	
12	0.8421	1.0841	-0.9310	0.4695	
13	0.8098	1.1787	-0.9546	0.3579	
14	0.7738	1.2720	-0.9844	0.2323	
15	0.7339	1.3637	-1.0009	0.0911	
16	0.6897	1.4534	-1.0025	0.0671	
17	0.7341	1.3638	-1.0012	0.0913	
18	0.6897	1.4534	-1.0026	0.0671	
19	0.7341	1.3638	-1.0012	0.0913	
20	0.6897	1.4534	-1.0026	0.0671	
21	0.6920	1.4437	-0.9991	0.0540	
22	0.6966	1.4348	-0.9996	0.0374	
23	0.7012	1.4259	-0.9999	0.0211	
24	0.7057	1.4170	-1.0000	0.0050	0.01
25	0.7101	1.4080	-1.0000	0.0109	
26	0.7057	1.4170	-1.0000	0.0050	
27	0.7101	1.4080	-1.0000	0.0109	
28	0.7057	1.4170	-1.0000	0.0050	
29	0.7101	1.4080	-1.0000	0.0109	
30	0.7097	1.4089	-0.9999	0.0093	
31	0.7092	1.4098	-0.9999	0.0077	
32	0.7088	1.4107	-0.9999	0.0061	
33	0.7083	1.4116	-0.9999	0.0045	
34	0.7079	1.4125	-0.9999	0.0029	0.001
35	0.7074	1.4134	-0.9999	0.0013	
36	0.7070	1.4143	-1.0000	0.0002	
37	0.7075	1.4133	-0.9999	0.0014	
38	0.7070	1.4142	-1.0000	0.0000	

computing μ from Eq. (17) and substituting into Eq. (18) one finds dx_1 and dx_2. The second approximation to the minimizing solution is then given by Eq. (4). The question arises as to whether to compute a third approximation. Let us assume that using $\mathbf{x}^{(2)}$ the constraint of Eq. (2) is satisfied with sufficient accuracy. Then in determining the change $d\mathbf{x}$ to be made in $\mathbf{x}^{(2)}$ to form $\mathbf{x}^{(3)}$, should it be desired, Eq. (20) would require $d\psi = 0$. With this condition Eq. (19) then yields

$$\left| \frac{d\text{IP}}{dS} \right|_{d\psi=0} = \left[x_1^2 + x_2^2 - \frac{x_1^2 x_2^2 [(1/a^2) + (1/b^2)]^2}{(x_1^2/a^4) + (x_2^2/b^4)} \right]^{1/2}_{\mathbf{x}^{(2)}} \tag{21}$$

Equation (21) provides an expression for the rate of change of the IP in the direction of steepest descent consistent with maintaining $\psi = 0$. If $|d\text{IP}/dS|_{d\psi=0}$ is not small, it is apparent that the minimum of the IP has not been reached, that is, that $\mathbf{x}^{(2)}$ does not correspond to the minimizing solution. The approximation must then be improved by calculating $\mathbf{x}^{(3)}$. The computations continue until (1) the nth approximation is reached, where $\psi(\mathbf{x}^{(n)}) \simeq 0$, requiring $d\psi = 0$ in the calculation of $\mathbf{x}^{(n+1)}$ and (2) $|d\text{IP}/dS|_{d\psi=0}$ is below some acceptable level.

To illustrate the computational procedure the specific case $a = 1$, $b = 2$ is chosen. The exact solution to the problem yields $x_1^* = 1/\sqrt{2}$, $x_2^* = \sqrt{2}$, and a corresponding performance index of $\text{IP}^* = -1$. The approach to this solution by the steepest-descent technique is shown for the starting condition $x_1 = 1$, $x_2 = 0$ in Table 6-3(1). With $dS = 0.1$, it is seen that $|d\text{IP}/dS|_{d\psi=0}$ begins to oscillate after some 15 iterations. The value of dS is then reduced to 0.01 and the process continued. When $|d\text{IP}/dS|_{d\psi=0}$ once again begins to oscillate, the value of dS is further reduced to 0.001. The excellent agreement reached with the exact solution is illustrated in the table. In all cases the constant k in Eq. (20) was chosen to be unity, so that at each step the solution attempted to nullify the deviation of ψ from zero. If with this choice the quantity $dS^2 - d\psi^2/4[(x_1^2/a^4) + (x_2^2/b^4)]$ in Eq. (17) was negative, the value of $d\psi$ was reduced to make the quantity equal zero.

6-4   CONTROL FORMULATION

As previously discussed, designers in working with a system usually subdivide the system variables into two categories. These are (1) those variables which define the configuration of the system, termed *state variables*, and (2) those variables which are available to change the state of the system, termed *control variables*. The problem of Section 6-3 as presented and solved took no note of this distinction between the variables. It is assumed there that changes in the state variables could be effected directly without working through control variables. Let us now reformulate the example of Section 6-3 so as to introduce control variables.

The problem is taken to be that of determining the control variables u_1 and u_2 which cause the state variables x_1 and x_2 to minimize the performance index

$$\text{IP} = -x_1 x_2 \tag{1}$$

subject to the restrictions that x_1 and x_2 satisfy the state-variable constraint

$$\psi(x_1, x_2) = \frac{x_1^2}{a^2} + \frac{x_2^2}{b^2} - 1 = 0 \tag{2}$$

and the two system equations

$$f_1(x_1, x_2, u_1, u_2) = 0$$
$$f_2(x_1, x_2, u_1, u_2) = 0 \tag{3}$$

To start the procedure, an initial estimate is made that the optimal control values are $\mathbf{u}^{(1)} = (u_1^{(1)}, u_2^{(1)})$. Then from Eq. (3) the corresponding values of the state variables are calculated to be $\mathbf{x}^{(1)} = (x_1^{(1)}, x_2^{(1)})$. In general, $\mathbf{x}^{(1)}$ will neither satisfy the state-variable constraint of Eq. (2) nor minimize the IP. Thus it is necessary to determine how to change the control $\mathbf{u}^{(1)}$ in forming the second approximation $\mathbf{u}^{(2)}$ so as to obtain state variables $\mathbf{x}^{(2)}$ which better satisfy the constraint of Eq. (2) while providing a lower value of the IP.

As the state and control variables are interrelated through Eq. (3), the relation between differential changes in these quantities is given by the differential of Eq. (3), which is

$$df_1 = 0 = \frac{\partial f_1}{\partial x_1}\bigg|_{(1)} dx_1 + \frac{\partial f_1}{\partial x_2}\bigg|_{(1)} dx_2 + \frac{\partial f_1}{\partial u_1}\bigg|_{(1)} du_1 + \frac{\partial f_1}{\partial u_2}\bigg|_{(1)} du_2$$

$$df_2 = 0 = \frac{\partial f_2}{\partial x_1}\bigg|_{(1)} dx_1 + \frac{\partial f_2}{\partial x_2}\bigg|_{(1)} dx_2 + \frac{\partial f_2}{\partial u_1}\bigg|_{(1)} + du_1 \frac{\partial f_2}{\partial u_2}\bigg|_{(1)} du_2 \tag{4}$$

The notation $|_{(1)}$ or $[\ \ ]_{(1)}$ indicates that the quantity is to be evaluated at $\mathbf{x}^{(1)}$ and $\mathbf{u}^{(1)}$. In matrix notation Eq. (4) may be written

$$\begin{bmatrix} df_1 \\ df_2 \end{bmatrix} = \begin{bmatrix} 0 \\ 0 \end{bmatrix} = \begin{bmatrix} \dfrac{\partial f_1}{\partial x_1} & \dfrac{\partial f_1}{\partial x_2} \\ \dfrac{\partial f_2}{\partial x_1} & \dfrac{\partial f_2}{\partial x_2} \end{bmatrix}_{(1)} \begin{bmatrix} dx_1 \\ dx_2 \end{bmatrix} + \begin{bmatrix} \dfrac{\partial f_1}{\partial u_1} & \dfrac{\partial f_1}{\partial u_2} \\ \dfrac{\partial f_2}{\partial u_1} & \dfrac{\partial f_2}{\partial u_2} \end{bmatrix}_{(1)} \begin{bmatrix} du_1 \\ du_2 \end{bmatrix} \tag{5}$$

Defining the matrices

$$\mathbf{f} = \begin{bmatrix} f_1 \\ f_2 \end{bmatrix} \qquad \mathbf{x} = \begin{bmatrix} x_1 \\ x_2 \end{bmatrix} \qquad \mathbf{u} = \begin{bmatrix} u_1 \\ u_2 \end{bmatrix} \tag{6a}$$

and

$$\mathbf{F} = \begin{bmatrix} \dfrac{\partial f_1}{\partial x_1} & \dfrac{\partial f_1}{\partial x_1} \\[2ex] \dfrac{\partial f_2}{\partial x_1} & \dfrac{\partial f_2}{\partial x_2} \end{bmatrix}_{(1)} \qquad \mathbf{G} = \begin{bmatrix} \dfrac{\partial f_1}{\partial u_1} & \dfrac{\partial f_1}{\partial u_2} \\[2ex] \dfrac{\partial f_2}{\partial u_1} & \dfrac{\partial f_2}{\partial u_2} \end{bmatrix}_{(1)} \tag{6b}$$

Equation (4) becomes simply

$$d\mathbf{f} = 0 = \mathbf{F}\,d\mathbf{x} + \mathbf{G}\,d\mathbf{u} \tag{7}$$

Consider next the change in the value of $\boldsymbol{\psi}(\mathbf{x}^{(1)})$ resulting from changing $\mathbf{u}^{(1)}$. In terms of the state-variable change, one has

$$d\psi = \left.\frac{\partial \psi}{\partial x_1}\right|_{x^{(1)}} dx_1 + \left.\frac{\partial \psi}{\partial x_2}\right|_{x^{(2)}} dx_2 \tag{8}$$

which on defining the matrix

$$\frac{\partial \boldsymbol{\psi}}{\partial \mathbf{x}} = \begin{bmatrix} \dfrac{\partial \psi}{\partial x_1} & \dfrac{\partial \psi}{\partial x_2} \end{bmatrix}_{(1)} = \begin{bmatrix} \dfrac{2x_1}{a^2} & \dfrac{2x_2}{b^2} \end{bmatrix}_{(1)} \tag{9}$$

becomes†

$$d\boldsymbol{\psi} = \frac{\partial \boldsymbol{\psi}}{\partial \mathbf{x}}\,d\mathbf{x} \tag{10}$$

However, the state variables are not altered directly but through changes in the control variables. To determine the effect on $\boldsymbol{\psi}(\mathbf{x}^{(1)})$ of changes in $\mathbf{u}^{(1)}$, the constraint of Eq. (7) must be used. Rather than solving for $d\mathbf{u}$ in terms of $d\mathbf{x}$ from Eq. (7) these relations are introduced using undetermined multipliers. On forming the product of $df_1 = 0$ with the multiplier λ_{ψ_1} and $df_2 = 0$ with the multiplier λ_{ψ_2} and adding these products to $d\psi$ in Eq. (10) one obtains the augmented function $d\boldsymbol{\Psi}$. Define the matrix $\boldsymbol{\lambda}_\psi$ by

$$\boldsymbol{\lambda}_\psi = \begin{bmatrix} \lambda_{\psi_1} \\ \lambda_{\psi_2} \end{bmatrix} \tag{11}$$

Then in matrix notation, using Eq. (7) for $d\mathbf{f}$, the equation for $d\boldsymbol{\Psi}$ is

$$d\boldsymbol{\Psi} = d\boldsymbol{\psi} = \frac{\partial \boldsymbol{\psi}}{\partial \mathbf{x}}\,d\mathbf{x} + \boldsymbol{\lambda}_\psi'[\mathbf{F}\,d\mathbf{x} + \mathbf{G}\,d\mathbf{u}] \tag{12}$$

where $\boldsymbol{\lambda}_\psi'$ is the transpose of $\boldsymbol{\lambda}_\psi$. Collecting the coefficients of $d\mathbf{x}$ together, Eq. (12) may be written

$$d\boldsymbol{\Psi} = \left[\frac{\partial \boldsymbol{\psi}}{\partial \mathbf{x}} + \boldsymbol{\lambda}_\psi'\mathbf{F}\right] d\mathbf{x} + \boldsymbol{\lambda}_\psi'\mathbf{G}\,d\mathbf{u} \tag{13}$$

† In this problem $d\psi$ is a 1×1 matrix, that is, a scalar.

Since the stated purpose of introducing $d\mathbf{f}$ from Eq. (7) into Eq. (10) was to express the effect of $d\mathbf{u}$ on $d\boldsymbol{\psi}$ directly, let us choose $\boldsymbol{\lambda}_\psi$, which is at our disposal, so that the coefficient of $d\mathbf{x}$ vanishes in Eq. (13). Thus λ_{ψ_1} and λ_{ψ_2} are determined from the two equations

$$\frac{\partial \boldsymbol{\psi}}{\partial \mathbf{x}} + \boldsymbol{\lambda}_\psi' \mathbf{F} = 0 \tag{14}$$

Multiplying Eq. (14) from the right by the inverse of the matrix $\mathbf{F}$ and taking the transpose of the equation one obtains as the solution†

$$\boldsymbol{\lambda}_\psi = -(\mathbf{F}^{-1})' \frac{\partial \boldsymbol{\psi}'}{\partial \mathbf{x}} \tag{15}$$

With λ_{ψ_1} and λ_{ψ_2} given by Eq. (15), Eq. (13) becomes

$$d\boldsymbol{\Psi} = d\boldsymbol{\psi} = \boldsymbol{\lambda}_\psi' \mathbf{G}\, d\mathbf{u} \tag{16}$$

expressing the change in $\boldsymbol{\psi}$ to be expected from differential changes in the control variables.

In a similar manner, the change in the IP resulting from differential change in the control $\mathbf{u}^{(1)}$ may be found. First form

$$d\text{IP} = \frac{\partial \text{IP}}{\partial \mathbf{x}}\, d\mathbf{x} \tag{17}$$

where the matrix $\partial \text{IP}/\partial \mathbf{x}$ is defined by

$$\frac{\partial \text{IP}}{\partial \mathbf{x}} = \left[\frac{\partial \text{IP}}{\partial x_1} \quad \frac{\partial \text{IP}}{\partial x_2} \right]_{(1)} = [-x_2 \quad -x_1]_{(1)} \tag{18}$$

Then to express the change in the IP in terms of the control change, the differential system equations of (7) are used. Introducing the two multipliers λ_{IP_1} and λ_{IP_2}, their respective product with df_1 and df_2 from Eq. (7) is appended to $d\text{IP}$ to form the augmented function

$$d\mathscr{IP} = d\text{IP} = \frac{\partial \text{IP}}{\partial \mathbf{x}}\, d\mathbf{x} + \boldsymbol{\lambda}_{\text{IP}}'[\mathbf{F}\, d\mathbf{x} + \mathbf{G}\, d\mathbf{u}] \tag{19}$$

where

$$\boldsymbol{\lambda}_{\text{IP}} = \begin{bmatrix} \lambda_{\text{IP}_1} \\ \lambda_{\text{IP}_2} \end{bmatrix} \tag{20}$$

---

† The transpose of the product of several matrices is equal to the product of the transposed matrices taken in reverse order.

To eliminate $d\mathbf{x}$ from Eq. (19) the multipliers $\boldsymbol{\lambda}_{\mathrm{IP}}$ are chosen so that the coefficient of $d\mathbf{x}$ in Eq. (19) vanishes. Thus

$$\frac{\partial \mathrm{IP}}{\partial \mathbf{x}} + \boldsymbol{\lambda}_{\mathrm{IP}}'\mathbf{F} = 0 \tag{21}$$

from which one finds

$$\boldsymbol{\lambda}_{\mathrm{IP}} = -(\mathbf{F}^{-1})'\,\frac{\partial \mathrm{IP}'}{\partial \mathbf{x}} \tag{22}$$

With $\boldsymbol{\lambda}_{\mathrm{IP}}$ determined in this manner, Eq. (19) becomes

$$d\mathscr{IP} = \boldsymbol{\lambda}_{\mathrm{IP}}'\mathbf{G}\,d\mathbf{u} \tag{23}$$

In line with the discussion of Section 6-3, the control change $d\mathbf{u}$ is to be determined so as to move in the direction of steepest descent of $\mathscr{IP}$ consistent with achieving a given value of $d\boldsymbol{\psi}$. However, as also noted before, if too large a change is made at one time, the relations of Eqs. (16) and (23) cannot be expected to hold. Thus some limitation must be placed on the distance moved in control space. Analogous to Eq. 6-3(10) let us require that the distance moved in control space be the specified value dS.

$$dS^2 = W_{11}\,du_1^2 + 2W_{12}\,du_1\,du_2 + W_{22}\,du_2^2 \tag{24}$$

where dS is the admissible change at one time. In matrix notation

$$dS^2 = d\mathbf{u}'\mathbf{W}\,d\mathbf{u} \tag{25}$$

The symmetric matrix $\mathbf{W}$, which can be intrepreted as the metric in control space, is given by

$$\mathbf{W} = \begin{bmatrix} W_{11} & W_{12} \\ W_{21} & W_{22} \end{bmatrix} \qquad W_{12} = W_{21} \tag{26}$$

It must be chosen so that the quadratic form of Eq. (25) is positive definite. Aside from this restriction the elements of $\mathbf{W}$ are at the disposal of the designer to aid convergence by allowing different emphasis to be placed on the changes du_1 and du_2. If no preference exists, or one's intuition is poor, $\mathbf{W}$ would be chosen as the identity matrix ($W_{11} = W_{22} = 1$, $W_{12} = 0$), in which case Eq. (24) reduces to

$$dS^2 = du_1^2 + du_2^2 \tag{27}$$

Having determined $\boldsymbol{\lambda}_{\psi}$ and $\boldsymbol{\lambda}_{\mathrm{IP}}$ from Eqs. (15) and (22) the problem becomes that of determining $d\mathbf{u}$ so as to minimize

$$d\mathrm{IP} = \boldsymbol{\lambda}_{\mathrm{IP}}'\mathbf{G}\,d\mathbf{u} \tag{28}$$

while satisfying

$$d\boldsymbol{\psi} = \boldsymbol{\lambda}'_{\psi}\mathbf{G}\,d\mathbf{u} \tag{29}$$

and

$$dS^2 = d\mathbf{u}'\mathbf{W}\,d\mathbf{u} \tag{30}$$

with specified values for $d\boldsymbol{\psi}$ and dS. Rather than solve the problem as formulated here for the case of only two state variables, let us extend the formulation to treat the general problem of n state variables. This is done in Section 6-5.

However, before going on, for purposes of comparison with the work of Section 6-3, consider that the system equations of Eq. (3) are given by

$$f_1 = x_1 - u_1 = 0$$
$$f_2 = x_2 - u_2 = 0 \tag{31}$$

In this case one finds from Eq. (6b) that

$$\mathbf{F} = \begin{bmatrix} 1 & 0 \\ 0 & 1 \end{bmatrix} \qquad \mathbf{G} = \begin{bmatrix} -1 & 0 \\ 0 & -1 \end{bmatrix} \tag{32}$$

Consequently, using this result Eqs. (15) and (22) become

$$\boldsymbol{\lambda}_{\psi} = \begin{bmatrix} \lambda_{\psi_1} \\ \lambda_{\psi_2} \end{bmatrix} = -\begin{bmatrix} 1 & 0 \\ 0 & 1 \end{bmatrix}\begin{bmatrix} \dfrac{2x_1^{(1)}}{a^2} \\ \dfrac{2x_2^{(1)}}{b^2} \end{bmatrix}$$

and

$$\boldsymbol{\lambda}_{\mathrm{IP}} = \begin{bmatrix} \lambda_{\mathrm{IP}_1} \\ \lambda_{\mathrm{IP}_2} \end{bmatrix} = -\begin{bmatrix} 1 & 0 \\ 0 & 1 \end{bmatrix}\begin{bmatrix} -x_2^{(1)} \\ -x_1^{(1)} \end{bmatrix}$$

from which one has

$$\lambda_{\psi_1} = \frac{-2x_1^{(1)}}{a^2} \qquad \lambda_{\psi_2} = \frac{-2x_2^{(1)}}{b^2}$$

$$\lambda_{\mathrm{IP}_1} = x_2^{(1)} \qquad \lambda_{\mathrm{IP}_2} = x_1^{(1)}$$

Substituting into Eqs. (28), (29), and (30) the problem is seen to be that of determining $d\mathbf{u}$ so as to minimize

$$d\mathrm{IP} = -x_2^{(1)}\,du_1 - x_1^{(1)}\,du_2 \tag{33}$$

while satisfying

$$d\psi = \frac{2x_1^{(1)}}{a^2}\,du_1 + \frac{2x_2^{(1)}}{b^2}\,du_2 \tag{34}$$

and, after taking **W** in Eq. (26) to be the identity matrix, one has

$$dS^2 = du_1^2 + du_2^2 \tag{35}$$

The problem contained in Eqs. (33), (34), and (35) is just that given by Eqs. 6-3(8), 6-3(9), and 6-3(10) with du_1 and du_2 replacing dx_1 and dx_2 as is to be expected from the system equations of (31).

6-5 THE CASE OF n STATE VARIABLES AND m CONTROL VARIABLES — FORMULATION†

Our problem is now that of optimizing a system defined by n state variables $x_1, \cdots, x_n$ which are subject to m control variables $u_1, \cdots, u_m$. In particular, our task is taken to be that of minimizing the system performance index

$$\text{IP} = \text{IP}(x_1, x_2, \cdots, x_n) \tag{1}$$

while satisfying the p state-variable constraints

$$\psi_i(x_1, x_2, \cdots, x_n) = 0 \qquad i = 1, \cdots, p < n \tag{2}$$

and the n system equations

$$f_i(x_1, \cdots, x_n, u_1, \cdots, u_m) = 0 \qquad i = 1, \cdots, n \tag{3}$$

Let $\mathbf{u}^{(1)}$ be the matrix defining the initial estimate of the optimal controls operating on the system and let the matrix $\mathbf{x}^{(1)}$ be the resulting state of the system obtained on using $\mathbf{u}^{(1)}$ in Eq. (3). In matrix notation,

$$\underset{\substack{n \times 1 \text{ matrix of}\\ \text{state variables}}}{\mathbf{x} =} \begin{bmatrix} x_1 \\ x_2 \\ \vdots \\ x_n \end{bmatrix} \qquad\qquad \underset{\substack{m \times 1 \text{ matrix of}\\ \text{control variables}}}{\mathbf{u} =} \begin{bmatrix} u_1 \\ u_2 \\ \vdots \\ u_m \end{bmatrix} \tag{4a}$$

and

$$\underset{\substack{p \times 1 \text{ matrix of}\\ \text{known constraint}\\ \text{functions}}}{\boldsymbol{\psi}(\mathbf{x}) =} \begin{bmatrix} \psi_1 \\ \psi_2 \\ \vdots \\ \psi_p \end{bmatrix} \qquad\qquad \underset{\substack{n \times 1 \text{ matrix of}\\ \text{known system}\\ \text{functions}}}{\mathbf{f}(\mathbf{x}, \mathbf{u}) =} \begin{bmatrix} f_1 \\ f_2 \\ \vdots \\ f_n \end{bmatrix} \tag{4b}$$

† The procedure used here as well as in Section 6-8 follows that developed by A. E. Bryson and W. F. Denham in their paper, "A Steepest-Ascent Method for Solving Optimum Programming Problems," *J. Appl. Mech.*, June 1962, pp. 247–257. An equivalent treatment has also been developed by H. J. Kelley in "Gradient Theory of Optimal Flight Paths," *J. Am. Rocket Soc.*, Oct. 1960, pp. 947–954.

As $\mathbf{x}^{(1)}$ is determined by satisfying Eq. (3), one has

$$\mathbf{f}(\mathbf{x}^{(1)}, \mathbf{u}^{(1)}) = 0 \tag{5}$$

However, as the choice of $\mathbf{u}^{(1)}$ is quite arbitrary, the constraints of Eq. (2) will, in general, not be satisfied; that is, $\boldsymbol{\psi}(\mathbf{x}^{(1)}) \neq 0$. The control set $\mathbf{u}^{(1)}$ must be changed so as to drive $\boldsymbol{\psi}$ closer to zero while lowering the value of the IP.

Forming the differential of each of the system equations of (3) one finds that changes in the state and control variables about $\mathbf{x}^{(1)}$, $\mathbf{u}^{(1)}$ are related through the perturbation equations

$$d\mathbf{f} = 0 = \mathbf{F}\,d\mathbf{x} + \mathbf{G}\,d\mathbf{u} \qquad (n \text{ equations}) \tag{6}$$

where

$$
\begin{aligned}
\mathbf{F} &= \\
n \times n & \\
\text{matrix} &
\end{aligned}
\begin{bmatrix}
\dfrac{\partial f_1}{\partial x_1} & \dfrac{\partial f_1}{\partial x_2} & \cdots & \dfrac{\partial f_1}{\partial x_n} \\
\vdots & & & \\
\dfrac{\partial f_n}{\partial x_1} & \dfrac{\partial f_n}{\partial x_2} & \cdots & \dfrac{\partial f_n}{\partial x_n}
\end{bmatrix}_{(1)}
$$

$$
\begin{aligned}
\mathbf{G} &= \\
n \times m & \\
\text{matrix} &
\end{aligned}
\begin{bmatrix}
\dfrac{\partial f_1}{\partial u_1} & \dfrac{\partial f_1}{\partial u_2} & \cdots & \dfrac{\partial f_1}{\partial u_m} \\
\vdots & & & \\
\dfrac{\partial f_n}{\partial u_1} & \dfrac{\partial f_n}{\partial u_2} & \cdots & \dfrac{\partial f_n}{\partial u_m}
\end{bmatrix}_{(1)}
\tag{7}
$$

The matrices $\mathbf{F}$ and $\mathbf{G}$, being evaluated at $\mathbf{x}^{(1)}$, $\mathbf{u}^{(1)}$, are made up of known elements.

Consider next the change in each of the state constraints about its value at $\mathbf{x}^{(1)}$ resulting from changing $\mathbf{u}^{(1)}$. The objective is to express $d\boldsymbol{\psi}$ in terms of $d\mathbf{u}$ directly. One forms first

$$d\boldsymbol{\psi} = \frac{\partial \boldsymbol{\psi}}{\partial \mathbf{x}}\,d\mathbf{x} \qquad (p \text{ equations}) \tag{8}$$

where

$$
\begin{aligned}
\frac{\partial \boldsymbol{\psi}}{\partial \mathbf{x}} &= \\
p \times n & \\
\text{matrix} &
\end{aligned}
\begin{bmatrix}
\dfrac{\partial \psi_1}{\partial x_1} & \dfrac{\partial \psi_1}{\partial x_2} & \cdots & \dfrac{\partial \psi_1}{\partial x_n} \\
\vdots & & & \\
\dfrac{\partial \psi_p}{\partial x_1} & \dfrac{\partial \psi_p}{\partial x_2} & \cdots & \dfrac{\partial \psi_p}{\partial x_n}
\end{bmatrix}_{(1)}
\tag{9}
$$

is made up of known elements. Next, to replace $d\mathbf{x}$ in Eq. (8) by $d\mathbf{u}$ the relations of Eq. (6) are used. To $d\psi_i$ in Eq. (8) is appended the sum of the products of each of the relations of Eq. (6) with an undetermined multiplier. Denoting these multipliers by $\lambda_{\psi_i1}, \cdots, \lambda_{\psi_in}$ one obtains

$$d\Psi_i = d\psi_i = \frac{\partial \psi_i}{\partial \mathbf{x}} d\mathbf{x} + \lambda'_{\psi_i}[\mathbf{F}\,d\mathbf{x} + \mathbf{G}\,d\mathbf{u}] \tag{10}$$

where

$$\frac{\partial \psi_i}{\partial \mathbf{x}} = \left[\frac{\partial \psi_i}{\partial x_1} \quad \cdots \quad \frac{\partial \psi_i}{\partial x_n}\right]_{(1)} \qquad \lambda_{\psi_i} = \begin{bmatrix} \lambda_{\psi_i1} \\ \vdots \\ \lambda_{\psi_in} \end{bmatrix} \tag{11}$$

What was just done with $d\psi_i$ can be done with each of the other quantities in Eq. (8). Defining

$$\lambda_\psi = [\lambda_{\psi_1} \quad \lambda_{\psi_2} \quad \cdots \quad \lambda_{\psi_p}] \tag{12a}$$

$$\underset{\substack{n \times p \\ \text{matrix}}}{\lambda_\psi} = \begin{bmatrix} \lambda_{\psi_11} & \lambda_{\psi_21} & \cdots & \lambda_{\psi_p1} \\ \lambda_{\psi_12} & & & \\ \vdots & & & \\ \lambda_{\psi_1n} & \lambda_{\psi_2n} & \cdots & \lambda_{\psi_pn} \end{bmatrix} \tag{12b}$$

the general equation corresponding to Eq. (10) is

$$d\Psi = d\psi = \frac{d\psi}{\partial \mathbf{x}} d\mathbf{x} + \lambda'_\psi[\mathbf{F}\,d\mathbf{x} + \mathbf{G}\,d\mathbf{u}] \qquad (p \text{ equations}) \tag{13}$$

Then, regrouping and using the multipliers λ_ψ to eliminate the coefficient of $d\mathbf{x}$ in Eq. (13) one has that

$$d\psi = \lambda'_\psi \mathbf{G}\,d\mathbf{u} \qquad (p \text{ equations}) \tag{14}$$

where the multipliers λ_ψ are determined so as to satisfy the equations

$$\frac{\partial \psi}{\partial \mathbf{x}} + \lambda'_\psi \mathbf{F} = 0 \tag{15}$$

Equation (15) is actually p sets of equations for the p sets of multipliers $\lambda_{\psi_1}, \lambda_{\psi_2}, \cdots, \lambda_{\psi_p}$. For example, the n multipliers in the ith set, $\lambda_{\psi_i1} \cdots \lambda_{\psi_in}$ must satisfy the n equations

$$\frac{\partial \psi_i}{\partial \mathbf{x}} + \lambda'_{\psi_i} \mathbf{F} = 0 \tag{16}$$

yielding

$$\lambda_{\psi_i} = -(\mathbf{F}^{-1})' \frac{\partial \psi'_i}{\partial \mathbf{x}} \tag{17}$$

Equations (16) and (17) determine λ_{ψ_i} for $i = 1, 2, \cdots, p$.

In a similar manner, the change in the IP about $\mathbf{x}^{(1)}$, $\mathbf{u}^{(1)}$ is found. Forming

$$dIP = \frac{\partial IP}{\partial \mathbf{x}}\, d\mathbf{x} \tag{18}$$

where

$$\frac{\partial IP}{\partial \mathbf{x}} = \left[\frac{\partial IP}{\partial x_1} \quad \cdots \quad \frac{\partial IP}{\partial x_n}\right]_{(1)} \tag{19}$$

$$1 \times n \text{ matrix}$$

the sum of the product of each relation in Eq. (6) with an undetermined multiplier is appended to Eq. (18). Denoting the multipliers by

$$\boldsymbol{\lambda}_{IP} = \begin{bmatrix} \lambda_{IP_1} \\ \vdots \\ \lambda_{IP_n} \end{bmatrix} \tag{20}$$

$$n \times 1 \text{ matrix}$$

one obtains upon collecting coefficients of $d\mathbf{x}$,

$$d\mathscr{IP} = dIP = \left[\frac{\partial IP}{\partial \mathbf{x}} + \boldsymbol{\lambda}'_{IP}\mathbf{F}\right] d\mathbf{x} + \boldsymbol{\lambda}'_{IP}\mathbf{G}\, d\mathbf{u} \tag{21}$$

The multipliers $\boldsymbol{\lambda}_{IP}$ are determined so that the coefficient of $d\mathbf{x}$ vanishes, Thus

$$\boldsymbol{\lambda}_{IP} = -(\mathbf{F}^{-1})'\frac{\partial IP'}{\partial \mathbf{x}} \tag{22}$$

Using Eq. (22), the change in the performance index resulting from changing the control about $\mathbf{u}^{(1)}$ is seen to be

$$dIP = \boldsymbol{\lambda}'_{IP}\mathbf{G}\, d\mathbf{u} \tag{23}$$

Finally, a restriction must be imposed on the magnitude of the allowable change in control space. Denoting the admissible change by dS the condition is imposed that

$$dS^2 = d\mathbf{u}'\mathbf{W}\, d\mathbf{u} \tag{24}$$

where $\mathbf{W}$ is a symmetric, positive definite, $m \times m$ matrix representing the metric in control space.

6-6   THE CASE OF n STATE VARIABLES AND m CONTROL VARIABLES — SOLUTION

Having made the initial estimate that the optimal control matrix is $\mathbf{u}^{(1)}$, solution of Eqs. 6-5(3) provides knowledge of the resulting state $\mathbf{x}^{(1)}$. Inserting $\mathbf{x}^{(1)}$ into Eqs. 6-5(1) and 6-5(2) the corresponding values of the $IP(\mathbf{x}^{(1)})$ and the constraint matrix $\boldsymbol{\psi}(\mathbf{x}^{(1)})$ are found. The second approximation to the control, $\mathbf{u}^{(2)}$,

$$\mathbf{u}^{(2)} = \mathbf{u}^{(1)} + d\mathbf{u} \tag{1}$$

is to be determined so that the values of the state variables $\mathbf{x}^{(2)}$ found from Eq. 6-5(3) using $\mathbf{u}^{(2)}$ provide a lower value of the IP than $\mathrm{IP}(\mathbf{x}^{(1)})$ while moving $\boldsymbol{\psi}$ closer to zero.

To achieve the ultimate goal of reaching the minimum of the IP with $\boldsymbol{\psi} = 0$ may require several steps. However, to most expeditiously utilize the allowable control change dS, at each step $d\mathbf{u}$ will be determined so as to minimize

$$d\mathrm{IP} = \boldsymbol{\lambda}'_{\mathrm{IP}}\mathbf{G}\,d\mathbf{u} \tag{2}$$

while satisfying the $p + 1$ constraints

$$d\boldsymbol{\psi} = \boldsymbol{\lambda}'_{\psi}\mathbf{G}\,d\mathbf{u} \qquad (p \text{ equations}) \tag{3}$$

$$dS^2 = d\mathbf{u}'\mathbf{W}\,d\mathbf{u} \tag{4}$$

where the values of the p elements of the matrix $d\boldsymbol{\psi}$ and the magnitude of the admissible control change dS are specified.

The minimization of $d\mathrm{IP}$ under the $p + 1$ constraints of Eqs. (3) and (4) is accomplished by introducing the $p + 1$ multipliers $\nu_1, \cdots, \nu_p$ and μ. The product is taken of ν_1 with the first constraint of Eq. (3), ν_2 with the second, and so on through the pth constraint. In addition, the product of the multiplier μ is taken with the remaining constraint of Eq. (4) and each of these $p + 1$ products is then added to $d\mathrm{IP}$ in Eq. (2). Thus one obtains the augmented quantity [see Eq. 6-3(11)]

$$d\mathrm{IP} = \boldsymbol{\lambda}'_{\mathrm{IP}}\mathbf{G}\,d\mathbf{u} + \boldsymbol{\nu}'[d\boldsymbol{\psi} - \boldsymbol{\lambda}'_{\psi}\mathbf{G}\,d\mathbf{u}] + \mu[dS^2 - d\mathbf{u}'\mathbf{W}\,d\mathbf{u}] \tag{5}$$

For $d\mathbf{u}$ to minimize $d\mathrm{IP}$ requires that $d\mathbf{u}$ satisfy $d^2\mathrm{IP} = 0$. From Eq. (5) one finds

$$d^2\mathrm{IP} = 0 = \boldsymbol{\lambda}'_{\mathrm{IP}}\mathbf{G}\,d^2\mathbf{u} - \boldsymbol{\nu}'\boldsymbol{\lambda}'_{\psi}\mathbf{G}\,d^2\mathbf{u} - \mu\,d\mathbf{u}'\mathbf{W}\,d^2\mathbf{u} - \mu\,d^2\mathbf{u}'\mathbf{W}\,d\mathbf{u} \tag{6}$$

Since $\mathbf{W}$ is a symmetric matrix, the last term in Eq. (6) is equal to the term preceding it.† That is,

$$d^2\mathbf{u}'\mathbf{W}\,d\mathbf{u} = d\mathbf{u}'\mathbf{W}\,d^2\mathbf{u} \tag{7}$$

Combining the last two terms in Eq. (6) then yields

$$d^2\mathrm{IP} = 0 = [\boldsymbol{\lambda}'_{\mathrm{IP}}\mathbf{G} - \boldsymbol{\nu}'\boldsymbol{\lambda}'_{\psi}\mathbf{G} - 2\mu\,d\mathbf{u}'\mathbf{W}]\,d^2\mathbf{u} \tag{8}$$

---

† This can be seen by noting that $d^2\mathbf{u}'\mathbf{W}\,d\mathbf{u}$ is a scalar quantity. Thus $d^2\mathbf{u}'\mathbf{W}\,d\mathbf{u} = (d^2\mathbf{u}'\mathbf{W}\,d\mathbf{u})' = d\mathbf{u}'\mathbf{W}'\,d^2\mathbf{u} = d\mathbf{u}'\mathbf{W}\,d^2\mathbf{u}$. The fact that $\mathbf{W}$ is symmetric has been used to replace $\mathbf{W}'$ by $\mathbf{W}$.

For Eq. (8) to be satisfied for arbitrary changes $d^2\mathbf{u}$ in $d\mathbf{u}$ requires that the coefficient of $d^2\mathbf{u}$ vanish in Eq. (8). Applying this condition, one finds

$$\boldsymbol{\lambda}'_{\mathrm{IP}}\mathbf{G} - \boldsymbol{\nu}'\boldsymbol{\lambda}'_{\psi}\mathbf{G} - 2\mu\, d\mathbf{u}'\mathbf{W} = 0 \tag{9}$$

From Eq. (9), one may solve for the required value of $d\mathbf{u}$ to minimize $d\mathrm{IP}$. Thus

$$d\mathbf{u}' = \frac{1}{2\mu}\,[\boldsymbol{\lambda}'_{\mathrm{IP}} - \boldsymbol{\nu}'\boldsymbol{\lambda}'_{\psi}]\mathbf{G}\mathbf{W}^{-1} \tag{10a}$$

or

$$d\mathbf{u} = \frac{\mathbf{W}^{-1}\mathbf{G}'}{2\mu}\,[\boldsymbol{\lambda}_{\mathrm{IP}} - \boldsymbol{\lambda}_{\psi}\boldsymbol{\nu}] \tag{10b}$$

where the $p + 1$ multipliers $\nu_1, \cdots, \nu_p$ and μ remain to be determined so as to satisfy the $p + 1$ constraints of Eqs. (3) and (4).

Before doing so, let us determine what is required for the solution of Eq. (10) to be the minimizing solution. For $d\mathrm{IP}$ to be minimized by $d\mathbf{u}$ requires

$$d^2\mathrm{IP} = 0 \tag{11a}$$

$$d^3\mathrm{IP} > 0 \tag{11b}$$

Forming $d^3\mathrm{IP}$ from Eq. (8) one has on using Eq. (9) that

$$d^3\mathrm{IP} = -2\mu\, d^2\mathbf{u}'\mathbf{W}\, d^2\mathbf{u} \tag{12}$$

Since $d^2\mathbf{u}'\mathbf{W}\, d^2\mathbf{u}$ is positive definite by definition of $\mathbf{W}$, Eq. (11b) requires that for our solution to minimize $d\mathrm{IP}$,

$$\mu < 0 \tag{13}$$

Continuing with the determination of the multipliers $\boldsymbol{\nu}$ and μ, the solution for $d\mathbf{u}$ from Eq. (10b) is substituted into the p constraint equations of Eq. (3), yielding

$$d\boldsymbol{\psi} = \frac{\boldsymbol{\lambda}'_{\psi}\mathbf{G}\mathbf{W}^{-1}\mathbf{G}'}{2\mu}\,[\boldsymbol{\lambda}_{\mathrm{IP}} - \boldsymbol{\lambda}_{\psi}\boldsymbol{\nu}] \tag{14}$$

For ease of manipulation in what follows, let us define the following matrices:

$$\mathbf{I}_{\psi\,\mathrm{IP}} = \boldsymbol{\lambda}'_{\psi}\mathbf{G}\mathbf{W}^{-1}\mathbf{G}'\boldsymbol{\lambda}_{\mathrm{IP}} \qquad (p \times 1 \text{ matrix})$$

$$\mathbf{I}_{\psi\psi} = \boldsymbol{\lambda}'_{\psi}\mathbf{G}\mathbf{W}^{-1}\mathbf{G}'\boldsymbol{\lambda}_{\psi} \qquad (p \times p \text{ matrix}) \tag{15}$$

$$\mathbf{I}_{\mathrm{IP\,IP}} = \boldsymbol{\lambda}'_{\mathrm{IP}}\mathbf{G}\mathbf{W}^{-1}\mathbf{G}'\boldsymbol{\lambda}_{\mathrm{IP}} \qquad (1 \times 1 \text{ matrix})$$

where $\mathbf{I}_{\psi\psi}$ and $\mathbf{I}_{\mathrm{IP\,IP}}$ are seen to be symmetric; that is, $\mathbf{I}_{\psi\psi} = \mathbf{I}'_{\psi\psi}$, $\mathbf{I}_{\mathrm{IP\,IP}} = \mathbf{I}'_{\mathrm{IPIP}}$. Equation (14) then becomes

$$d\boldsymbol{\psi} = \frac{1}{2\mu}\,[\mathbf{I}_{\psi\,\mathrm{IP}} - \mathbf{I}_{\psi\psi}\boldsymbol{\nu}] \tag{16}$$

Solving Eq. (16) for the p constants $\mathbf{v}$ in terms of μ, one finds

$$\mathbf{v} = -2\mu \mathbf{I}_{\psi\psi}^{-1}\, d\boldsymbol{\psi} + \mathbf{I}_{\psi\psi}^{-1}\mathbf{I}_{\psi\,\mathrm{IP}}$$

or   (17)

$$\mathbf{v}' = -2\mu\, d\boldsymbol{\psi}'\mathbf{I}_{\psi\psi}^{-1} + \mathbf{I}_{\psi\mathrm{IP}}'\mathbf{I}_{\psi\psi}^{-1}$$

The only constraint that remains to be satisfied is Eq. (4) governing the admissible step size in control space. Substituting $d\mathbf{u}$ and $d\mathbf{u}'$ from Eq. (10) into Eq. (4) gives

$$dS^2 = \frac{1}{4\mu^2}\,[\boldsymbol{\lambda}_{\mathrm{IP}}' - \mathbf{v}'\boldsymbol{\lambda}_{\psi}']\mathbf{GW}^{-1}\mathbf{WW}^{-1}\mathbf{G}'[\boldsymbol{\lambda}_{\mathrm{IP}} - \boldsymbol{\lambda}_{\psi}\mathbf{v}] \tag{18}$$

which, using the definitions of Eq. (15) and the fact that $\mathbf{W}^{-1}\mathbf{W} = \mathbf{I}$, may be written

$$dS^2 = \frac{1}{4\mu^2}\,[\mathbf{I}_{\mathrm{IP\,IP}} - \mathbf{I}_{\psi\mathrm{IP}}'\mathbf{v} - \mathbf{v}'I_{\psi\,\mathrm{IP}} + \mathbf{v}'\mathbf{I}_{\psi\psi}\mathbf{v}] \tag{19}$$

Equations (17) and (19) are $p + 1$ equations for the p multipliers $\mathbf{v}$ and the single multiplier μ. Using Eq. (17) to eliminate $\mathbf{v}$ from Eq. (19) the remaining multiplier μ is given by†

$$2\mu = \left[\frac{\mathbf{I}_{\mathrm{IP\,IP}} - \mathbf{I}_{\psi\mathrm{IP}}'\mathbf{I}_{\psi\psi}^{-1}\mathbf{I}_{\psi\,\mathrm{IP}}}{dS^2 - d\boldsymbol{\psi}'\mathbf{I}_{\psi\psi}^{-1}\,d\boldsymbol{\psi}}\right]^{1/2} \tag{20}$$

where Eq. (13) has been used to select the negative sign for the square root.

Using the results of Eqs. (17) and (20) to eliminate $\mathbf{v}$ and μ, substitution of $d\mathbf{u}$ from Eq. (10) into Eq. (2) yields for the expected change in the IP,

$$d\mathrm{IP} = -[dS^2 - d\boldsymbol{\psi}'\mathbf{I}_{\psi\psi}^{-1}\,d\boldsymbol{\psi}]^{1/2}[\mathbf{I}_{\mathrm{IP\,IP}} - \mathbf{I}_{\psi\mathrm{IP}}'\mathbf{I}_{\psi\psi}^{-1}\mathbf{I}_{\psi\,\mathrm{IP}}]^{1/2} + \mathbf{I}_{\psi\mathrm{IP}}'\mathbf{I}_{\psi\psi}^{-1}\,d\boldsymbol{\psi} \tag{21}$$

If the constraint $\boldsymbol{\psi} = 0$ has been satisfied so that one can set $d\boldsymbol{\psi} = 0$, Eq. (21) yields

$$\left.\left|\frac{d\mathrm{IP}}{dS}\right|\right|_{d\boldsymbol{\psi}=0} = [\mathbf{I}_{\mathrm{IP\,IP}} - \mathbf{I}_{\psi\mathrm{IP}}'\mathbf{I}_{\psi\psi}^{-1}\mathbf{I}_{\psi\,\mathrm{IP}}]^{1/2} \tag{22}$$

which provides a measure of the deviation of the solution $\mathbf{x}^{(1)}$ from the minimizing solution.

For convenience $d\mathbf{u}$ as given by Eq. (10) is presented below with $\mathbf{v}$ and μ replaced by their values from Eqs. (17) and (20)

$$d\mathbf{u} = -\mathbf{W}^{-1}\mathbf{G}'[\boldsymbol{\lambda}_{\mathrm{IP}} - \boldsymbol{\lambda}_{\psi}\mathbf{I}_{\psi\psi}^{-1}\mathbf{I}_{\psi\,\mathrm{IP}}]\left[\frac{dS^2 - d\boldsymbol{\psi}'\mathbf{I}_{\psi\psi}^{-1}\,d\boldsymbol{\psi}}{\mathbf{I}_{\mathrm{IP\,IP}} - \mathbf{I}_{\psi\mathrm{IP}}'\mathbf{I}_{\psi\psi}^{-1}\mathbf{I}_{\psi\,\mathrm{IP}}}\right]^{1/2}$$
$$+ \mathbf{W}^{-1}\mathbf{G}'\boldsymbol{\lambda}_{\psi}\mathbf{I}_{\psi\psi}^{-1}\,d\boldsymbol{\psi} \tag{23}$$

† All the linear terms in μ in Eq. (19) cancel after substituting for $\mathbf{v}$ and $\mathbf{v}'$ from Eq. (17). Solving for $4\mu^2$ and taking the square root one obtains the result of Eq. (20).

If $\psi(\mathbf{x}^{(1)})$ is not zero, the choice of $d\psi$ remains to be made so that $\psi(\mathbf{x}^{(2)})$ will be closer to zero. Generally, one chooses

$$d\psi = -k\psi(\mathbf{x}^{(1)}) \qquad 0 < k \leqslant 1$$

with the parameter k selected so that the value of $d\psi$ used is within the range for which Eq. (3) can be expected to be valid. An upper limit is placed on the magnitude of $d\psi$ by the requirement that the numerator† under the square root in (23) be positive so that $d\mathbf{u}$ will be real.

With this determination, the first step in the procedure is complete. To the initial estimate of the control $\mathbf{u}^{(1)}$, the change $d\mathbf{u}$ given by Eq. (23) is added, forming the second approximation to the optimal control

$$\mathbf{u}^{(2)} = \mathbf{u}^{(1)} + d\mathbf{u}$$

Using $\mathbf{u}^{(2)}$, the corresponding state of the system is obtained by solving for $\mathbf{x}^{(2)}$ from Eq. 6-5(3). If $\mathbf{x}^{(2)}$ is not such that $\psi(\mathbf{x}^{(2)})$ is sufficiently close to zero [see Eq. 6-5(2)], the procedure followed in finding $\mathbf{u}^{(2)}$ and $\mathbf{x}^{(2)}$ must be repeated to obtain $\mathbf{u}^{(3)}$ and $\mathbf{x}^{(3)}$. Suppose that for the nth approximation $\mathbf{u}^{(n)}$, $\mathbf{x}^{(n)}$ one finds that $\psi(\mathbf{x}^{(n)}) = 0$ is satisfied with sufficient accuracy. The question remains of how close this solution is to the minimizing solution. To provide an answer, the rate of change of the IP in the direction of steepest descent in control space consistent with maintaining $\psi = 0$ is calculated from Eq. (22). If this slope is below some acceptably small value, the procedure is terminated and the nth approximation $\mathbf{x}^{(n)}$, $\mathbf{u}^{(n)}$ accepted as the minimizing solution.‡ If using $\mathbf{x}^{(n)}$ and $\mathbf{u}^{(n)}$ $|d\text{IP}/dS|_{d\psi = 0}$ is too large, the procedure is continued until a solution is reached for which $|d\text{IP}/dS|_{d\psi = 0}$ does fall below the chosen level.

The reader is urged to complete the solution of the problem started in Section 6-4 using the general methodology developed here. The results can be checked against those obtained in Section 6-3.

6-7 THE ADJOINT EQUATIONS

In Section 6-8 the theory just developed to find the minimum of a function is extended to treat the problem of minimizing a functional. In preparation, the present section introduces the concept of the adjoint differential form. For the purposes of this book it is only necessary to deal with first-order ordinary differential forms.

---

† The denominator under the square root in Eq. (23) is positive by its construction, representing as it does $[d\text{IP}/ds]^2$ holding ψ constant [see Eq. (22)].

‡ Recall the discussion of Section 6-2 which points out the necessity of the minimizing point lying in a well with steep sides if solution accuracy is to be obtained.

Consider the system of n linear differential forms $L_i(\mathbf{Z})$ involving the n variables $Z_1, \cdots, Z_n$ and given by

$$\mathbf{L}(\mathbf{Z}) = \begin{bmatrix} L_1(\mathbf{Z}) \\ \vdots \\ L_n(\mathbf{Z}) \end{bmatrix} = \frac{d\mathbf{Z}}{dt} - \mathbf{A}(t)\mathbf{Z} \tag{1}$$

$n \times 1$ matrix
of differential
forms

where

$$\mathbf{Z} = \begin{bmatrix} Z_1 \\ \vdots \\ Z_n \end{bmatrix} \qquad \mathbf{A}(t) = \begin{bmatrix} A_{11}(t) & \cdots & A_{1n}(t) \\ \vdots & & \\ A_{n1}(t) & \cdots & A_{nn}(t) \end{bmatrix} \tag{2}$$

$n \times 1$ $n \times n$
matrix   matrix

Let there be n adjoint variables λ_i, $i = 1, \cdots, n$. Then the set of n differential forms $M_i(\boldsymbol{\lambda})$ adjoint to $L_i(\mathbf{Z})$ is defined by the condition that

$$\sum_{i=1}^{n} [\lambda_i L_i(\mathbf{Z}) - Z_i M_i(\boldsymbol{\lambda})] = \boldsymbol{\lambda}'\mathbf{L}(\mathbf{Z}) - \mathbf{Z}'\mathbf{M}(\boldsymbol{\lambda})$$

be integrable. This requires that

$$\boldsymbol{\lambda}'\mathbf{L}(\mathbf{Z}) - \mathbf{Z}'\mathbf{M}(\boldsymbol{\lambda}) = \frac{dQ}{dt} \tag{3}$$

where

$$\boldsymbol{\lambda} = \begin{bmatrix} \lambda_1 \\ \vdots \\ \lambda_n \end{bmatrix} \qquad \mathbf{M}(\boldsymbol{\lambda}) = \begin{bmatrix} M_1(\boldsymbol{\lambda}) \\ \vdots \\ M_n(\boldsymbol{\lambda}) \end{bmatrix} \tag{4}$$

$n \times 1$ matrix $n \times 1$ matrix
of adjoint   of adjoint
variables   differential forms

and Q is a single as yet unspecified function.

Now from Eq. (1),

$$\boldsymbol{\lambda}'\mathbf{L}(\mathbf{Z}) = \boldsymbol{\lambda}'\frac{d\mathbf{Z}}{dt} - \boldsymbol{\lambda}'\mathbf{A}(t)\mathbf{Z} \tag{5}$$

Rewriting the first term on the right, one has

$$\boldsymbol{\lambda}'\mathbf{L}(\mathbf{Z}) = \frac{d}{dt}(\boldsymbol{\lambda}'\mathbf{Z}) - \frac{d\boldsymbol{\lambda}'}{dt}\mathbf{Z} - \boldsymbol{\lambda}'\mathbf{A}(t)\mathbf{Z} \tag{6}$$

Since the terms in Eq. (6) are scalar quantities, the tranpose of each one equals the value of the term itself. Transposing the last two terms on the right, Eq. (6) may be written

$$\lambda' \mathbf{L}(\mathbf{Z}) = \frac{d}{dt}(\lambda'\mathbf{Z}) - \mathbf{Z}'\left[\frac{d\lambda}{dt} + \mathbf{A}'(t)\lambda\right] \tag{7}$$

Defining

$$\mathbf{M}(\lambda) = -\frac{d\lambda}{dt} - \mathbf{A}'(t)\lambda \tag{8}$$

Eq. (7) may be written in the desired form of Eq. (3), yielding

$$\lambda'\mathbf{L}(\mathbf{Z}) - \mathbf{Z}'\mathbf{M}(\lambda) = \frac{d}{dt}(\lambda'\mathbf{Z}) \tag{9}$$

Thus by our definition the differential form $\mathbf{M}(\lambda)$ given by Eq. (8) is adjoint to $\mathbf{L}(\mathbf{Z})$. The adjoint equations are obtained by setting $\mathbf{M}(\lambda)$ equal to zero. From Eq. (8) the n adjoint equations are seen to be

$$\frac{d\lambda}{dt} = -\mathbf{A}'(t)\lambda \qquad (n \text{ equations}) \tag{10}$$

For the adjoint variables satisfying the equations of (10), integration of Eq. (9) between t_i and t_f gives

$$\lambda'\mathbf{Z}\bigg|_{t_f} - \lambda'\mathbf{Z}\bigg|_{t_i} - \int_{t_i}^{t_f} \lambda'\mathbf{L}(\mathbf{Z})\,dt \tag{11}$$

Equation (11) provides a relation between the values of the $2n$ variables $\mathbf{Z}$ and λ at t_i and t_f and the behavior over the interval of the adjoint variables $\lambda(t)$ and the differential forms $\mathbf{L}(\mathbf{Z})$.

Example   The Two-point Boundary-value Problem Again. Besides application to the direct techniques of this chapter the results just developed can be used to solve two-point boundary-value problems, (see Sections 4-2 and 4-3). As an example, consider the problem of determining the four variables $Z_1(t), \cdots,$ $Z_4(t)$ subject to the nonlinear differential constraints

$$\begin{aligned}
\dot{Z}_1 &= f_1(t, \mathbf{Z}) \\
\dot{Z}_2 &= f_2(t, \mathbf{Z}) \\
\dot{Z}_3 &= f_3(t, \mathbf{Z}) \\
\dot{Z}_4 &= f_4(t, \mathbf{Z})
\end{aligned} \tag{12}$$

and the end conditions

$$\begin{array}{cc}
\text{at } t = t_i \text{ (specified)} & \text{at } t = t_f \text{ (specified)} \\
Z_1(t_i) = Z_{1_0} & Z_3(t_f) = Z_{3_f} \\
Z_2(t_i) = Z_{2_0} & Z_4(t_f) = Z_{4_f}
\end{array} \tag{13}$$

As the four equations of (12) are nonlinear, in general it will not be possible to integrate them analytically. Resort must be had to numerical integration. To start the numerical integration of Eqs. (12) at t_i requires that the values $Z_1(t_i), \cdots, Z_4(t_i)$ be known. Since only Z_{1_0} and Z_{2_0} are specified by Eq. (13), let us make an estimate of the correct values of $Z_3(t_i)$ and $Z_4(t_i)$, say $Z_{3_0}^{(1)}$ and $Z_{4_0}^{(1)}$, and integrate Eq. (12) from t_i to t_f using them. However, since $Z_{3_0}^{(1)}$ and $Z_{4_0}^{(1)}$ will be chosen rather arbitrarily, it is unlikely that at t_f the values of $Z_3^{(1)}(t_f)$ and $Z_4^{(1)}(t_f)$ resulting will satisfy the terminal conditions of Eq. (13). Let us see how to use the results of this section to improve the degree to which the terminal conditions are satisfied.

Perturbations about the nominal solution obtained using the estimated initial conditions are governed by the first variation of the equations of Eq. (12). Thus

$$\delta \dot{Z}_i = \sum_{j=1}^{4} \frac{\partial f_i}{\partial Z_j} \delta Z_j \qquad i = 1, \cdots, 4 \tag{14}$$

In matrix notation one has

$$\mathbf{L}(\delta \mathbf{Z}) = \delta \dot{\mathbf{Z}} - \mathbf{A}(t)\, \delta \mathbf{Z} = 0 \tag{15}$$

where

$$\delta \mathbf{Z} = \begin{bmatrix} \delta Z_1 \\ \vdots \\ \delta Z_4 \end{bmatrix} \qquad \mathbf{A}(t) = \begin{bmatrix} \dfrac{\partial f_1}{\partial Z_1} & \cdots & \dfrac{\partial f_1}{\partial Z_4} \\ \vdots & & \\ \dfrac{\partial f_4}{\partial Z_1} & \cdots & \dfrac{\partial f_4}{\partial Z_4} \end{bmatrix} \tag{16}$$

$$\underset{\text{matrix}}{4 \times 1} \qquad\qquad \underset{\text{matrix}}{4 \times 4}$$

with the elements of $\mathbf{A}$ evaluated along the nominal solution.

Comparing the form of Eq. (15) with that of Eq. (1) it is seen that they are identical. Thus the equations adjoint to the differential form of Eq. (15) are analogous to those of Eq. (10) and given by

$$\dot{\boldsymbol{\lambda}} = -\mathbf{A}'(t)\boldsymbol{\lambda} \tag{17}$$

where the variables $\boldsymbol{\lambda}' = [\lambda_1, \cdots, \lambda_4]$ are the adjoint variables. Then from Eq. (11) since $\mathbf{L}(\delta \mathbf{Z})$ equals zero, one has that

$$\boldsymbol{\lambda}' \, \delta \mathbf{Z} \Big|_{t_f} = \boldsymbol{\lambda}' \, \delta \mathbf{Z} \Big|_{t_i} \tag{18}$$

The end conditions on $\lambda_1, \cdots, \lambda_4$ are at our disposal and remain to be chosen. Equation (18) provides a means of relating changes $\delta \mathbf{Z}(t_i)$ in the initial values $\mathbf{Z}(t_i)$ to changes $\delta \mathbf{Z}(t_f)$ in the terminal values $\mathbf{Z}(t_f)$.

The information desired from Eq. (18) is the change required in the initial values so that the terminal conditions of Eq. (13) can be better satisfied. Now at t_i the values of $Z_1(t_i)$ and $Z_2(t_i)$ are specified by Eq. (13), hence one has

$$\delta Z_1(t_i) \;=\; \delta Z_2(t_i) \;=\; 0 \tag{19}$$

Similarly, since the values of Z_3 and Z_4 are specified at t_f, the changes $\delta Z_3(t_f)$ and $\delta Z_4(t_f)$ desired are given by the difference between the specified values and the values obtained from the nominal solution. Thus

$$\delta Z_3(t_f) \;=\; Z_{3_f} - Z_3^{(1)}(t_f)$$

$$\delta Z_4(t_f) \;=\; Z_{4_f} - Z_4^{(1)}(t_f) \tag{20}$$

If Eq. (18) is to be useful, the quantities $\delta Z_1(t_f)$ and $\delta Z_2(t_f)$, which are not known, must be eliminated from it. To do this the terminal values of the corresponding adjoint variables which are at our disposal are chosen to be

$$\lambda_1(t_f) = 0$$

$$\lambda_2(t_f) = 0 \tag{21a}$$

Then to complete the specification of the terminal values of the adjoint variables let us choose

$$\lambda_3(t_f) = 1$$

$$\lambda_4(t_f) = 0 \tag{21b}$$

Using the terminal values of Eq. (21), the equations of (17) can be integrated backward to the initial time. Denoting the solution thus obtained by $\bar{\lambda}(t)$, on using Eqs. (19) and (21), Eq. (18) yields

$$\bar{\lambda}_3(t_i)\,\delta Z_3(t_i) + \bar{\lambda}_4(t_i)\,\delta Z_4(t_i) \;=\; \delta Z_3(t_f) \tag{22}$$

Equation (22) is a single equation for the required changes in the initial values $\delta Z_3(t_i)$ and $\delta Z_4(t_i)$ in terms of the value $\delta Z_3(t_f)$ given by Eq. (20). To obtain a second relation, let us specify a second independent set of terminal values of the adjoint variables by

$$\lambda_1(t_f) = 0$$

$$\lambda_2(t_f) = 0$$

$$\lambda_3(t_f) = 0$$

$$\lambda_4(t_f) = 1 \tag{23}$$

and once again integrate Eq. (17) backward to t_i. Denoting this solution by $\bar{\bar{\boldsymbol{\lambda}}}(t)$, Eq. (18) yields

$$\bar{\bar{\lambda}}_3(t_i)\,\delta Z_3(t_i) + \bar{\bar{\lambda}}_4(t_i)\,\delta Z_4(t_i) = \delta Z_4(t_f) \tag{24}$$

Equations (22) and (24) are solved simultaneously for the change in the initial values $Z_3^{(1)}(t_i)$ and $Z_4^{(1)}(t_i)$ required to meet the specified terminal conditions. One obtains

$$\delta Z_3(t_i) = \frac{\delta Z_3(t_f)\bar{\bar{\lambda}}_4(t_i) - \delta Z_4(t_f)\bar{\lambda}_4(t_i)}{\bar{\lambda}_3(t_i)\bar{\bar{\lambda}}_4(t_i) - \bar{\lambda}_4(t_i)\bar{\bar{\lambda}}_3(t_i)}$$

$$\delta Z_4(t_i) = \frac{\delta Z_4(t_f)\bar{\lambda}_3(t_i) - \delta Z_3(t_f)\bar{\bar{\lambda}}_3(t_i)}{\bar{\lambda}_3(t_i)\bar{\bar{\lambda}}_4(t_i) - \bar{\lambda}_4(t_i)\bar{\bar{\lambda}}_3(t_i)} \tag{25}$$

After forming

$$Z_3^{(2)}(t_i) = Z_3^{(1)}(t_i) + \delta Z_3(t_i)$$

$$Z_4^{(2)}(t_i) = Z_4^{(1)}(t_i) + \delta Z_4(t_i) \tag{26}$$

Equation (12) is integrated using the specified initial values Z_{1_0} and Z_{2_0} from Eq. (13) and the improved estimates $Z_3^{(2)}(t_i)$ and $Z_4^{(2)}(t_i)$ from Eq. (26). The resulting solution can be expected to satisfy the terminal conditions on Z_3 and Z_4 given by Eq. (13) as long as the use of Eq. (14) is valid. If the changes $\delta Z_3(t_f)$ and $\delta Z_4(t_f)$ required by Eq. (20) are too large, Eq. (14), which considers only first-order perturbations, cannot be expected to hold. If this occurs, the new initial values will be inaccurate and the procedure must be repeated. To aid convergence the change in the terminal values required at one time is usually limited to the range in which Eq. (14) may be expected to be valid.

The reader is urged to apply the technique just developed to the problem first discussed in Sections 4-2 and 4-3. In that problem analogous to the relations of Eq. (12) one has Eq. 4-2(12) with the relation of Eq. 4-2(13) used to eliminate the control a. The end conditions for the problem are given by Eq. 4-2(14).

6-8 APPLICATION TO THE MINIMIZATION OF A FUNCTIONAL (THE PROBLEM OF MAYER IN THE CALCULUS OF VARIATIONS)—FORMULATION

Stating the problem in matrix notation let our task be that of determining the control matrix $\mathbf{u}(t)$ in the interval t_i to t_f so as to minimize the performance index

$$\mathrm{IP} = \mathrm{IP}[\mathbf{x}(t_f), t_f] \tag{1}$$

while satisfying the p terminal constraints,

$$\boldsymbol{\Psi}[\mathbf{x}(t_f), t_f] = \begin{bmatrix} \psi_1 \\ \psi_2 \\ \vdots \\ \psi_p \end{bmatrix} = 0 \qquad (2)$$

$p \times 1$ matrix of known
functions of $\mathbf{x}(t_f)$ and t_f

the single stopping condition,

$$\Omega[\mathbf{x}(t_f), t_f] = 0 \qquad (3)$$

a known function of $\mathbf{x}(t_f)$ and t_f, and the n system equations,

$$\frac{d\mathbf{x}}{dt} = \mathbf{f}(\mathbf{x}, \mathbf{u}, t) \qquad (4)$$

where

$$\mathbf{x}(t) = \begin{bmatrix} x_1(t) \\ \vdots \\ x_n(t) \end{bmatrix} \qquad\qquad \mathbf{u}(t) = \begin{bmatrix} u_1(t) \\ \vdots \\ u_m(t) \end{bmatrix}$$

$n \times 1$ matrix $m \times 1$ matrix
of state variables   of control variables

$$\qquad (5)$$

$$\mathbf{f}(\mathbf{x}, \mathbf{u}, t) = \begin{bmatrix} f_1 \\ \vdots \\ f_n \end{bmatrix}$$

$n \times 1$ matrix of known
functions of $\mathbf{x}(t)$, $\mathbf{u}(t)$, and t

It will be assumed that all the initial values are specified at $t_i = 0$, although the method could be extended to include the case where some of the initial values are unknown.†

$$t_i = 0 \qquad \mathbf{x}(t_i) \text{ specified} \qquad (6)$$

To start the procedure of finding the minimizing solution an initial estimate must be made of the control function in the interval between t_i and the time at which the stopping condition of Eq. (3) is satisfied. Let this estimate of the optimal control be denoted by $\mathbf{u}^{(1)}(t)$. Inserting $\mathbf{u}^{(1)}(t)$ into Eq. (4) the

---

† The reader is referred to the references that follow for treatment of this aspect as well as the more general problems which involve control- and state-variable inequality constraints. W. F. Denham and A. E. Bryson, "Optimal Programming Problems with Inequality Constraints, II: Solution by Steepest-Ascent," *AIAA J.*, Vol. 2, No. 1, Jan, 1964, pp. 25–34; W. F. Denham, "Steepest-Ascent Solution of Optimal Programming Problems," *Raytheon Rept. BR-2393*, April 1963.

system equations are integrated† from the known initial state of Eq. (6) until the stopping condition $\Omega[\mathbf{x}(t_f), t_f] = 0$ is reached. The time at which this occurs is denoted by $t_f^{(1)}$. By storing the values of the state variables between $t_i = 0$ and $t_f^{(1)}$ knowledge is obtained of the state-variable behavior $\mathbf{x}^{(1)}(t)$ which corresponds to using $\mathbf{u}^{(1)}(t)$.

Now since the initial choice of the control vector $\mathbf{u}^{(1)}(t)$ is quite arbitrary, it cannot be expected that the values of $\mathbf{x}(t_f^{(1)})$ and $t_f^{(1)}$ obtained from the integration of Eq. (4) will satisfy the constraints of Eq. (2) or minimize the IP of Eq. (1). Thus it is necessary to determine how to change $\mathbf{u}^{(1)}(t)$ so as to better satisfy the constraints of Eq. (2) while obtaining a lower value of the IP.

Let the second approximation to the optimal control vector be denoted by

$$\mathbf{u}^{(2)}(t) = \mathbf{u}^{(1)}(t) + \delta\mathbf{u}(t) \tag{7}$$

where $\delta\mathbf{u}(t)$ is to be determined to meet the above objectives. Our immediate problem is to determine what effect changing the control $\mathbf{u}^{(1)}(t)$ by $\delta\mathbf{u}(t)$ has on the terminal values of the state variables and through them on the values of the end constraints and the IP.

Consider first the variation of the system equations of (4) which provide the equations governing the variation in $\mathbf{x}^{(1)}(t)$ resulting from changing $\mathbf{u}^{(1)}(t)$. Taking the variation of Eq. (4) one obtains the perturbation equations

$$\frac{d}{dt}\delta\mathbf{x} = \mathbf{F}\,\delta\mathbf{x} + \mathbf{G}\,\delta\mathbf{u} \tag{8}$$

where

$$\mathbf{F}(t) = \begin{bmatrix} \dfrac{\partial f_1}{\partial x_1} & \dfrac{\partial f_1}{\partial x_2} & \cdots & \dfrac{\partial f_1}{\partial x_n} \\ \vdots & & & \\ \dfrac{\partial f_n}{\partial x_1} & \dfrac{\partial f_n}{\partial x_2} & \cdots & \dfrac{\partial f_n}{\partial x_n} \end{bmatrix}_{(1)}$$

$n \times n$ matrix of known functions of time

$$\mathbf{G}(t) = \begin{bmatrix} \dfrac{\partial f_1}{\partial u_1} & \dfrac{\partial f_1}{\partial u_2} & \cdots & \dfrac{\partial f_1}{\partial u_m} \\ \vdots & & & \\ \dfrac{\partial f_n}{\partial u_1} & \dfrac{\partial f_n}{\partial u_2} & \cdots & \dfrac{\partial f_n}{\partial u_m} \end{bmatrix}_{(1)}$$

$n \times m$ matrix of known functions of time

$$\tag{9}$$

† In the sense of this chapter the term "integration" refers to numerical computer integration.

The elements of $\mathbf{F}$ and $\mathbf{G}$ are known functions of time in the interval between $t_i = 0$ and $t_f^{(1)}$ since the partial derivatives are evaluated along the nominal solution $\mathbf{x}^{(1)}(t)$, $\mathbf{u}^{(1)}(t)$.

In the notation of Section 6-7, Eq. (8) may be written [see Eq. 6-7(1)]

$$\mathbf{L}(\delta x) = \mathbf{G}\,\delta\mathbf{u} \tag{10}$$

Let us introduce the n adjoint variables λ_i, $i = 1, \cdots, n$ which satisfy the adjoint equations

$$\frac{d\boldsymbol{\lambda}}{dt} = -\mathbf{F}'\boldsymbol{\lambda} \tag{11}$$

that result from setting the differential form adjoint to $\mathbf{L}(\delta\mathbf{x})$ equal to zero. Then over the interval $0 \leqslant t \leqslant t_f^{(1)}$, the analogous relation to Eq. 6-7(11) is

$$\boldsymbol{\lambda}'\,\delta\mathbf{x}\Big|_{t_f^{(1)}} = \boldsymbol{\lambda}'\,\delta\mathbf{x}\Big|_0 + \int_0^{t_f^{(1)}} \boldsymbol{\lambda}'\mathbf{L}(\delta\mathbf{x})\,dt \tag{12}$$

At t_i and t_f the total change in $\mathbf{x}$ is made up of two parts: one part occurring from the direct variation of $\mathbf{x}$ holding t_i or t_f fixed and the other resulting from changing the end time. Thus [see Eq. 3-2(10),

$$d\mathbf{x}(t_i) = \delta\mathbf{x}(t_i) + \dot{\mathbf{x}}(t_i)\,dt_i \tag{13a}$$

$$d\mathbf{x}(t_f) = \delta\mathbf{x}(t_f) + \dot{\mathbf{x}}(t_f)\,dt_f \tag{13b}$$

As for this problem $\mathbf{x}(t_i)$ and t_i are specified, it follows that $d\mathbf{x}(t_i) = dt_i = 0$. Equation (13) then gives

$$\delta\mathbf{x}(t_i) = 0 \tag{14}$$

Equation (12) can now be rewritten using the result of Eq. (14) and the relation of Eq. (10), yielding

$$\boldsymbol{\lambda}'\,\delta\mathbf{x}\Big|_{t_f^{(1)}} = \int_0^{t_f^{(1)}} \boldsymbol{\lambda}'\mathbf{G}\,\delta\mathbf{u}\,dt \tag{15}$$

Recall that it is our desire to determine the change $d\mathrm{IP}[\mathbf{x}(t_f^{(1)}), t_f^{(1)}]$ in the performance index as well as the change $d\boldsymbol{\psi}[\mathbf{x}(t_f^{(1)}), t_f^{(1)}]$ in the p terminal constraints resulting from changing $\mathbf{u}^{(1)}(t)$ by $\delta\mathbf{u}(t)$ over the interval $0 \leqslant t \leqslant t_f^{(1)}$. Equation (15) is seen to provide a relation between the behavior of $\delta\mathbf{u}(t)$ over the interval and the terminal value of the product $\boldsymbol{\lambda}'\,\delta\mathbf{x}$. Thus if $\boldsymbol{\lambda}'(t_f^{(1)})\,\delta\mathbf{x}(t_f^{(1)})$ can be made to equal, in turn, $d\mathrm{IP}$ and $d\boldsymbol{\psi}$, the desired information will have been obtained. Consider first, for the sake of definiteness, having Eq. (15) represent the change in the jth constraint $d\psi_j[\mathbf{x}(t_f^{(1)}), t_f^{(1)}]$. Now

$$d\psi_j = \frac{\partial\psi_j}{\partial\mathbf{x}_f}\,d\mathbf{x}(t_f^{(1)}) + \frac{\partial\psi_j}{\partial t_f}\Big|_{t_f^{(1)}}\,dt_f \tag{16}$$

where

$$\frac{\partial \psi_j}{\partial \mathbf{x}_f} = \left[\frac{\partial \psi_j}{\partial x_1(t_f)} \quad \cdots \quad \frac{\partial \psi_j}{\partial x_n(t_f)} \right] \Bigg|_{t_f^{(1)}} \tag{17}$$

$1 \times n$
matrix

Since our solution $\mathbf{x}^{(1)}$, $\mathbf{u}^{(1)}$ satisfies the stopping condition of Eq. (3) one also has

$$d\Omega = 0 = \frac{\partial \Omega}{\partial \mathbf{x}_f} d\mathbf{x}(t_f^{(1)}) + \frac{\partial \Omega}{\partial t_f}\Bigg|_{t_f^{(1)}} dt_f \tag{18}$$

where

$$\frac{\partial \Omega}{\partial \mathbf{x}_f} = \left[\frac{\partial \Omega}{\partial x_1(t_f)} \quad \cdots \quad \frac{\partial \Omega}{\partial x_n(t_f)} \right] \Bigg|_{t_f^{(1)}} \tag{19}$$

$1 \times n$
matrix

Then using the relation

$$d\mathbf{x}(t_f) = \delta\mathbf{x}(t_f) + \dot{\mathbf{x}}(t_f)\, dt_f$$

to eliminate $d\mathbf{x}(t_f)$ from Eqs. (16) and (18) one obtains

$$d\psi_j = \frac{\partial \psi_j}{\partial \mathbf{x}_f} \delta\mathbf{x}_f + \dot{\psi}_j\, dt_f \tag{20a}$$

$$d\Omega = 0 = \frac{\partial \Omega}{\partial \mathbf{x}_f} \delta\mathbf{x}_f + \dot{\Omega}\, dt_f \tag{20b}$$

where $\dot{\psi}_j$ and $\dot{\Omega}$ are defined by

$$\dot{\psi}_j = \left[\frac{\partial \psi_j}{\partial t_f} + \frac{\partial \psi_j}{\partial \mathbf{x}_f} \dot{\mathbf{x}}_f \right] \Bigg|_{t_f^{(1)}} \tag{21a}$$

$$\dot{\Omega} = \left[\frac{\partial \Omega}{\partial t_f} + \frac{\partial \Omega}{\partial \mathbf{x}_f} \dot{\mathbf{x}}_f \right] \Bigg|_{t_f^{(1)}} \tag{21b}$$

Equation (20b) can be used to eliminate dt_f from Eq. (20a). Solving Eq. (20b) for dt_f and substituting into Eq. (20a) one obtains

$$d\psi_j = \left[\frac{\partial \psi_j}{\partial \mathbf{x}_f} - \frac{\dot{\psi}_j}{\dot{\Omega}} \frac{\partial \Omega}{\partial \mathbf{x}_f} \right] \delta\mathbf{x}(t_f^{(1)}) \tag{22}$$

A comparison of Eq. (22) with the left side of Eq. (15) then shows how the values of the multipliers at $t_f^{(1)}$ are to be chosen to obtain the desired result.

Denoting by $\boldsymbol{\lambda}_{\psi_j}(t)$ the set of n multipliers which satisfy at $t_f^{(1)}$ the condition

$$\boldsymbol{\lambda}'_{\psi_j}(t_f^{(1)}) = \frac{\partial \psi_j}{\partial \mathbf{x}_f} - \frac{\psi_j}{\Omega} \frac{\partial \Omega}{\partial \mathbf{x}_f} \tag{23}$$

Eq. (15) can be written

$$d\psi_j = \int_0^{t_f^{(1)}} \boldsymbol{\lambda}'_{\psi_j}(t)\mathbf{G}(t)\, \delta\mathbf{u}(t)\, dt \tag{24}$$

The multipliers $\boldsymbol{\lambda}_{\psi_j}(t)$ are completely specified by Eq. (23). Integrating the adjoint equations of Eq. (11), that is,

$$\frac{d\boldsymbol{\lambda}_{\psi_j}}{dt} = -\mathbf{F}'\boldsymbol{\lambda}_{\psi_j} \qquad (n \text{ equations}) \tag{25}$$

backward from the known terminal values of $\boldsymbol{\lambda}_{\psi_j}(t_f^{(1)})$ given by Eq. (23) one obtains $\boldsymbol{\lambda}_{\psi_j}(t)$ in the interval $0 \leqslant t \leqslant t_f^{(1)}$. As the jth constraint was chosen arbitrarily it is apparent that the results obtained hold for each one of the constraints $j = 1, \cdots, p$ and that there are p sets of n multipliers. Denoting by $\boldsymbol{\lambda}_\psi$ the matrix

$$\boldsymbol{\lambda}_\psi = [\boldsymbol{\lambda}_{\psi_1} \quad \boldsymbol{\lambda}_{\psi_2} \quad \cdots \quad \boldsymbol{\lambda}_{\psi_p}] \tag{26}$$
$$1 \times p \text{ matrix}$$

Eq. (24) may be written

$$d\boldsymbol{\psi} = \int_0^{t_f^{(1)}} \boldsymbol{\lambda}'_\psi \mathbf{G}\, \delta\mathbf{u}\, dt \tag{27}$$

In a similar manner one finds that the change in the performance index is given by

$$d\mathrm{IP} = \int_0^{t_f^{(1)}} \boldsymbol{\lambda}'_{\mathrm{IP}}\mathbf{G}\, \delta\mathbf{u}\, dt \tag{28}$$

where the set of n multipliers $\boldsymbol{\lambda}_{\mathrm{IP}}$ satisfy the terminal condition

$$\boldsymbol{\lambda}'_{\mathrm{IP}}(t_f^{(1)}) = \frac{\partial \mathrm{IP}}{\partial \mathbf{x}_f} - \frac{\dot{\mathrm{IP}}}{\dot{\Omega}} \frac{\partial \Omega}{\partial \mathbf{x}_f} \tag{29}$$

and the adjoint equations

$$\frac{d\boldsymbol{\lambda}_{\mathrm{IP}}}{dt} = -\mathbf{F}'\boldsymbol{\lambda}_{\mathrm{IP}} \qquad (n \text{ equations}) \tag{30}$$

The notation

$$\frac{\partial \text{IP}}{\partial \mathbf{x}_f} = \left[\frac{\partial \text{IP}}{\partial x_1(t_f)} \quad \cdots \quad \frac{\partial \text{IP}}{\partial x_n(t_f)} \right] \Big|_{t_f^{(1)}}$$

$1 \times n$ matrix

$$\dot{\text{IP}} = \left[\frac{\partial \text{IP}}{\partial t_f} + \frac{\partial \text{IP}}{\partial \mathbf{x}_f} \dot{\mathbf{x}}_f \right] \Big|_{t_f^{(1)}} \tag{31}$$

corresponds to that used in Eqs. (19) and (21).

As stated earlier, the control change $\delta\mathbf{u}(t)$ is at our disposal so that the terminal values $\mathbf{x}(t_f^{(2)})$ obtained using $\mathbf{u}^{(2)}(t)$ provide a lower value of the IP than that obtained using the first approximation $\mathbf{u}^{(1)}(t)$ while better satisfying the constraints of Eq. (2). However, if the linearized equations of (8) are to be valid, the change to be made must be limited. To this end the restriction is imposed that

$$dS^2 = \int_0^{t_f^{(1)}} \delta\mathbf{u}' \mathbf{W} \, \delta\mathbf{u} \, dt \tag{32}$$

where $\mathbf{W}$ is a symmetric, positive definite $m \times m$ matrix. Beyond this the choice to be made for the elements of $\mathbf{W}$ is at the disposal of the designer to aid convergence to the optimal solution.

6-9   SOLUTION OF THE PROBLEM

Having made the initial estimate $\mathbf{u}^{(1)}(t)$, our problem is seen to be that of determining $\delta\mathbf{u}(t)$ so that $\mathbf{u}^{(2)}(t)$ will provide a still better approximation to the optimal control. In particular, using the notation of Section 6-8, let us determine $\delta\mathbf{u}(t)$ so as to minimize

$$d\text{IP} = \int_0^{t_f^{(1)}} \boldsymbol{\lambda}'_{\text{IP}} \mathbf{G} \, \delta\mathbf{u} \, dt \tag{1}$$

while satisfying the p terminal constraints

$$d\boldsymbol{\psi} = \int_0^{t_f^{(1)}} \boldsymbol{\lambda}'_\psi \mathbf{G} \, \delta\mathbf{u} \, dt \tag{2}$$

and the single constraint on the magnitude of the control change over the interval

$$dS^2 = \int_0^{t_f^{(1)}} \delta\mathbf{u}' \mathbf{W} \, \delta\mathbf{u} \, dt \tag{3}$$

where the p elements in the matrix $d\boldsymbol{\psi}$ and the magnitude of dS are specified.

Introducing the p multipliers, $\mathbf{v}' = [\nu_1, \cdots, \nu_p]$, the product is taken of each of the p constraints of Eq. (2) with the corresponding numbered multiplier and the results added to Eq. (1). Similarly, the product is taken of the constraint of Eq. (3) with the single multiplier μ and the result also added to Eq. (1), yielding

$$d\text{IP} = \int_0^{t_f^{(1)}} \boldsymbol{\lambda}_{\text{IP}}'\mathbf{G}\,\delta\mathbf{u}\,dt + \mathbf{v}'\left[d\boldsymbol{\psi} - \int_0^{t_f^{(1)}} \boldsymbol{\lambda}_\psi'\mathbf{G}\,\delta\mathbf{u}\,dt\right]$$

$$+\mu\left[dS^2 - \int_0^{t_f^{(1)}} \delta\mathbf{u}'\mathbf{W}\,\delta\mathbf{u}\,dt\right] \quad (4)$$

Equation (4) is directly analogous to Eq. 6-6(5) and the procedure to be followed here will parallel the developments of Section 6-6.

For $\delta\mathbf{u}(t)$ to minimize $d\text{IP}$ as given by Eq. (4) requires that it be such that $d^2\text{IP}$ vanish. From Eq. (4) one obtains

$$d^2\text{IP} = 0 = \int_0^{t_f^{(1)}} [\boldsymbol{\lambda}_{\text{IP}}'\mathbf{G} - \mathbf{v}'\boldsymbol{\lambda}_\psi'\mathbf{G} - 2\mu\,\delta\mathbf{u}'\mathbf{W}]\,\delta^2\mathbf{u}\,dt \quad (5)$$

Since each element of $\delta^2\mathbf{u}(t)$ may be treated as being independent of the others because of the $p + 1$ multipliers $\mathbf{v}$ and μ, for Eq. (5) to be satisfied requires that the coefficient of $\delta^2\mathbf{u}(t)$ vanish. Thus one has

$$\boldsymbol{\lambda}_{\text{IP}}'\mathbf{G} - \mathbf{v}'\boldsymbol{\lambda}_\psi'\mathbf{G} - 2\mu\,\delta\mathbf{u}'\mathbf{W} = 0 \quad (6)$$

which on solving for the control change yields

$$\delta\mathbf{u}'(t) = \frac{1}{2\mu}[\boldsymbol{\lambda}_{\text{IP}}' - \mathbf{v}'\boldsymbol{\lambda}_\psi']\mathbf{G}\mathbf{W}^{-1} \quad (7)$$

or

$$\delta\mathbf{u}(t) = \frac{\mathbf{W}^{-1}\mathbf{G}'}{2\mu}[\boldsymbol{\lambda}_{\text{IP}} - \boldsymbol{\lambda}_\psi\mathbf{v}]$$

Equation (7) is seen to be identical in form to the corresponding relations of Eq. 6-6(10). The $p + 1$ multipliers $\nu_1, \cdots, \nu_p$ and μ remain to be determined so that the $p + 1$ constraints of Eqs. (2) and (3) are satisfied. In Section 6-6 the $p + 1$ multipliers $\nu_1, \cdots, \nu_p$ and μ are determined so as to satisfy Eqs. 6-6(3) and 6-6(4). Comparing the procedure followed in Section 6-6 with our present requirements it is seen that if one defines [see Eq. 6-6(15)],

$$\mathbf{I}_{\psi\,\text{IP}} = \int_0^{t_f^{(1)}} \boldsymbol{\lambda}_\psi'\mathbf{G}\mathbf{W}^{-1}\mathbf{G}'\boldsymbol{\lambda}_{\text{IP}}\,dt \qquad (p \times 1 \text{ matrix})$$

$$\mathbf{I}_{\psi\psi} = \int_0^{t_f^{(1)}} \boldsymbol{\lambda}_\psi'\mathbf{G}\mathbf{W}^{-1}\mathbf{G}'\boldsymbol{\lambda}_\psi\,dt \qquad (p \times p \text{ matrix}) \quad (15)$$

$$\mathbf{I}_{\text{IP IP}} = \int_0^{t_f^{(1)}} \boldsymbol{\lambda}_{\text{IP}}'\mathbf{G}\mathbf{W}^{-1}\mathbf{G}'\boldsymbol{\lambda}_{\text{IP}}\,dt \qquad (1 \times 1 \text{ matrix})$$

that the results of Section 6-6 may be transcribed directly.

The desired control change is therefore [see Eq. 6-6(23)]

$$\delta\mathbf{u}(t) = -\mathbf{W}^{-1}\mathbf{G}'[\boldsymbol{\lambda}_{\mathrm{IP}} - \boldsymbol{\lambda}_\psi \mathbf{I}_{\psi\psi}^{-1}\mathbf{I}_{\psi\,\mathrm{IP}}]\left[\frac{dS^2 - d\boldsymbol{\psi}'\mathbf{I}_{\psi\psi}^{-1}\,d\boldsymbol{\psi}}{\mathbf{I}_{\mathrm{IP\,IP}} - \mathbf{I}_{\psi\mathrm{IP}}'\mathbf{I}_{\psi\psi}^{-1}\mathbf{I}_{\psi\,\mathrm{IP}}}\right]^{1/2}$$

$$+ \mathbf{W}^{-1}\mathbf{G}'\boldsymbol{\lambda}_\psi \mathbf{I}_{\psi\psi}^{-1}\,d\boldsymbol{\psi} \quad (16)$$

where $d\boldsymbol{\psi}$ is selected so that

$$d\boldsymbol{\psi} = -k\boldsymbol{\psi}[\mathbf{x}(t_f^{(1)}), t_f^{(1)}] \qquad 0 < k \leqslant 1 \quad (17)$$

so as to obtain for $\boldsymbol{\psi}[\mathbf{x}(t_f^{(2)}), t_f^{(2)}]$ a value closer to zero. In addition, the magnitude of $d\boldsymbol{\psi}$ is constrained by the condition that the numerator under the square root in Eq. (16) be positive. Using $\delta\mathbf{u}(t)$ from Eq. (16) one finds [see Eq. 6-6(21)]

$$d\mathrm{IP} = -[dS^2 - d\boldsymbol{\psi}'\mathbf{I}_{\psi\psi}^{-1}\,d\boldsymbol{\psi}]^{1/2}[\mathbf{I}_{\mathrm{IP\,IP}} - \mathbf{I}_{\psi\mathrm{IP}}'\mathbf{I}_{\psi\psi}^{-1}\mathbf{I}_{\psi\,\mathrm{IP}}]^{1/2} + \mathbf{I}_{\psi\mathrm{IP}}'\,\mathbf{I}_{\psi\psi}^{-1}d\boldsymbol{\psi} \quad (18)$$

Hence the magnitude of the rate of change of the IP in the direction of steepest descent consistent with maintaining $d\boldsymbol{\psi} = 0$ is given by

$$\left|\frac{d\mathrm{IP}}{dS}\right|_{d\psi=0} = [\mathbf{I}_{\mathrm{IP\,IP}} - \mathbf{I}_{\psi\mathrm{IP}}'\mathbf{I}_{\psi\psi}^{-1}\mathbf{I}_{\psi\,\mathrm{IP}}]^{1/2} \quad (19)$$

The development of the tools required for our solution procedure is now complete. Listed below are the steps to be undertaken so as to reach the optimal solution.

1. Estimate the elements of the control vector $\mathbf{u}^{(1)}(t)$ and integrate Eq. 6-8(4) from the known initial state until the stopping condition of Eq. 6-8(3) is satisfied at $t_f^{(1)}$. Store the results for $\mathbf{x}^{(1)}(t)$.

2. Integrate the adjoint equations of Eq. 6-8(25) and Eq. 6-8(30) backward from the known terminal conditions of Eqs. 6-8(23) and 6-8(29) to find the $p + 1$ sets of multipliers $\boldsymbol{\lambda}_{\psi_1}, \cdots, \boldsymbol{\lambda}_{\psi_p}$ and $\boldsymbol{\lambda}_{\mathrm{IP}}$. Store the results.

3. Along with step 2, carry out the integrations required to obtain $\mathbf{I}_{\psi\,\mathrm{IP}}$, $\mathbf{I}_{\mathrm{IP\,IP}}$, and $\mathbf{I}_{\psi\psi}$ as given by Eq. (15).

4. Select dS^2 so as to obtain a reasonable value for the mean-square control change in the interval as given by $dS^2/t_f^{(1)}$.

5. Examine the values at $t_f^{(1)}$ of the p terminal constraint functions $\psi_1, \cdots, \psi_p$ resulting from using $\mathbf{u}^{(1)}(t)$ and select the desired changes $d\psi_1, \cdots, d\psi_p$ within the limitation that $[dS^2 - d\boldsymbol{\psi}'\mathbf{I}_{\psi\psi}^{-1}\,d\boldsymbol{\psi}] \geqslant 0$. If the choice of $d\boldsymbol{\psi}$ makes this quantity negative, reduce $d\boldsymbol{\psi}$ by a constant factor so that $[dS^2 - d\boldsymbol{\psi}'\mathbf{I}_{\psi\psi}^{-1}\,d\boldsymbol{\psi}]$ equals zero.

6. From Eq. (16) determine the control change $\delta\mathbf{u}(t)$ and add it to $\mathbf{u}^{(1)}(t)$ to form the next approximation $\mathbf{u}^{(2)}(t)$ in the interval $0 \leqslant t \leqslant t_f^{(1)}$.

7. If the elements of $\boldsymbol{\psi}[\mathbf{x}(t_f^{(1)}), t_f^{(1)}]$ in step 5 are sufficiently close to zero, examine $|d\mathrm{IP}/dS|_{d\psi=0}$ from Eq. (19) to see if the solution found is sufficiently close to the minimizing solution.

8. If with the constraint $\boldsymbol{\psi} = 0$ satisfied the value of $|d\mathrm{IP}/dS|_{d\psi\,=\,0}$ is not sufficiently small, return to step 1 and using $\mathbf{u}^{(2)}(t)$ in place of $\mathbf{u}^{(1)}(t)$ repeat the process. Should $t_f^{(2)}$ be greater than $t_f^{(1)}$ extend the range of definition of $\mathbf{u}^{(2)}(t_f^{(1)})$ and use it for the control in the interval between $t_f^{(1)}$ and $t_f^{(2)}$.

9. Terminate the process when the solution obtained satisfies the constraint $\boldsymbol{\psi} = 0$ with sufficient accuracy and gives a value for $|d\mathrm{IP}/dS|_{d\psi\,=\,0}$ below some acceptable level. The solution thus found is our approximation to the minimizing solution.

6-10 APPLICATION OF THE TECHNIQUE TO THE LUNAR ASCENT PROBLEM

To illustrate the use of the technique just developed, let us return to the example of the lunar ascent vehicle first discussed in Section 3-9. Recall that the problem is that of determining the trajectory to be followed by a space ship initially at rest on the lunar surface to enable it to reach specified orbital terminal conditions in minimum time. Utilizing a flat moon approximation, the problem is given in Section 3-9 as that of minimizing

$$\mathrm{IP} = t_f \tag{1}$$

subject to the system equations

$$\dot{x}_1 = x_3$$

$$\dot{x}_2 = x_4$$

$$\dot{x}_3 = A \cos \theta \tag{2}$$

$$\dot{x}_4 = A \sin \theta - g_m$$

and the end conditions

$$
\begin{array}{ll}
\text{at } t = t_i = 0 & \text{at } t = t_f \text{ (unspecified)} \\
x_1 = 0 & x_2 = h = 50{,}000 \text{ ft} \\
x_2 = 0 & x_3 = U = 5444 \text{ ft/sec} \\
x_3 = 0 & x_4 = 0 \\
x_4 = 0 &
\end{array}
\tag{3}
$$

To put the problem in the form given in Section 6-9, one of the terminal conditions of Eq. (3) must be selected as the stopping condition. The choice made is to let the terminal time be defined by attainment of the horizontal velocity U. Thus [see Eq. 6-8(2) and 6-8(3)],

$$\Omega[\mathbf{x}(t_f), t_f] = x_3(t_f) - U = 0 \tag{4}$$

and

$$\boldsymbol{\psi}[\mathbf{x}(t_f), t_f] = \begin{bmatrix} \psi_1 \\ \psi_2 \end{bmatrix} = \begin{bmatrix} x_2(t_f) - h \\ x_4(t_f) \end{bmatrix} = 0 \tag{5}$$

The reason for this selection is that it is easy to visualize cases in which the nominal control program (thrust-angle behavior) is such that the trajectory never reaches either of the other two terminal conditions. Trajectories of this type are shown in Fig. 6-2. Thus if either the condition $x_2(t_f) = h$ or the con-

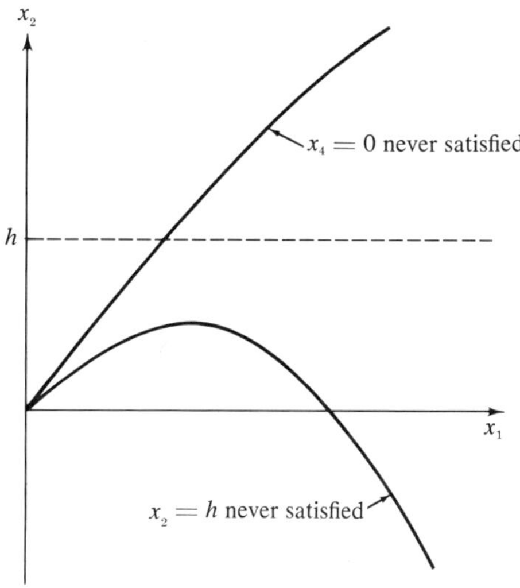

FIGURE 6-2   Illustration of the need for care
in the choice of the stopping condition

dition $x_4(t_f) = 0$ is used as the stopping condition, the estimated control behavior may be such that the condition cannot be reached. In this case integration of the system equations of (1) from the known initial state using the estimated optimal control behavior would continue indefinitely. On the other hand, the stopping condition selected will eventually be satisfied no matter what the thrust angle behavior so long as it is positive.

The equations analogous to Eq. 6-8(8) governing the behavior of perturbations in the system are obtained by taking the variation of the system equations of (1). One finds

$$\frac{d}{dt}\,\delta x_1 = \delta x_3$$

$$\frac{d}{dt}\,\delta x_2 = \delta x_4$$

$$\frac{d}{dt}\,\delta x_3 = -A \sin \theta \, \delta\theta$$

$$\frac{d}{dt}\,\delta x_4 = A \cos \theta \, \delta\theta$$

(6)

The matrices $\mathbf{F}(t)$ and $\mathbf{G}(t)$ defined by Eq. 6-8(9) are thus seen to be

$$\mathbf{F}(t) = \begin{bmatrix} 0 & 0 & 1 & 0 \\ 0 & 0 & 0 & 1 \\ 0 & 0 & 0 & 0 \\ 0 & 0 & 0 & 0 \end{bmatrix} \qquad \mathbf{G}(t) = \begin{bmatrix} 0 \\ 0 \\ -A \sin \theta \\ A \cos \theta \end{bmatrix}$$

(7)

The elements of $\mathbf{F}(t)$ and $\mathbf{G}(t)$ are evaluated along the nominal trajectory resulting from using the control function $\theta(t)$ whose improvement is desired.

The multipliers $\boldsymbol{\lambda}_\psi$ and $\boldsymbol{\lambda}_{\mathrm{IP}}$ may now be introduced. Let

$$\boldsymbol{\lambda}_\psi(t) = [\boldsymbol{\lambda}_{\psi_1}(t) \quad \boldsymbol{\lambda}_{\psi_2}(t)] = \begin{bmatrix} \lambda_{\psi_1 1}(t) & \lambda_{\psi_2 1}(t) \\ \lambda_{\psi_1 2}(t) & \lambda_{\psi_2 2}(t) \\ \lambda_{\psi_1 3}(t) & \lambda_{\psi_2 3}(t) \\ \lambda_{\psi_1 4}(t) & \lambda_{\psi_2 4}(t) \end{bmatrix}$$

(8a)

and

$$\boldsymbol{\lambda}_{\mathrm{IP}} = \begin{bmatrix} \lambda_{\mathrm{IP}_1} \\ \lambda_{\mathrm{IP}_2} \\ \lambda_{\mathrm{IP}_3} \\ \lambda_{\mathrm{IP}_4} \end{bmatrix}$$

(8b)

Then using Eq. (7) it is seen from Eqs. 6-8(25) and 6-8(30) that the sets of multipliers $\boldsymbol{\lambda}_\psi$ satisfy the adjoint equations

$$\dot{\lambda}_{\psi_i 1} = 0$$

$$\dot{\lambda}_{\psi_i 2} = 0$$

$$\dot{\lambda}_{\psi_i 3} = -\lambda_{\psi_i 1} \qquad i = 1, 2$$

$$\dot{\lambda}_{\psi_i 4} = -\lambda_{\psi_i 2}$$

(9)

while the set of multipliers $\boldsymbol{\lambda}_{\mathrm{IP}}$ satisfy the corresponding equations

$$
\begin{aligned}
\dot{\lambda}_{\mathrm{IP}_1} &= 0 \\
\dot{\lambda}_{\mathrm{IP}_2} &= 0 \\
\dot{\lambda}_{\mathrm{IP}_3} &= -\lambda_{\mathrm{IP}_1} \\
\dot{\lambda}_{\mathrm{IP}_4} &= -\lambda_{\mathrm{IP}_2}
\end{aligned}
\tag{10}
$$

The terminal conditions on the multipliers are given by Eqs. 6-8(23) and 6-8(29) in conjunction with the specific functions of this example as provided by Eqs. (1), (4), and (5). Since

$$
\frac{\partial \psi_1}{\partial \mathbf{x}_f} = [0 \quad 1 \quad 0 \quad 0] \qquad \frac{\partial \mathrm{IP}}{\partial \mathbf{x}_f} = [0 \quad 0 \quad 0 \quad 0]
$$

$$
\frac{\partial \psi_2}{\partial \mathbf{x}_f} = [0 \quad 0 \quad 0 \quad 1] \qquad \frac{\partial \Omega}{\partial \mathbf{x}_f} = [0 \quad 0 \quad 1 \quad 0]
\tag{11}
$$

while

$$
\frac{\partial \Omega}{\partial t_f} = \frac{\partial \psi_1}{\partial t_f} = \frac{\partial \psi_2}{\partial t_f} = 0 \qquad \frac{\partial \mathrm{IP}}{\partial t_f} = 1
\tag{12}
$$

one has from Eqs. 6-8(21) and 6-8(31) that

$$
\begin{aligned}
\dot{\psi}_1 &= \dot{x}_2(t_f) & \dot{\mathrm{IP}} &= 1 \\
\dot{\psi}_2 &= \dot{x}_4(t_f) & \dot{\Omega} &= \dot{x}_3(t_f)
\end{aligned}
\tag{13}
$$

With these results the terminal conditions on the multipliers become

$$
\boldsymbol{\lambda}_{\psi_1}(t_f) = \begin{bmatrix} 0 \\ 1 \\ -\dot{x}_2(t_f) \\ \dot{x}_3(t_f) \\ 0 \end{bmatrix}
\qquad
\boldsymbol{\lambda}_{\psi_2}(t_f) = \begin{bmatrix} 0 \\ 0 \\ -\dot{x}_4(t_f) \\ \dot{x}_3(t_f) \\ 1 \end{bmatrix}
\qquad
\boldsymbol{\lambda}_{\mathrm{IP}}(t_f) = \begin{bmatrix} 0 \\ 0 \\ -1 \\ \dot{x}_3(t_f) \\ 0 \end{bmatrix}
\tag{14}
$$

As there is only a single control variable, the thrust angle $\theta(t)$, the matrix $\mathbf{W}$ in Eq. 6-9(3) consists of only one element $W_{11}(t)$. Since there exists no a priori reason to unevenly weight the control changes over the interval we choose $W_{11}(t)$ to be unity everywhere in $0 \leqslant t \leqslant t_f$.

Let us consider two initial estimates of the optimal control. To start the process of finding the optimal control in case 1 the assumption is made that $\theta(t) = 10°$ in the interval from $t = 0$ until the stopping condition of Eq. (4) is

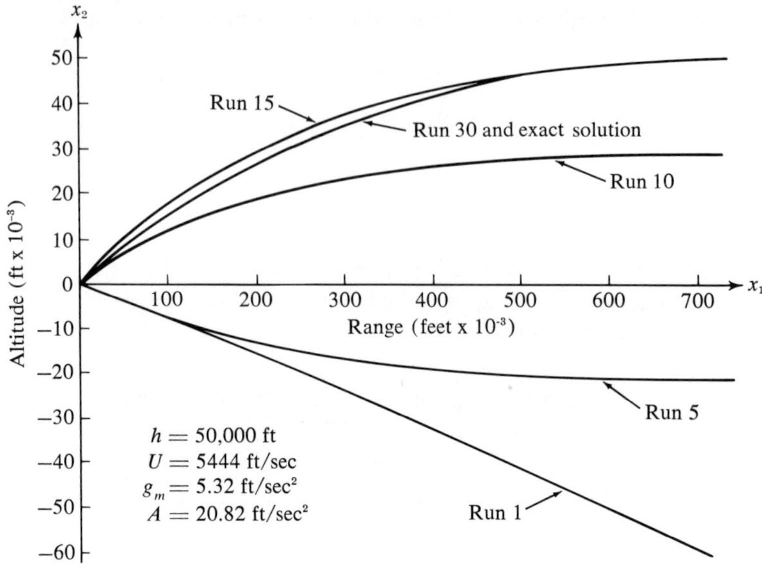

FIGURE 6-3   Variation of trajectory profile under successive iterations —initial control assumption $\theta(t) = 10°$

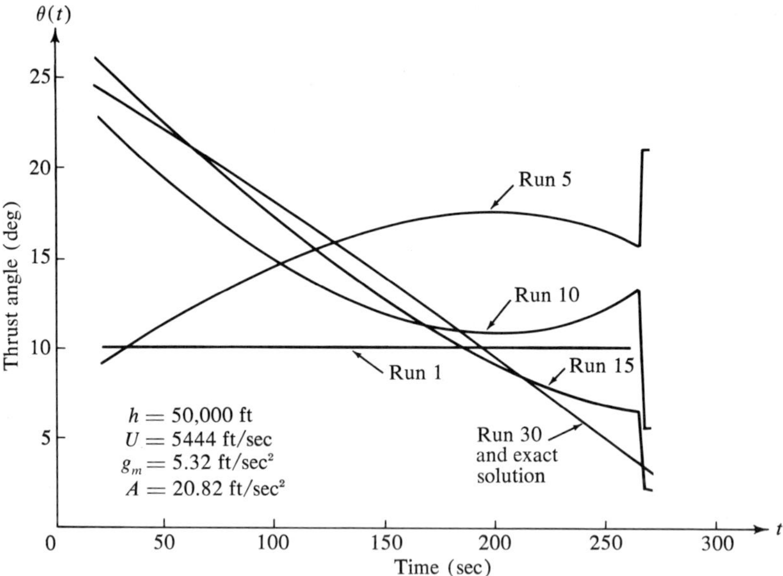

FIGURE 6-4   Variation of thrust angle under successive iterations—initial control assumption $\theta(t) = 10°$

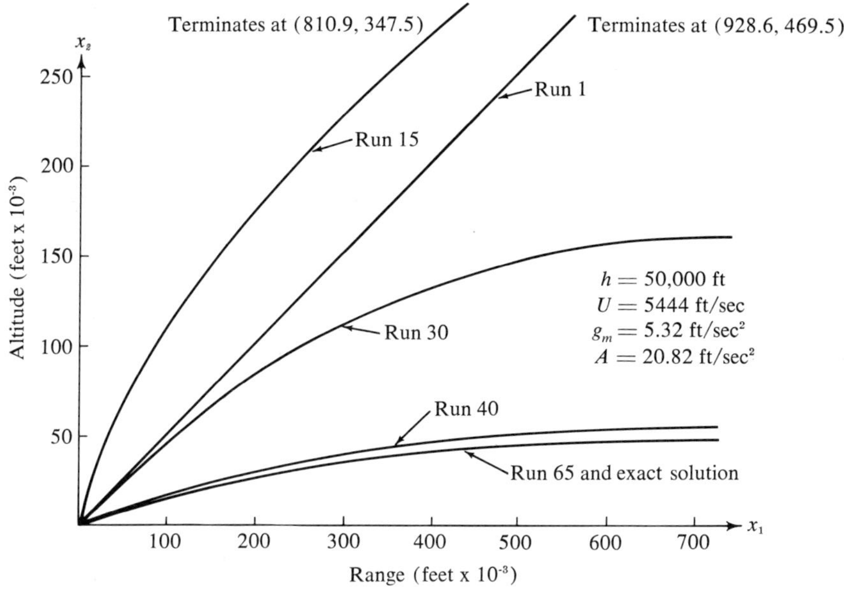

FIGURE 6-5   Variation of trajectory profile under successive iterations—initial control assumption $\theta(t) = 40°$

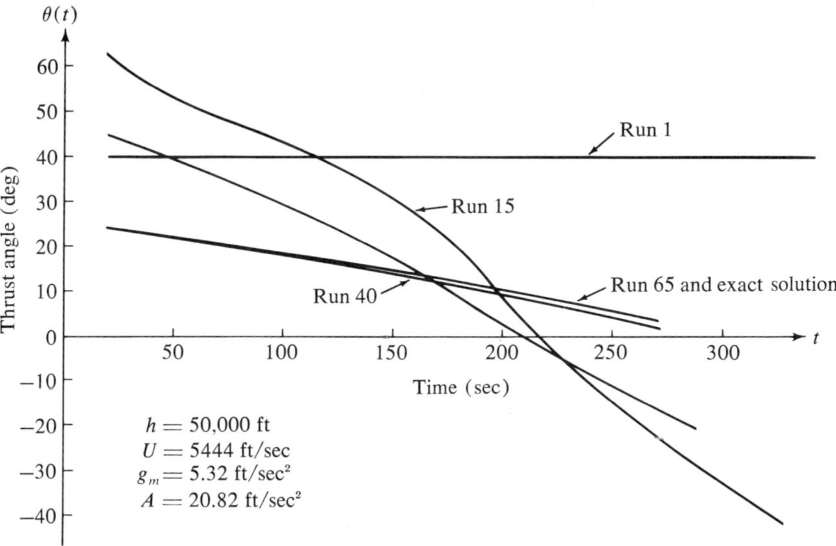

FIGURE 6-6   Variation of thrust angle under successive iterations—initial control assumption $\theta(t) = 40°$

satisfied. The results of successively improving the solution thus found using the method of Section 6-9† is illustrated in Figs. 6-3 and 6-4. Comparison is made there of the approximate solutions with the exact solution obtained in Section 3-9. For case 2 the initial estimate of the optimal control is taken to be $\theta(t) = 40°$ from $t = 0$ until the stopping condition of Eq. (4) is reached. The results are given in Figs. 6-5 and 6-6.

For both cases the runs were made with the limits $|d\psi_1| \leqslant 10,000$ ft and $|d\psi_2| \leqslant 200$ ft/sec. The solutions started with $dS = 1.0$. Once the required changes $d\psi_1$ and $d\psi_2$ were less than 10,000 ft and 200 ft/sec, the value of dS was reduced by a factor of 10 every 10 iterations down to $dS = 0.001$. When the values of $d\psi_1$ and $d\psi_2$ first fell within the limiting values, the value of $|dIP/dS|_{d\psi=0}$ in both case 1 and case 2 was on the order of unity. At the termination of the process the value of $|dIP/dS|_{d\psi=0}$ was on the order of 10^{-2}.

6-11   THE PENALTY-FUNCTION APPROACH

An alternative scheme to the method developed here of handling terminal conditions on the state variables is the penalty-function approach. The technique can also be used to treat problems involving inequality constraints.

The idea is a very simple one. Imagine that $\psi = 0$ is a terminal constraint to be satisfied by the problem solution. The penalty function approach is to adjoin a positive definite function of ψ, for example, $k\psi^2$, where $k > 0$, to the IP to be minimized,

$$\mathscr{IP} = IP + k\psi^2$$

To illustrate, in the lunar ascent problem of Section 6-10 the augmented performance index including the terminal penalty functions might be

$$\mathscr{IP} = t_f + k_1[x_2(t_f) - 50,000]^2 + k_2[x_3(t_f) - 5444]^2 + k_3 x_4^2(t_f)$$

The value of the performance index is thus penalized for deviations of ψ from zero. If the optimal control is found so as to minimize $\mathscr{IP}$, the solution will also attempt to drive ψ to the neighborhood of zero. The larger the value of k, the greater the penalty paid for deviations of ψ from zero, and hence the closer the solution will attempt to attain ψ equal zero.

The advantage of the technique is that it eliminates the need for directly satisfying the constraints; hence the computational solution is simpler. However, as the value of k is increased in an attempt to refine the solution, instabilities can result. They occur because of the effort by the control to drive ψ to

† Because the adjoint equations of (9) and (10) are quite simple, their integration subject to the terminal conditions of Eq. (14) can be carried out without resort to the computer.

zero in $\mathscr{IP}$ at the expense of minimizing IP. Thus the penalty-function approach might best be used to quickly bring the solution to the neighborhood of the desired result. For an accurate solution direct treatment of the constraints is usually necessary.

6-12 CONCLUDING REMARKS

The techniques presented in this chapter are first-order schemes and as such conceptually among the simplest available. Once having mastered the material in the chapter the reader may be interested in the more sophisticated second-order techniques that have been developed (1–4).

An additional scheme of great promise is the conjugate gradient method (5). The convergence properties of this technique appear to be excellent and computational simplicity is retained.

REFERENCES

1. S. Mitter, "Successive Approximation Methods for the Solution of Optimal Control Problems," *Automatica*, Vol. 3, 1966, pp. 133–149.
2. S. R. McReynolds and A. E. Bryson, "A Successive Sweep Method for Solving Optimal Programming Problems," Preprint Volume, *Joint Automatic Control Conference*, 1965, pp. 551–555.
3. J. V. Breakwell, J. L. Speyer, and A. E. Bryson, "Optimization and Control of Non-Linear Systems Using the Second Variation," *SIAM J. Control*, Ser. A, Vol. 1, No. 2, 1963.
4. H. J. Kelley, "Guidance Theory and Extremal Fields," *IRE Trans. Automat. Control*, Oct. 1962, pp. 75–82.
5. L. S. Lasdon, S. K. Mitter, and A. D. Warren, "The Conjugate Gradient Method for Optimal Control Problems," *IEEE Trans. Automat. Control*, April 1967, pp. 132–138.

7

DYNAMIC PROGRAMMING
AND OPTIMAL CONTROL

7-1 THE OPTIMAL ALLOCATION PROBLEM

We shall introduce the dynamic programming concepts and terminology by considering a rather simple problem called the optimal allocation problem. Suppose one initially has $\$c_1$ which one wants to invest in the following simple situation. At the beginning of every year there are two choices for investment. An investment of $\$x_1$ in the first choice yields a return of $g(x_1)$ at the end of the year, where g is a known function. An investment of the balance, $\$(c_1 - x_1)$ in the second choice yields a return of $h(c_1 - x_1)$ at the end of the year. It is not necessary to consider g and h in terms of dollars. They may, for example, represent the output of some production process. As a cost of obtaining $g(x_1)$, x_1 is reduced to ax_1, where $0 \leqslant a < 1$ and as a cost of obtaining $h(c_1 - x_1)$, $(c_1 - x_1)$ is reduced to $b(c_1 - x_1)$, where $0 \leqslant b < 1$.

We are required to review the investment situation at the end of each year and divide and reinvest the total available capital at the end of each year between the two choices in such a manner that the total return is maximized over a period of n years. If $n = 1$, the problem reduces to maximizing the function

$$R_1(c_1, x_1) = g(x_1) + h(c_1 - x_1) \tag{1}$$

over all values of x_1 such that $0 \leqslant x_1 \leqslant c_1$.

If $n = 2$, we shall have a return given by Eq. (1) and a capital of $ax_1 + b(c_1 - x_1)$ to reinvest at the end of the first year.†

† For simplicity a and b will be held constant, although they could be made a function of the stage.

210

Let $c_2 = ax_1 + b(c_1 - x_1)$. The return for a two-year process is then

$$R_2(c_1, x_1, x_2) = g(x_1) + h(c_1 - x_1) + g(x_2) + h(c_2 - x_2)$$

The maximization is over the two variables x_1, x_2, where $0 \leqslant x_1 \leqslant c_1$ and $0 \leqslant x_2 \leqslant c_2$. Similarly, if we have an n-year process, the total return from the n-stage process is

$$R_n(c_1, x_1, x_2, \cdots, x_n) = g(x_1) + h(c_1 - x_1) + g(x_2) + h(c_2 - x_2)$$
$$+ \cdots + g(x_n) + h(c_n - x_n) \quad (2)$$

where

$$c_2 = ax_1 + b(c_1 - x_1) \qquad 0 \leqslant x_1 \leqslant c_1$$

$$c_3 = ax_2 + b(c_2 - x_2) \qquad 0 \leqslant x_2 \leqslant c_2$$

$$\vdots$$

$$c_n = ax_{n-1} + b(c_{n-1} - x_{n-1}) \qquad 0 \leqslant x_{n-1} \leqslant c_{n-1}$$

$$0 \leqslant x_n \leqslant c_n$$

The maximization is over the n variables $x_1, x_2, \cdots, x_n$.

Let us compare some methods of finding a solution to the problem posed. One method of solving the n-dimensional maximization problem is to equate to zero each of the partial derivatives of the right side of Eq. (2) with respect to the n variables. This will yield n simultaneous equations in n variables of the form

$$g'(x_n) - h'(c_n - x_n) = 0$$
$$g'(x_{n-1}) - h'(c_{n-1} - x_{n-1}) + (a - b)h'(c_n - x_n) = 0 \quad (3)$$
$$\vdots$$
$$g'(x_1) - h'(c_1 - x_1) + (a - b)h'(c_2 - x_2) = 0$$

However, even if these equations can be solved, the solution is not correct unless the absolute maximum occurs at an interior point and not on the boundary of the set of allowable x. Thus to find the absolute maximum the solutions obtained from Eq. (3) must be compared with those found allowing the maximum to occur on the boundary. We can see that the computation involved for any reasonable-sized problem is large.

Alternatively, consider solving this problem numerically by direct enumeration of all possibilities. For example, for a 10-stage process if each of the intervals $(0, c_1)$, $(0, c_2)$, $\cdots$, $(0, c_{10})$ is divided into 10 parts and the set $(x_1, x_2, \cdots, x_{10})$ which maximizes the total return sought, some 10^{10} comparisons would be necessary to determine the maximum value of R_{10}. This means a computing time on the order of 2780 hr, assuming 1 millisec for each computation. It is also not difficult to see that the computing time increases rapidly with an increase in the number of stages.

With this in mind we now attempt to solve this problem using a different technique. Notice that the maximum total return depends only on the number of stages n and the initial quantity c_1. Thus let us formally define:

$$f_n(c_1) = \text{the maximum total return from an } n\text{-stage} \atop \text{process starting with an initial quantity } c_1 \tag{4}$$

Notice that in the original problem n and c_1 are fixed quantities, whereas in Eq. (4), n and c_1 are treated as variables. We are now attempting to solve the problem for an arbitrary n and c_1. The original problem has been embedded in a more general problem. If the solution to this general problem can be found, then the solution of the original problem immediately follows as it is just a special case of the general problem.

From Eqs. (4) and (2) we have

$$f_n(c_1) = \max_{(x_1, x_2 \cdots x_n)} R_n(c_1, x_1, x_2, \cdots, x_n) \qquad n = 2, 3, \cdots \tag{5}$$

with

$$f_1(c_1) = \max_{\mathscr{R}} \left[g(x_1) + h(c_1 - x_1) \right] \tag{5a}$$

where $\mathscr{R}$ is the admissible range of the optimizing variables. We seek now an expression for $f_2(c_1)$ in terms of $f_1(c_1)$. The total return for the two-stage process consists of the return from the first stage plus the return from the second stage in which $ax_1 + b(c_1 - x_1)$ is allocated. Thus

$$f_2(c_1) = \max_{\mathscr{R}} \left[g(x_1) + h(c_1 - x_1) + R_1(c_2, x_2) \right] \tag{6}$$

where

$$c_2 = ax_1 + b(c_1 - x_1)$$

However, with an initial allocation of x_1 in the first stage, the return from the second stage if x_2 is chosen optimally is, by Eq. (4), $f_1[ax_1 + b(c_1 - x_1)]$. Using the optimal choice it follows that Eq. (6) can be written

$$f_n(c_1) = \max_{\mathscr{R}} \left[g(x_1) + h(c_1 - x_1) + f_1(ax_1 + b(c_1 - x_1)) \right] \tag{7}$$

The replacement of $R_1(c_2, x_2)$ in Eq. (6) by $f_1[ax_1 + b(c_1 - x_1)]$ is most important in reducing the number of alternative decisions that must be considered. In place of considering all possible values x_2 at the start of the second year, as in Eq. (6), the choice has been limited to only those which are optimal over the remaining stage. Using the same arguments we see that in general

$$f_n(c_1) = \max_{\mathscr{R}} \left[g(x_1) + h(c_1 - x_1) + f_{n-1}(ax_1 + b(c_1 - x_1)) \right] \tag{8}$$

The recursive relationship of (8) may be used to compute $f_k(c_1)$ as well as $x_k(c_1)$, the optimal allocation to be made at the beginning of a k-stage process starting with an initial amount c_1. The final result is a sequence $x_k(c_1)$ and $f_k(c_1)$ for $c_1 > 0$, $k = 1, 2, \cdots$.

If we are now given a particular c_1 and n, the allocation is as follows:

$$\begin{aligned}
x_1 &= x_n(c_1) \\
x_2 &= x_{n-1}[ax_1 + b(c_1 - x_1)] \\
x_3 &= x_{n-2}[ax_2 + b(c_2 - x_2)] \\
&\ \ \vdots \\
x_n &= x_1[ax_{n-1} + b(c_{n-1} - x_{n-1})]
\end{aligned} \tag{9}$$

where $x_1 = x_n(c_1)$ is the allocation at the start of the process with n decisions still to be m.de; $x_2 = x_{n-1}[ax_1 + b(c_1 - x_1)]$ is the allocation when there are $n - 1$ decisions still to be made and so on to the nth decision where $x_n = x_1[ax_{n-1} + b(c_{n-1} - x_{n-1})]$ is the optimum allocation when there is only one decision left to be made before the end of the process.

The n-dimensional maximization problem of Eq. (3) has been reduced through the recursive relationship of Eq. (8) to the solution of n one-dimensional maximization problems. The digital computer is ideally suited to compute such recursive relationships. Let us examine the number of comparisons required to find the solutions using this method. Dividing $(0, c_1)$ into 10 parts, it is seen that 10 comparisons are required to find $f_1(c_1)$ and the optimum allocation x_1 from Eq. (5a) for each c_1. If this is done for 10 values of c_1 some 100 comparisons are required to find $f_1(c_1)$ for the 10 values of c_1 selected. Similarly, working with $f_2(c_1)$ from Eq. (6), once again dividing $(0, c_1)$ into 10 parts, it is seen that 10 comparisons are required to find $f_2(c_1)$ and the optimum allocation x_2 for each value of c_1 chosen. Then to determine $f_2(c_1)$ for 10 values of c_1 once $f_1(c_1)$ is known, some 100 comparisons are again required. Thus, given a 10-stage process, some 10^3 comparisons are required to find $f_{10}(c_1)$ and the desired solution. This contrasts with the 10^{10} comparisons required for this case by direct enumeration of all possibilities.

Notice from Eq. (8) that the n-stage process has been treated by building up the solution recursively. It is necessary to consider one-stage, two-stage, $\cdots$, $(n - 1)$-stage, and n-stage processes and to decide on the *optimal policy* x for each of these processes to solve the original n-stage problem. The *optimal return function* is defined as the return from the optimum policy. The solution of the multistage decision process using the recursive method presented is termed *dynamic programming*.

Finally, the argument used in going from Eq. (6) to Eq. (7) is an example of the application of a general principle termed the *principle of optimality*, which states:

An optimal policy has the property that whatever the initial state and initial decision are, the remaining decisions must constitute an optimal policy with regard to the state resulting from the first decision.

Example 7-1(1)   Consider the problem of allocating in an optimal manner the initial resource $c_1 = 100$ in the three-stage process *(1)*, where

$$g(x) = 0.4\sqrt{x}$$
$$h(x) = 0.2\sqrt{x} \tag{10}$$

and

$$a = 0.6 \qquad b = 0.8$$

The value $f_3(100)$ and the allocation policy to be followed to achieve it are to be found.

From Eq. (8), determination of $f_3(100)$ is seen to depend on knowledge of $f_2(c_1)$, which itself cannot be found until the behavior of $f_1(c_1)$ is known. Thus to start we must find

$$f_1(c_1) = \max_{x_1} [0.4\sqrt{x_1} + 0.2\sqrt{c_1 - x_1}] \tag{11}$$

over the range of argument c_1 required to find $f_2(c_1)$. Since at any stage in the process after the first, the resource c_1 available will be less than the initial amount, it is sufficient to consider $0 \leqslant c_1 \leqslant 100$.

Although the form of Eq. (11) is simple enough to yield an analytic solution, this is generally not the case. Thus let us imagine that the behavior of $f_1(c_1)$ must be found numerically. We shall assume that knowledge of $f_1(c_1)$ at the 10 values $c_1 = 10, 20, \cdots, 100$ over the range $0 \leqslant c_1 \leqslant 100$ provides sufficient accuracy. Then for each value of c_1 the corresponding x_1 which maximizes $f_1(c_1)$, where $0 \leqslant x_1 \leqslant c_1$ is found by dividing the interval into 10 parts, inserting each of the 10 values of x_1 into Eq. (11) in turn and choosing that which maximizes $f_1(c_1)$. The results are shown in Table 7-1(1) under the heading One-stage process.

TABLE 7-1(1)   *Three-Stage Allocation Problem*

| | I | | | II | | | III | |
| | Three-stage process | | | Two-stage process | | | One-stage process | |
c_1	x_1	$f_3(c_1)$	x_1	$f_2(c_1)$	x_1	$f_1(c_1)$
10	6	3.484	7	2.551	8	1.413
20	12	4.928	14	3.608	16	1.998
30	18	6.035	21	4.419	24	2.448
40	24	6.969	28	5.103	32	2.828
50	30	7.792	35	5.706	40	3.160
60	36	8.536	42	6.251	48	3.462
70	42	9.220	49	6.752	56	3.740
80	48	9.856	56	7.217	64	3.997
90	54	10.454	63	7.656	72	4.240
100	60	11.02	70	8.07	80	4.47

Continuing, with $f_1(c_1)$ known, $f_2(c_1)$ is found from

$$f_2(c_1) = \max_{x_1} \left[0.4\sqrt{x_1} + 0.2\sqrt{c_1 - x_1} + f_1(0.6x_1 + 0.8(c_1 - x_1)) \right] \tag{12}$$

The behavior of $f_2(c_1)$, as with $f_1(c_1)$, is determined for each of the 10 values $c_1 = 10, 20, \cdots, 100$. For each c_1, the maximizing x_1 is found by dividing c_1 into 10 parts and testing each value of x_1 in Eq. (12). The maximizing x_1 and the corresponding value of $f_2(c_1)$ found are given in Table 7-1(1) under the heading Two-stage process. In the determination of $f_2(c_1)$, values of $f_1(c_1)$ not known are found by interpolation of the values contained in Table 7-1(1).

Finally the desired quantity $f_3(100)$ can be obtained from

$$f_3(c_1) = \max_{x_1} \left[0.4\sqrt{x_1} + 0.2\sqrt{c_1 - x_1} + f_2(0.6x_1 + 0.8(c_1 - x_1)) \right] \tag{13}$$

by the same procedure described for $f_2(c_1)$. Although only $f_3(100)$ is required, for completeness results for all the values of c_1 for which $f_2(c_1)$ and $f_1(c_1)$ are given are included in Table 7-1(1) under the heading Three-stage process. Again, when determination of $f_3(c_1)$ requires knowledge of $f_2(c_1)$ at a point not provided by the table the known values are interpolated.

Referring to column I, Table 7-1(1) shows that optimal allocation of the initial resource $c_1 = 100$ over the three-stage process will yield a return $f_3(100) = 11.02$. It is also shown that the initial allocation should be

$$x_3(100) = x_1 = 60 \tag{14}$$

With $x_1 = 60$, the resource available at the beginning of the second stage is found from

$$c_2 = ax_1 + b(c_1 - x_1) = 0.6(60) + 0.8(100 - 60) = 68 \tag{15}$$

Referring to column II of Table 7-1(1) it is found by interpolation of the listed values that optimal allocation of the 68 units of resource available at the start of the remaining two-stage process is given by choosing

$$x_2(68) = x_2 = 47.6 \tag{16}$$

The resource then left to be allocated over the remaining stage is given by

$$c_3 = ax_2 + b(c_2 - x_2) = 0.6(47.6) + 0.8(68 - 47.6) = 44.9 \tag{17}$$

Interpolation of the values listed under column III yields that for optimal allocation of the resource $c_3 = 44.9$ over the remaining stage, the final allocation must be

$$x_1(44.9) = x_3 = 35.9 \tag{18}$$

This concludes the solution of the problem stated. Although a numerical treatment was used to illustrate the usual method of attack, as noted before, the form of the problem is simple enough that an analytic solution can be found. The reader may want to verify analytically that

$$f_1(c_1) = 0.447\sqrt{c_1} \qquad x_1(c_1) = 0.8c_1$$

$$f_2(c_1) = 0.807\sqrt{c_1} \qquad x_2(c_1) = 0.7c_1 \tag{19}$$

$$f_3(c_1) = 1.102\sqrt{c_1} \qquad x_3(c_1) = 0.6c_1$$

7-2 APPLICATION TO THE CONTROL PROBLEM—THEORY

We now employ the dynamic programming approach to formulate the solution of a fairly general control problem. Consider a system with n state variables and m control variables,

$$\mathbf{x}(t) \Rightarrow x_1(t), \cdots, x_n(t) \qquad \text{state variables}$$

$$\mathbf{u}(t) \Rightarrow u_1(t), \cdots, u_m(t) \qquad \text{control variables}$$

The problem is that of choosing the control vector $\mathbf{u}(t)$ so as to minimize the performance index

$$\mathrm{IP} = g(\mathbf{x}_f, t_f) + \int_{t_i}^{t_f} \bar{f}[\tau, \mathbf{x}(\tau), \mathbf{u}(\tau)]\, d\tau \tag{1}$$

subject to the differential constraints

$$\dot{x}_j = f_j[t, \mathbf{x}(t), \mathbf{u}(t)] \qquad j = 1, \cdots, n \tag{2}$$

and the end conditions

$$t_i \text{ and } x_j(t_i) \qquad j = 1, \cdots, n \qquad \text{specified} \tag{3a}$$

$$x_j(t_f) \qquad j = 1, \cdots, p \leqslant n \qquad \text{specified}$$

$$t_f \qquad\qquad\qquad \text{specified or unspecified} \tag{3b}$$

To start the dynamic programming approach let us embed the original problem in a more general one. Removing the restriction of Eq. (3a) that the system start at the specified values $x_j(t_i)$ at t_i, we define

$$\mathrm{IP}[\mathbf{x}(t), t] = \text{minimum performance index (optimal return) for} \tag{4}$$
$$\text{the process starting at state } \mathbf{x}(t) \text{ at time } t$$

Therefore, subject to the terminal conditions of Eq. (3b) and the differential constraints of Eq. (2) which govern the process, we have

$$\mathrm{IP}[\mathbf{x}(t), t] = \min_{\{\mathbf{u}\}} \left\{ g[\mathbf{x}(t_f), t_f] + \int_{t}^{t_f} \bar{f}[\tau, \mathbf{x}(\tau), \mathbf{u}(\tau)]\, d\tau \right\} \tag{5}$$

The symbol $\{\mathbf{u}\}$ on the right side is meant to indicate that the minimization is to be carried out over the range of admissible controls in the interval t to t_f.

Proceeding, the integral in Eq. (5) is divided into an integral over an initial period Δt followed by one over the remaining interval from $t + \Delta t$ to t_f. Thus

$$\mathrm{IP}[\mathbf{x}(t),\, t] = \min_{\{\mathbf{u}\}} \left\{ g[\mathbf{x}(t_f),\, t_f] + \int_t^{t+\Delta t} \bar{f}[\tau,\, \mathbf{x}(\tau),\, \mathbf{u}(\tau)]\, d\tau \right.$$

$$\left. + \int_{t+\Delta t}^{t_f} \bar{f}[\tau,\, \mathbf{x}(\tau),\, \mathbf{u}(\tau)]\, d\tau \right\} \quad (6)$$

where, since the range of the first integral is small, it can be represented by

$$\int_t^{t+\Delta t} \bar{f}[\tau,\, \mathbf{x}(\tau),\, \mathbf{u}(\tau)]\, d\tau = \bar{f}[t,\, \mathbf{x}(t),\, \mathbf{u}(t)]\, \Delta t + O(\Delta t^2)$$

yielding

$$\mathrm{IP}[\mathbf{x}(t),\, t] = \min_{\{\mathbf{u}\}} \left\{ \bar{f}[t,\, \mathbf{x}(t),\, \mathbf{u}(t)]\, \Delta t + O(\Delta t^2) + g[\mathbf{x}(t_f),\, t_f] \right.$$

$$\left. + \int_{t+\Delta t}^{t_f} \bar{f}[\tau,\, \mathbf{x}(\tau),\, \mathbf{u}(\tau)]\, d\tau \right\} \quad (7)$$

Now $g[\mathbf{x}(t_f),\, t_f]$ plus the integral in Eq. (7) is the return in going to t_f from the state $\mathbf{x}(t) + \Delta\mathbf{x}$ existing at time $t + \Delta t$; the increment $\Delta\mathbf{x}$ being that resulting from using the control $\mathbf{u}(t)$ over the interval Δt starting at time t. However, whatever the control decision $\mathbf{u}(t)$ at time t, the *principle of optimality* states that the control policy over the remaining interval must be optimal if the policy over the entire interval is to be optimal.

By Eq. (4), starting at state $\mathbf{x}(t) + \Delta\mathbf{x}$ at time $t + \Delta t$, the return from the optimal policy is given by $\mathrm{IP}[\mathbf{x}(t) + \Delta\mathbf{x},\, t + \Delta t]$. Substituting this quantity for the integral plus $g[\mathbf{x}(t_f),\, t_f]$ in Eq. (7) one then obtains

$$\mathrm{IP}[\mathbf{x}(t),\, t] = \min_{\mathbf{u}(t)} \{ \bar{f}[t,\, x(t),\, u(t)]\, \Delta t + O(\Delta t^2)$$

$$+ \mathrm{IP}[\mathbf{x}(t) + \mathbf{f}(t,\, \mathbf{x}(t),\, \mathbf{u}(t))\, \Delta t + O(\Delta t^2),\, t + \Delta t]\} \quad (8)$$

where with the aid of Eq. (2), $\Delta\mathbf{x}$ has been replaced by

$$\Delta\mathbf{x} = \mathbf{f}[t,\, \mathbf{x}(t),\, \mathbf{u}(t)]\, \Delta t + O(\Delta t^2)$$

Note in Eq. (8) that the minimization is now done over the control $\mathbf{u}(t)$ to be applied at time t rather than, as in Eq. (7), over the control in the interval t to t_f.

Expanding $\mathrm{IP}[\mathbf{x}(t) + \Delta\mathbf{x},\, t + \Delta t]$ about its value at $\mathbf{x}(t)$ and t, Eq. (8) yields

$$\mathrm{IP}[x(t),\, t] = \min_{\mathbf{u}(t)} \left\{ \bar{f}[t,\, \mathbf{x}(t),\, \mathbf{u}(t)]\, \Delta t + \mathrm{IP}[\mathbf{x}(t),\, t] \right.$$

$$+ \sum_{j=1}^{n} \frac{\partial \mathrm{IP}}{\partial x_j}\, [\mathbf{x}(t),\, t] f_j[t,\, \mathbf{x}(t),\, \mathbf{u}(t)]\, \Delta t$$

$$\left. + \frac{\partial \mathrm{IP}}{\partial t}\, [\mathbf{x}(t),\, t]\, \Delta t + O(\Delta t^2) \right\} \tag{9}$$

Since $\mathrm{IP}[\mathbf{x}(t),\, t]$ and $\partial \mathrm{IP}[\mathbf{x}(t),\, t]/\partial t$ are not functions of $\mathbf{u}(t)$, they can be taken out from under the minimization on the right side of Eq. (9), the term $\mathrm{IP}[\mathbf{x}(t),\, t]$ canceling the same term on the left. Then dividing Eq. (9) by Δt and taking the limit as $\Delta t \to 0$ one obtains

$$0 = \frac{\partial \mathrm{IP}}{\partial t}\, [\mathbf{x}(t),\, t] + \min_{\mathbf{u}(t)} \left\{ \bar{f}[t,\, \mathbf{x}(t),\, u(t)] \right.$$

$$\left. + \sum_{j=1}^{n} \frac{\partial \mathrm{IP}}{\partial \mathbf{x}_j}\, [\mathbf{x}(t),\, t] f_j[t,\, \mathbf{x}(t),\, u(t)] \right\} \tag{10}$$

Equation (10) is to be solved for $\mathrm{IP}[\mathbf{x}(t),\, t]$ and the optimal control policy in the interval t to t_f.

To aid in the solution, Eq. (10) may be written differently. Assuming no control limitations, minimization of the bracketed term in Eq. (10) requires that its partial derivative with respect to each of the control variables vanish. On using controls in Eq. (10) which satisfy this necessary condition, as well as sufficiency conditions for the minimum, the minimization notation can be removed. Thus for these controls, Eq. (10) can be replaced† by the set of equations

$$0 = \frac{\partial \mathrm{IP}}{\partial t}\, (\mathbf{x},\, t) + \bar{f}(t,\, \mathbf{x},\, \mathbf{u}) + \sum_{j=1}^{n} \frac{\partial \mathrm{IP}}{\partial x_j}\, (\mathbf{x},\, t) f_j(t,\, \mathbf{x},\, \mathbf{u}) \tag{11a}$$

$$0 = \frac{\partial \bar{f}}{\partial u_k}\, (t,\, \mathbf{x},\, \mathbf{u}) + \sum_{j=1}^{n} \frac{\partial \mathrm{IP}}{\partial x_j}\, (\mathbf{x},\, t) \frac{\partial f_j}{\partial u_k}\, (t,\, \mathbf{x},\, \mathbf{u}) \qquad k = 1,\, \cdots,\, m \tag{11b}$$

† The controls which satisfy Eq. (11b) must minimize

$$\left[\bar{f}(t,\, \mathbf{x},\, \mathbf{u}) + \sum_{j=1}^{n} \frac{\partial \mathrm{IP}}{\partial x_j}\, (\mathbf{x},\, t) f_j(t,\, \mathbf{x},\, u) \right]$$

not merely render the quantity stationary.

The condition to be satisfied at t_f by the solution of Eq. (11) remains to be stated. This is easily done by noting from Eq. (5) that

$$\text{IP}[\mathbf{x}(t_f), t_f] = g[\mathbf{x}(t_f), t_f] \tag{12}$$

The relation of Eq. (12) provides the required condition for the determination of $\text{IP}(\mathbf{x}, t)$, and with it, the optimal control policy from Eq. (11). Once the solution for the optimal return function is known for arbitrary starting points $\mathbf{x}(t)$ and t, the desired solution satisfying the particular starting point of Eq. (3a) is automatically determined.

7-3   A PERFECT INTEGRATOR

To illustrate use of the method, let us consider the optimization of a system which is simple enough to allow analytic solution. We assume a linear first-order plant which behaves as a perfect integrator,

$$\dot{x} = u \tag{1}$$

The problem is taken to be that of minimizing the performance index

$$\text{IP} = \int_0^{t_f} (x^2 + u^2)\, dt \tag{2}$$

subject to the initial conditions

$$t_i = 0$$
$$x(t_i) = x_0 \tag{3}$$

and, in turn, the two sets of terminal conditions

$$\left.\begin{array}{l} t_f \quad \text{specified} \\ x(t_f) \quad \text{unspecified} \end{array}\right\} \tag{4a}$$

$$\left.\begin{array}{l} t_f \quad \text{specified} \\ x(t_f) = x_f \quad \text{specified} \end{array}\right\} \tag{4b}$$

For this example Eq. 7-2(10) becomes

$$0 = \frac{\partial \text{IP}}{\partial t}(x, t) + \min_{u(t)} \left[x^2 + u^2 + \frac{\partial \text{IP}}{\partial x}(x, t)u \right] \tag{5}$$

which, written in the form of Eq. 7-2(11), is equivalent to the two equations

$$0 = \frac{\partial \text{IP}}{\partial t}(x, t) + x^2 + u^2 + \frac{\partial \text{IP}}{\partial x}(x, t)u \tag{6a}$$

$$0 = 2u + \frac{\partial \text{IP}}{\partial x}(x, t) \tag{6b}$$

Equation (6b) yields the relation between the optimal control law† and the performance index,

$$u^*(x, t) = -\frac{1}{2}\frac{\partial \text{IP}}{\partial x}(x, t) \tag{7}$$

Substituting this result into Eq. (6a) one then obtains the equation to be solved for determination of $\text{IP}(x, t)$,

$$0 = \frac{\partial \text{IP}}{\partial t}(x, t) + x^2 - \frac{1}{4}\left[\frac{\partial \text{IP}}{\partial x}(x, t)\right]^2 \tag{8}$$

The solution of Eq. (8) is to be found subject to the condition of Eq. 7-2(12). For this problem one has

$$\text{IP}[x(t_f), t_f] = 0 \tag{9}$$

Equation (8) is a nonlinear partial differential equation, a type that does not generally yield an analytic solution. However, in this case one can be found. Letting

$$\text{IP}(x, t) = x^2\varphi_2(t) + x\varphi_1(t) + \varphi_0(t) \tag{10}$$

and inserting Eq. (10) into Eq. (8) one finds

$$0 = x^2(\dot{\varphi}_2 + 1 - \varphi_2^2) + x(\dot{\varphi}_1 - \varphi_1\varphi_2) + \left(\dot{\varphi}_0 - \frac{\varphi_1^2}{4}\right) \tag{11}$$

Then since x^2, x, and 1 are linearly independent, for Eq. (11) to be satisfied for all x requires that the coefficient of each power vanish separately, yielding the equations

$$\dot{\varphi}_2 = \varphi_2^2 - 1 \tag{12a}$$

$$\dot{\varphi}_1 = \varphi_2\varphi_1 \tag{12b}$$

$$\dot{\varphi}_0 = \frac{\varphi_1^2}{4} \tag{12c}$$

---

† Checking the second-order terms, it is found that the control of Eq. (6b) does minimize the bracketed term in Eq. (5).

Equation (12a) for determination of $\varphi_2(t)$ is a Riccati equation. Once $\varphi_2(t)$ is known, $\varphi_1(t)$ is found as the solution of Eq. (12b), a linear equation, and $\varphi_0(t)$ results from direct quadrature (integration) of Eq. (12c).

The closed-form solution for a Riccati equation is not known in general. However, in this particular case the solution of Eq. (12a) can be found by separating the variables φ_2 and t. Depending on whether $|\varphi_2(t)| > 1$ or $|\varphi_2(t)| < 1$, one finds from Eq. (12),

$$|\varphi_2(t)| > 1$$

$$\varphi_2(t) = \operatorname{ctnh}(t_f - t + \alpha)$$

$$\varphi_1(t) = \frac{C_1}{\sinh(t_f - t + \alpha)} \tag{13a}$$

$$\varphi_0(t) = C_0 + \frac{C_1^2}{4}\operatorname{ctnh}(t_f - t + \alpha)$$

$$|\varphi_2(t)| < 1$$

$$\varphi_2(t) = \tanh(t_f - t + \alpha)$$

$$\varphi_1(t) = \frac{C_1}{\cosh(t_f - t + \alpha)} \tag{13b}$$

$$\varphi_0(t) = C_0 - \frac{C_1^2}{4}\tanh(t_f - t + \alpha)$$

The Case t_f Specified, $x(t_f)$ Unspecified. Applying the condition of Eq. (9) to the IP of Eq. (10) we have

$$0 = x^2(t_f)\varphi_2(t_f) + x(t_f)\varphi_1(t_f) + \varphi_0(t_f) \tag{14}$$

Since Eq. (14) is to be true for all $x(t_f)$ this requires

$$\varphi_2(t_f) = \varphi_1(t_f) = \varphi_0(t_f) = 0 \tag{15}$$

With Eq. (15), the results of Eq. (13b) for $|\varphi_2(t)| < 1$ yield

$$\alpha = 0 \qquad \varphi_2(t) = \tanh(t_f - t)$$

$$C_1 = 0 \qquad \varphi_1(t) = 0$$

$$C_0 = 0 \qquad \varphi_0(t) = 0$$

The generic solution to the problem is thus given by

$$\mathrm{IP}(x, t) = x^2 \tanh(t_f - t) \tag{16}$$

from which the optimal control law of Eq. (7) is found to be

$$u^*(x, t) = -x \tanh(t_f - t) \tag{17}$$

The minimum value of the performance index (optimal return function) for the specific initial conditions of Eq. (3) is then

$$IP[x(0), 0] = x_0^2 \tanh t_f \tag{18}$$

The Case t_f Specified, $x(t_f) = x_f$ Specified.   As before, condition (9) requires

$$0 = x_f^2 \varphi_2(t_f) + x_f \varphi_1(t_f) + \varphi_0(t_f) \tag{19}$$

Note that the optimal control of Eq. (7) must now have the capability of driving $x(t_f - t)$ to the specified value x_f as t approaches t_f no matter what the value of $x(t_f - t)$. With this in mind it is seen that the optimal control

$$u^*(x, t) = -\frac{1}{2} \frac{\partial IP}{\partial x} = -x\varphi_2(t) - \tfrac{1}{2}\varphi_1(t) \tag{20}$$

must grow unbounded as $t \to t_f$ for arbitrary $x(t)$. Applying this argument to the solution of Eq. (13a) where $|\varphi_2(t)| > 1$ yields

$$\alpha = 0$$

with which the control law of Eq. (20) becomes

$$u^*(x, t) = -x \operatorname{ctnh}(t_f - t) - \frac{1}{2} \frac{C_1}{\sinh(t_f - t)} \tag{21}$$

Using the solution of Eq. (13a), condition (19) may be written

$$0 = \lim_{t \to t_f} \left[x^2 \operatorname{ctnh}(t_f - t) + \frac{xC_1}{\sinh(t_f - t)} + \frac{C_1^2}{4} \operatorname{ctnh}(t_f - t) \right] + C_0 \tag{22}$$

For the limit in Eq. (22) to be finite requires

$$0 = x_f^2 + x_f C_1 + \frac{C_1^2}{4}$$

yielding

$$C_1 = -2x_f \tag{23}$$

Equation (22) then can be expressed in the form

$$0 = \lim_{\tau \to 0} \left[\frac{x^2 \cosh \tau - 2xx_f + x_f^2 \cosh \tau}{\sinh \tau} \right] + C_0 \tag{24}$$

where

$$\tau = t_f - t$$

Applying L'Hospital's rule and Eq. (1), Eq. (24) may be put in the form

$$0 = C_0 + \lim_{\tau \to 0} \left[\frac{-2xu \cosh \tau + x^2 \sinh \tau + 2x_f u + x_f^2 \sinh \tau}{\cosh \tau} \right]$$

from which one finds

$$C_0 = 0 \tag{25}$$

The generic optimal return function is then given by

$$IP(x, t) = x^2 \operatorname{ctnh}(t_f - t) - \frac{2xx_f}{\sinh(t_f - t)} + x_f^2 \operatorname{ctnh}(t_f - t) \qquad (26)$$

from which the optimal control law is found to be

$$u^*(x, t) = -x \operatorname{ctnh}(t_f - t) + \frac{x_f}{\sinh(t - t)} \qquad (27)$$

For the specific initial conditions of Eq. (3), the optimal return function of Eq. (27) becomes

$$IP(x_0, 0) = x_0^2 \operatorname{ctnh} t_f - \frac{2x_0 x_f}{\sinh t_f} + x_f^2 \operatorname{ctnh} t_f \qquad (28)$$

To conclude this section let us note that we have been rather fortunate in being able to solve the governing nonlinear partial differential equation of Eq. (8) as found from Eq. 7-2(11). It is only the extreme simplicity of the example that has made this possible. The practical importance of dynamic programming is not that it yields a nonlinear partial differential equation like Eq. (8) to be solved, but that it leads to an equation of the form of Eq. 7-2(8) that can be solved iteratively on a digital computer. How this is done is shown in Section 7-4.

7-4 THE COMPUTATIONAL SCHEME

To illustrate the main features of the computational method let us consider the problem of minimizing

$$IP = \int_0^{t_f} \bar{f}(t, x, u) \, dt \qquad (1)$$

where

$$\dot{x} = f(t, x, u) \qquad (2)$$

and

$$x(0) = x_0 \text{ and } t_f \qquad \text{specified} \qquad (3)$$

In addition we shall assume the control available is limited so that

$$|u(t)| \leqslant U \qquad (4)$$

Reviewing our earlier procedure, we embed the problem in a more general one by defining

$$IP[x(t), t] = \text{minimum performance index for the process} \qquad (5a)$$
$$\text{starting at state } x(t) \text{ at time } t$$

or

$$IP[x(t), t] = \min_{\{u\}} \int_t^{t_f} \bar{f}[\tau, x(\tau), u(\tau)] \, d\tau \qquad (5b)$$

Then the fundamental equation for determination of $IP[x(t), t]$ is that of Eq. 7-2(8). Rewriting this equation in slightly different form we have

$$IP[x(t), t] = \min_{u(t)} \{\bar{f}[t, x(t), u(t)]\, \Delta t + IP[x(t) + \Delta x, t + \Delta t] + O(\Delta t^2)\} \quad (6)$$

where

$$\Delta x = f[t, x(t), u(t)]\, \Delta t + O(\Delta t^2) \quad (7)$$

From this point our procedure differs from that followed in Section 7-2. Rather than expand $IP[x(t) + \Delta x, t + \Delta t]$ in Eq. (6) to find the governing partial differential equation, we ask how Eq. (6) can be solved directly. To do this it is necessary to consider the discrete form of the problem.

Subdividing the interval between t and t_f into n equal parts of size Δt so that $t_f - t = n\,\Delta t$, let us define

$$t_k = t + k\,\Delta t$$

$$x_k = x(t_k) \qquad k = 0, 1, \cdots, n$$

$$\dot{x}_k = \dot{x}(t_k) \qquad\qquad\qquad (8a)$$

as well as

$$u_k = u(t_{k-1}) \qquad k = 1, \cdots, n \qquad (8b)$$

The notation has been chosen so that the subscript on the control gives its position in the control sequence. Thus u_1 is the first control application, occurring at the starting point $t = t_0$, u_2 is the second, occurring at $t + \Delta t$, continuing through the nth control application u_n at $t_f - \Delta t$.

In this notation we let

$$IP_n(x_0) = IP(x_0, t_0) = IP(x, t) \quad (9)$$

The subscript n on $IP_n(x_0)$ indicates that this is the performance index at time $n\,\Delta t$ before t_f. Then writing

$$\bar{f}_k(t_k, x_k, u_{k+1}) = \bar{f}[t + k\,\Delta t, x(t + k\,\Delta t), u(t + k\,\Delta t)] \quad (10)$$

Eq. (5b) can be approximated by

$$IP_n(x_0) = \min_{\{u\}} \sum_{k=0}^{n-1} \bar{f}_k(t_k, x_k, u_{k+1})\, \Delta t \quad (11)$$

The minimization over $\{u\}$ now indicates a minimization over the n control values $u_1, \cdots, u_n$.

Using Eq. (11), the discrete form of Eq. (6) to order Δt is found to be

$$IP_n(x_0) = \min_{u_1}\{\bar{f}_0(t_0, x_0, u_1)\, \Delta t + IP_{n-1}[x_0 + f_0(t_0, x_0, u_1)\, \Delta t]\} \quad (12)$$

where use has been made of Eq. (2) to provide

$$\Delta x = f[t, x(t), u(t)] \, \Delta t = f_0(t_0, x_0, u_1) \, \Delta t \tag{13}$$

To review, in using Eq. (12) to determine $IP_n(x_0)$ the quantities t_0, x_0, and u_1 are those at time $n \, \Delta t$ before t_f, Also, note that the form of Eq. (12) is the same as that of Eq. 7-1(8) for the optimum allocation problem.

Equation (12) is a recursion relation for $IP_n(x_0)$. That is, if $IP_0(x_0)$ is known, setting $n = 1$ in Eq. (12) provides a relation for determination of $IP_1(x_0)$. With $IP_1(x_0)$ known, setting $n = 2$ in Eq. (12) provides a relation for $IP_2(x_0)$ and so on until once $IP_{n-1}(x_0)$ is known, $IP_n(x_0)$ can be found. To start the procedure note that $IP_0(x_0)$ is the value of the performance index at $t = t_f$, which by Eq. (5b) is zero.

$$IP_0(x_0) = 0 \tag{14}$$

Then

$$IP_1(x_0) = \min_{u_1} \bar{f}_0(t_f - \Delta t, x_0, u_1) \, \Delta t \tag{15}$$

The minimization in Eq. (15) must be carried out over all u_1 consistent with the constraint $|u_1| \leqslant U$. It usually is most convenient to do this by direct computation. Breaking the interval $-U \leqslant u_1 \leqslant U$ into m parts, and defining $\Delta u = 2U/m$, the values $u_1 = -U, -U + \Delta u, -U + 2 \, \Delta u, \cdots, U$ are inserted in turn into Eq. (15) and the minimizing value of u_1 for each desired x_0 recorded. The size of m thus determines the accuracy of the result.

In this procedure the inequality constraint on u_1 actually simplifies the work, because it limits the number of values u_1 that must be tested. This is in sharp contrast to the inclusion of control inequality constraints in the calculus of variations approach of Chapter 4.

With $IP_1(x_0)$ now tabulated along with the minimizing values $u_1(x_0)$, we find $IP_2(x_0)$ and the optimal control u_1 at time $2 \, \Delta t$ before t_f from

$$\begin{aligned} IP_2(x_0) = \min_{u_1} \{ &\bar{f}_0(t_f - 2 \, \Delta t, x_0, u_1) \, \Delta t \\ &+ IP_1[x_0 + f_0(t_f - 2 \, \Delta t, x_0, u_1) \, \Delta t] \} \end{aligned} \tag{16}$$

The process is continued until $IP_N(x_0)$ and the optimal control at this point are known, where N is defined by $t_f - N \, \Delta t = t_0 = 0$. Using this N, inserting the desired initial value of $x(0)$ as given by Eq. (3) into $IP_N(x_0)$, provides the minimum performance index for the problem specified.

The problem is seen to be an N-stage decision process. Denoting the optimal control at the beginning of a process starting $k \, \Delta t$ before t_f at state x_0 by $u_k(x_0)$, the result of the procedure is a sequence $u_k(x_0)$ and $IP_k(x_0)$ for $k = 1, 2, \cdots N$.

For our problem, then, the optimal control sequence is given by

$$
\begin{aligned}
u_1 &= u_N(x_0) \\
u_2 &= u_{N-1}(x_1) \\
u_3 &= u_{N-2}(x_2) \\
&\vdots \\
u_N &= u_1(x_{N-1})
\end{aligned}
\tag{17}
$$

where

$$
\begin{aligned}
x_1 &= x_0 + f(0, x_0, u_1)\,\Delta t \\
x_2 &= x_1 + f(\Delta t, x_1, u_2)\,\Delta t \\
&\vdots \\
x_{N-1} &= x_{N-2} + f(t_f - 2\,\Delta t, x_{N-2}, u_{N-1})\,\Delta t \\
x_N &= x_{N-1} + f(t_f - \Delta t, x_{N-1}, u_N)\,\Delta t
\end{aligned}
\tag{18}
$$

Example 7-4(1)   As a specific illustration consider the regulator problem requiring the minimization of

$$
\mathrm{IP} = \int_0^2 (tx^2 + u^2)\,dt
\tag{19}
$$

subject to the system equation,

$$
\dot{x} = u
\tag{20}
$$

the initial condition,

$$
x(0) = 1
\tag{21}
$$

and the control limitation,

$$
|u| \leqslant 0.5
\tag{22}
$$

The multiplier t of x^2 in the performance index reduces the penalty to the system for initial deviations of x from zero while maximizing the effect of terminal deviations as is sometimes desired.

Let us break the interval $0 \leqslant t \leqslant t_f = 2$ into 10 equal parts, $\Delta t = 0.2$, $N = 10$. Then $\mathrm{IP}_{10}(1)$ is desired along with the optimal control sequence $u_{10}(1), u_9(x_1), \cdots, u_1(x_9)$. From Eq. (15) we have

$$
\mathrm{IP}_1(x_0) = \min_{u_1}[((2 - \Delta t)x_0^2 + u_1^2)\,\Delta t]
\tag{23}
$$

Hence it is seen that

$$
u_1(x_0) = 0 \qquad \mathrm{IP}_1(x_0) = (2 - \Delta t)x_0^2\,\Delta t
\tag{24}
$$

where as before $u_k(x_0)$ denotes the optimal control at time $k\,\Delta t$ before t_f.

TABLE 7-4(1)   *Dynamic Programming Problem*

x_0	$IP_{10}(x_0)$	$u_{10}(x_0)$	$IP_9(x_0)$	$u_9(x_0)$	$IP_8(x_0)$	$u_8(x_0)$	$IP_7(x_0)$	$u_7(x_0)$	$IP_6(x_0)$	$u_6(x_0)$	$IP_5(x_0)$	$u_5(x_0)$	$IP_4(x_0)$	$u_4(x_0)$	$IP_3(x_0)$	$u_3(x_0)$	$IP_2(x_0)$	$u_2(x_0)$	$IP_1(x_0)$	$u_1(x_0)$
1.0	0.710	-0.5	0.829	-0.5	0.931	-0.5	1.00	-0.5	1.04	-0.5	1.04	-0.5	0.987	-0.5	0.862	-0.5	0.657	-0.3	0.360	0
0.9	0.567	-0.5	0.659	-0.5	0.739	-0.5	0.801	-0.5	0.837	-0.5	0.838	-0.5	0.796	-0.5	0.697	-0.5	0.532	-0.3	0.291	0
0.8	0.448	-0.5	0.517	-0.5	0.577	-0.5	0.624	-0.5	0.654	-0.5	0.657	-0.5	0.626	-0.5	0.551	-0.5	0.420	-0.3	0.230	0
0.7	0.348	-0.5	0.398	-0.5	0.441	-0.5	0.476	-0.5	0.498	-0.5	0.501	-0.5	0.479	-0.5	0.423	-0.4	0.322	-0.2	0.176	0
0.6	0.261	-0.4	0.298	-0.5	0.328	-0.5	0.352	-0.5	0.367	-0.5	0.369	-0.5	0.353	-0.5	0.312	-0.4	0.237	-0.2	0.129	0
0.5	0.188	-0.4	0.212	-0.4	0.233	-0.4	0.249	-0.4	0.259	-0.5	0.259	-0.4	0.247	-0.4	0.217	-0.3	0.165	-0.2	0.090	0
0.4	0.125	-0.3	0.141	-0.3	0.155	-0.3	0.164	-0.4	0.170	-0.4	0.169	-0.3	0.160	-0.3	0.140	-0.2	0.105	-0.1	0.057	0
0.3	0.076	-0.2	0.085	-0.2	0.092	-0.2	0.097	-0.3	0.099	-0.3	0.098	-0.2	0.091	-0.2	0.079	-0.2	0.059	-0.1	0.032	0
0.2	0.039	-0.1	0.042	-0.2	0.045	-0.2	0.047	-0.2	0.047	-0.2	0.046	-0.2	0.042	-0.1	0.036	-0.1	0.027	-0.1	0.014	0
0.1	0.013	-0.1	0.014	-0.1	0.014	-0.1	0.014	-0.1	0.014	-0.1	0.013	-0.1	0.012	0	0.009	0	0.006	0	0.003	0
0	0	0	0	0	0	0	0	0	0	0	0	0	0	0	0	0	0	0	0	0

Using Eq. (20) and the result of Eq. (24) the optimal return at $2\,\Delta t$ before $t = t_f = 2$ as expressed by Eq. (16) becomes

$$IP_2(x_0) = \min_{u_1}[((2 - 2\,\Delta t)x_0^2 + u_1^2)\,\Delta t + (2 - \Delta t)(x_0 + u_1\,\Delta t)^2\,\Delta t] \qquad (25)$$

The control that minimizes $IP_2(x_0)$ within the limitation of Eq. (22) is then

$$u_2(x_0) \qquad (26)$$

The solution for $u_2(x_0)$ is carried out numerically by direct search using the values $u_1 = -0.5, -0.4, -0.3, \cdots, 0.4, 0.5$ for $x_0 = 0, 0.1, \cdots, 1$. In a similar manner, the sets $u_3(x_0)$, $IP_3(x_0)$ through $u_{10}(x_0)$, $IP_{10}(x_0)$ are computed using Eq. (12). The results are shown in Table 7-4(1).

For the particular problem formulated the optimal return is given by

$$IP_{10}(1) = 0.710 \qquad (27)$$

Using Eqs. (17) and (18), Table 7-4(1) provides the optimal control sequence.

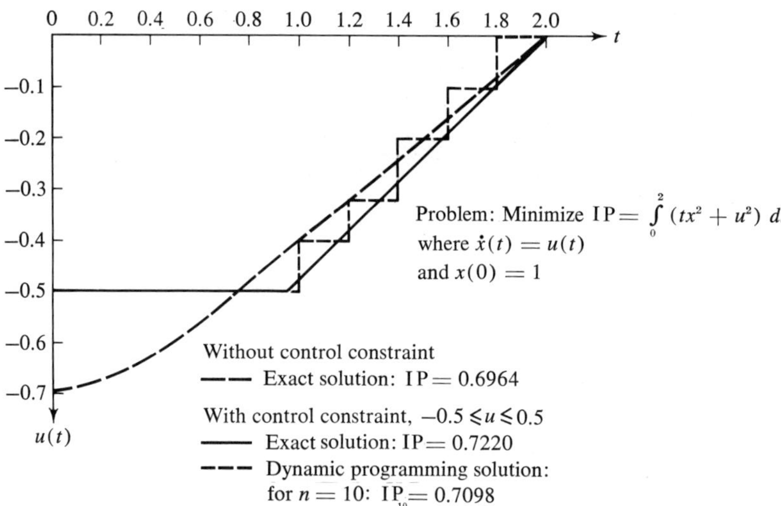

FIGURE 7-1   Control as a function of time for exact solution and dynamic programming solution

The result is presented in Fig. 7-1 and compared there with the calculus of variations solution to the problem both with and without the control constraint.

7-5 RELATION TO THE CALCULUS OF VARIATIONS

In this section the relation will be shown between the dynamic programming approach to the problem and that of the calculus of variations. By so doing, it is possible to provide a physical interpretation to the multipliers, $\boldsymbol{\lambda}(t)$, that have been used so extensively in our work. We shall consider the control problem as formulated in Section 7-2.

To review, given a system with n state variables and m control variables,

$$\mathbf{x}(t) \Rightarrow x_1(t), \cdots, x_n(t) \qquad \text{state variables}$$

$$\mathbf{u}(t) \Rightarrow u_1(t), \cdots, u_m(t) \qquad \text{control variables}$$

the problem is taken to be that of minimizing, by proper choice of the control variables, the performance index

$$\text{IP} = g(\mathbf{x}_f, t_f) + \int_{t_i}^{t_f} \bar{f}[\tau, \mathbf{x}(\tau), \mathbf{u}(\tau)] \, d\tau \tag{1}$$

subject to the differential constraints

$$\dot{x}_j = f_j[t, \mathbf{x}(t), \mathbf{u}(t)] \qquad j = 1, \cdots, n \tag{2}$$

and the end conditions

$$t_i \text{ and } x_j(t_i) \qquad j = 1, \cdots, n \qquad \text{specified} \tag{3a}$$

$$x_j(t_f) \qquad j = 1, \cdot \ \cdot, p \leqslant n \qquad \text{specified}$$

$$t_f \qquad\qquad\qquad\qquad \text{specified or unspecified} \tag{3b}$$

Solution of the problem, using the calculus of variations, requires introduction of the multipliers $\lambda_1(t), \cdot \ \cdot, \lambda_n(t)$. Forming the augmented integrand $F = \bar{f} + \sum_{j=1}^{n} \lambda_j(\dot{x}_j - f_j)$, assuming no control limitations, the Euler equations of 3-5(5) yield†

$$\dot{\lambda}_i = -\sum_{j=1}^{n} \lambda_j \frac{\partial f_j}{\partial x_i} + \frac{\partial \bar{f}}{\partial x_i} \qquad i = 1, \cdots, n \tag{4a}$$

$$\sum_{j=1}^{n} \lambda_j \frac{\partial f_j}{\partial u_k} - \frac{\partial \bar{f}}{\partial u_k} = 0 \qquad k = 1, \cdots, m \tag{4b}$$

while the transversality condition of Eq. 3-5(8) requires

$$\lambda_j(t_f) = -\frac{\partial g}{\partial x_j}(t_f) \qquad j = p + 1, \cdots, n \tag{5a}$$

† Equivalently, if the variable $x_{n+1}(t)$ is introduced satisfying the conditions $\dot{x}_{n+1} = \bar{f}$, $x_{n+1}(0) = 0$, then Eq. (1) becomes $\text{IP} = g(\mathbf{x}_f, t_f) + x_{n+1}(t_f)$, and the equations of 3-8 (5, 6, 11) would yield the results of Eqs. (4) and (5).

as well as, if t_f is not specified, the condition,

$$\left[\sum_{j=1}^{n} \lambda_j f_j - \bar{f}\right]\bigg|_{t_f} = \frac{\partial g}{\partial t_f} \qquad (t_f \text{ not specified}) \qquad (5b)$$

Equations (4) and (5) are presented here so that they may be compared with the conditions to be obtained for the problem using dynamic programming.

The solution of the problem by dynamic programming as presented in Section 7-2 requires that $IP(\mathbf{x}, t)$ satisfy Eq. 7-2(11) subject to the condition of Eq. 7-2(12). These are, respectively,

$$0 = \frac{\partial IP}{\partial t}(\mathbf{x}, t) + \bar{f}(t, \mathbf{x}, \mathbf{u}) + \sum_{j=1}^{n} \frac{\partial IP}{\partial x_j}(\mathbf{x}, t) f_j(t, \mathbf{x}, \mathbf{u}) \qquad (6a)$$

$$0 = \frac{\partial \bar{f}}{\partial u_k} + \sum_{j=1}^{n} \frac{\partial IP}{\partial x_j}(\mathbf{x}, t) \frac{\partial f_j}{\partial u_k}(t, \mathbf{x}, \mathbf{u}) \qquad k = 1, \cdots, m \qquad (6b)$$

and

$$IP(\mathbf{x}_f, t_f) = g(\mathbf{x}_f, t_f) \qquad (7)$$

One obtains from Eq. 7 that

$$dIP\bigg|_{t_f} = \frac{\partial g}{\partial t_f} dt_f + \sum_{j=1}^{n} \frac{\partial g}{\partial x_j(t_f)} dx_j(t_f) \qquad (8)$$

Note from Eq. (3b) that $dx_1(t_f)$ through $dx_p(t_f)$ all vanish, as well as dt_f when t_f is specified. Thus on expanding the left side of Eq. (8) and grouping the coefficients of the nonzero differentials from both sides together, one obtains

$$\sum_{j=p+1}^{n} \left[\frac{\partial IP}{\partial x_j}\bigg|_{t_f} - \frac{\partial g}{\partial x_j(t_f)}\right] dx_j(t_f) + \left[\frac{\partial IP}{\partial t}\bigg|_{t_f} - \frac{\partial g}{\partial t_f}\right] dt_f = 0 \qquad (9)$$

Then since the differentials remaining in Eq. (9) are arbitrary, the coefficient of each must vanish, giving

$$\frac{\partial IP}{\partial x_j}\bigg|_{t_f} = \frac{\partial g}{\partial x_j(t_f)} \qquad \begin{array}{l} j = p + 1, \cdots, n \\ (x_j(t_f) \text{ unspecified}) \end{array} \qquad (10a)$$

$$\frac{\partial IP}{\partial t}\bigg|_{t_f} = \frac{\partial g}{\partial t_f} \qquad (t_f \text{ unspecified}) \qquad (10b)$$

Note also, for later convenience, that on using Eq. (10b) the relation (6a) evaluated at t_f can be written

$$\left[\bar{f} + \sum_{j=1}^{n} \frac{\partial IP}{\partial x_j} f_j\right]\bigg|_{t_f} = -\frac{\partial g}{\partial t_f} \qquad (11)$$

To proceed to relate the two approaches, we form $\dfrac{d}{dt}\dfrac{\partial IP}{\partial x_i}(\mathbf{x}, t)$, obtaining

$$\frac{d}{dt}\frac{\partial IP}{\partial x_i} = \frac{\partial^2 IP}{\partial t\,\partial x_i} + \sum_{j=1}^{n}\frac{\partial^2 IP}{\partial x_j\,\partial x_i}f_j \tag{12}$$

where Eq. (2) has been used. Then taking $\partial/\partial x_i$ of Eq. (6a), on using Eq. (6b) one finds

$$\frac{\partial^2 IP}{\partial x_i\,\partial t} = -\frac{\partial \bar{f}}{\partial x_i} - \sum_{j=1}^{n}\frac{\partial^2 IP}{\partial x_i\,\partial x_j}f_j - \sum_{j=1}^{n}\frac{\partial IP}{\partial x_j}\frac{\partial f_j}{\partial x_i} \tag{13}$$

Assuming $IP(\mathbf{x}, t)$ has continuous second-order partial derivatives, it follows that

$$\frac{\partial^2 IP}{\partial t\,\partial x_i} = \frac{\partial^2 IP}{\partial x_i\,\partial t}$$

$$\frac{\partial^2 IP}{\partial x_i\,\partial x_j} = \frac{\partial^2 IP}{\partial x_j\,\partial x_i} \tag{14}$$

Using this result and Eq. (13) to eliminate $\partial^2 IP/\partial t\,\partial x_i$ from Eq. (12) there results

$$\frac{d}{dt}\frac{\partial IP}{\partial x_i} = -\sum_{j=1}^{n}\frac{\partial IP}{\partial x_j}\frac{\partial f_j}{\partial x_i} - \frac{\partial \bar{f}}{\partial x_i} \tag{15}$$

With Eq. (15) a direct comparison can be made between the results of Eqs. (4) and (5) from the calculus of variations and those obtained by dynamic programming. Defining

$$\lambda_i(t) = -\frac{\partial IP}{\partial x_i}(\mathbf{x}, t) \tag{16}$$

the relations of Eqs. (15), (6b), (10a), and (11) give, respectively, the adjoint equations, the control equation, and the transversality conditions of Eqs. (4) and (5) from the calculus of variations approach to the problem.

Thus the adjoint variables of the calculus of variations formulation are seen to equal the negative of the rate of change of the performance index with respect to the corresponding state variables.

One further comparison. The condition in Eq. 7-2(10) that the control minimize the bracketed term is equivalent to the requirement that $\mathbf{u}(t)$ maximize the negative of the term. In this form, using Eq. (16) and defining

$$H(t, \mathbf{x}^*, \mathbf{u}, \boldsymbol{\lambda}^*) = \sum_{j=1}^{n}\lambda_j^*(t)f_j[t, \mathbf{x}^*(t), \mathbf{u}(t)] - \bar{f}[t, \mathbf{x}^*(t), \mathbf{u}(t)] \tag{17}$$

the condition requires that the optimal control $\mathbf{u}^*$ maximize H.

$$H^* = H(t, \mathbf{x}^*, \mathbf{u}^*, \boldsymbol{\lambda}^*) \geqslant H(t, \mathbf{x}^*, \mathbf{u}, \boldsymbol{\lambda}^*) \tag{18}$$

The result of Eq. (18) is just that given by the maximum principle of Eq. 3-8(16) as found from the Weierstrass condition of Eq. 3-6(11).†

7-6   THE HAMILTON–JACOBI EQUATION

Let us now review, as well as generalize somewhat, the theoretical approach. In formulating the application of dynamic programming to the control problem in Section 7-2 the fundamental equation of Eq. 7-2(10) is to be solved subject to the boundary condition of Eq. 7-2(12). These are, respectively,

$$\frac{\partial \mathrm{IP}}{\partial t}(\mathbf{x}, t) + \min_{\mathbf{u}} \left[\bar{f}(t, \mathbf{x}, \mathbf{u}) + \sum_{j=1}^{n} \frac{\partial \mathrm{IP}}{\partial x_j}(\mathbf{x}, t) f_j(t, \mathbf{x}, \mathbf{u}) \right] = 0 \tag{1a}$$

$$\mathrm{IP}(\mathbf{x}, t) \bigg|_{t_f} = g(\mathbf{x}_f, t_f) \tag{1b}$$

where the minimization over $\mathbf{u}$ in Eq. (1a) must in general be carried out for $\mathbf{u}$ within some admissible region $\mathscr{R}$ of control space.

Let us define‡

$$\bar{H}\left(t, \mathbf{x}, \mathbf{u}, \frac{\partial \mathrm{IP}}{\partial \mathbf{x}}\right) = \bar{f}(t, \mathbf{x}, \mathbf{u}) + \sum_{j=1}^{n} \frac{\partial \mathrm{IP}}{\partial x_j}(\mathbf{x}, t) f_j(t, \mathbf{x}, \mathbf{u}) \tag{2}$$

Then Eq. (1a) becomes

$$\frac{\partial \mathrm{IP}}{\partial t} + \min_{\mathbf{u}} \bar{H}\left(t, \mathbf{x}, \mathbf{u}, \frac{\partial \mathrm{IP}}{\partial \mathbf{x}}\right) = 0 \tag{3}$$

Denote the $\mathbf{u}$ within $\mathscr{R}$ which instantaneously minimizes the Hamiltonian $\bar{H}$ of Eq. (2) by $\mathbf{u}^*$. From the definition of $\bar{H}$ it is seen that the minimization will yield $\mathbf{u}^*$ in the form

$$\mathbf{u}^* = \mathbf{u}^*\left(t, \mathbf{x}, \frac{\partial \mathrm{IP}}{\partial \mathbf{x}}\right) \tag{4}$$

---

† The apparent difference between the definitions of Eqs. (17) and 3-8(15) is not real. It results only because in the latter case the integral term in the performance index has been transformed into an end variable through introduction of an additional differential constraint.

‡ The quantity $\bar{H}$ is related to the negative of H as defined in Eq. 7-5(17).

Now define

$$\bar{H}^*\left(t, \mathbf{x}, \frac{\partial \mathrm{IP}}{\partial \mathbf{x}}\right) = \bar{H}\left(t, \mathbf{x}, \mathbf{u}, \frac{\partial \mathrm{IP}}{\partial \mathbf{x}}\right)\bigg|_{\mathbf{u}\,=\,\mathbf{u}^*(t,\mathbf{x},\partial \mathrm{IP}/\partial \mathbf{x})} \tag{5}$$

Thus $\bar{H}^*(t, \mathbf{x}, \partial \mathrm{IP}/\partial \mathbf{x})$ is the minimum value of the Hamiltonian, with respect to $\mathbf{u}$ within $\mathscr{R}$. In terms of this minimum value Eq. (3) becomes

$$\frac{\partial \mathrm{IP}}{\partial t}(\mathbf{x}, t) + \bar{H}^*\left(t, \mathbf{x}, \frac{\partial \mathrm{IP}}{\partial \mathbf{x}}\right) = 0 \tag{6}$$

Equation (6) is termed the Hamilton–Jacobi equation for the control problem. When Eq. (6) is solved with Eq. (1b) as a boundary condition one obtains $\mathrm{IP}(\mathbf{x}, t)$. Knowledge of $\mathrm{IP}(\mathbf{x}, t)$ implies that the IP is known for a process with arbitrary starting time and initial state.

The solution for $\mathrm{IP}(\mathbf{x}, t)$ may then be used to calculate $\partial \mathrm{IP}(\mathbf{x}, t)/\partial \mathbf{x}$ for substitution into Eq. (4). The result is $\mathbf{u}^*$ determined explicitly as a function of t, the current time, and $\mathbf{x}$, the current state.

$$\mathbf{u}^* = \mathbf{u}^*(\mathbf{x}, t) \tag{7}$$

This is the control-law or closed-loop solution to the optimization problem. That the closed-loop control is known once $\mathrm{IP}(\mathbf{x}, t)$ is obtained is an essential feature of the dynamic programming approach.

7-7 CONCLUSIONS

We have attempted in this chapter by means of exposition and example to illustrate the main features of dynamic programming. For a more complete treatment of the subject, the reader is referred to works by Bellman and Dreyfus (2–5) among others.

To review, the scheme has the computational advantage over that of enumerating all possible solutions, of eliminating those solutions which cannot be minimizing ones. However, even with this reduction in the number of cases which must be considered, difficulty is encountered.

The example treated in Section 7-4 had only a single stage variable. If the performance index is a function of n state variables $\mathrm{IP} = \mathrm{IP}[x_1(t), \cdots, x_n(t), t]$, the solution of the equation comparable to Eq. 7-4(12) requires that the IP be known for each time t on an n-dimensional grid of points defined by $0 \leqslant x_1 \leqslant N_1, 0 \leqslant x_2 \leqslant N_2, \cdots, 0 \leqslant x_n \leqslant N_n$. Not only does this increase the computational time required but, even more important, the amount of computer storage required increases rapidly. This increasing difficulty has been termed by Bellman the "curse of dimensionality." For $n \geqslant 3$, solution of the problem usually requires either use of a very coarse grid or polynomial approximation of one sort or another to replace the tabulated values in storage. This latter method is discussed in the references cited.

In contrast to the calculus of variations approach, inclusion of control inequality constraints actually simplifies the dynamic programming computational solution by limiting the range of control values which need be considered. Further, either analytically through solution of the Hamilton–Jacobi equation, or computationally through generation of the $u_k(c)$ as solutions of a recursion relation of the form of Eq. 7-4(12), a method of generating the optimal closed-loop control has been given.

In practice, however, analytic solution of the Hamilton-Jacobi equation in all but the most trivial cases is usually impossible. Thus, in general, the optimal control law $\mathbf{u}^*(\mathbf{x}, t)$ is found through the dynamic programming approach by computational solution of a recursion relation of the form of Eq. 7-4(12). Since the solution is carried out computationally, nonlinearities in the system equations do not introduce difficulties above those present if the equations are linear. The major drawback of this approach is the limitation already discussed on the dimension of the state vector of the system.

REFERENCES

1. Y. Fang and T. J. Higgins, "Optimal Design of Automatic Control Systems by Dynamic Programming," *Proceedings of the Dynamic Programming Workshop*, J. E. Gibson, ed., Purdue University, Lafayette, Ind., 1961, pp. 67–86.
2. Richard Bellman, *Dynamic Programming*, Princeton Univ. Press, Princeton, N.J., 1957.
3. Richard Bellman, *Adaptive Control Processes: A Guided Tour*, Princeton Univ. Press, Princeton, N.J., 1961.
4. Richard Bellman and Stuart Dreyfus, *Applied Dynamic Programming*, Princeton Univ. Press, Princeton, N.J., 1962.
5. Stuart Dreyfus, *Dynamic Programming and the Calculus of Variations*, Academic Press, New York, 1965.

8

OPTIMAL FEEDBACK CONTROL

8-1 INTRODUCTION

In Chapter 7 we have seen that the optimal feedback control or control vector law $\mathbf{u}^*(\mathbf{x}, t)$ can be found through dynamic programming in one of two ways; either by analytic solution of a Hamilton–Jacobi type equation or by computational solution of the relation from which the Hamilton–Jacobi equation was derived. The limitations of these two approaches are discussed in Chapter 7. Before going on to determine alternative schemes let us review why knowledge of the control law vector $\mathbf{u}^*(\mathbf{x}, t)$ is desired above that of the control function vector $\mathbf{u}^*(t)$.

The control function vector $\mathbf{u}^*(t)$ is determined as the optimal control to take the system between specified initial and terminal states (manifolds). Now there are problems of engineering interest, for example, regulator problems, for which the possible range of initial conditions is very large. To provide optimal system response over the set of initial conditions that might be encountered using the control function solution would require determining $\mathbf{u}^*(t)$ for each possible initial state. This obviously can be impractical. By contrast, if the control law $\mathbf{u}^*(\mathbf{x}, t)$ is known, the optimal control is determined at any time by knowledge of the current state $\mathbf{x}$.

At the other extreme consider a problem in which only one initial state is

235

to be expected. Here the control function solution $\mathbf{u}^*(t)$ would appear to have more meaning. However, even in this case feedback information is required to overcome the effects of errors and disturbances which change the system response from the nominal trajectory $\mathbf{x}^*(t)$. If $\mathbf{u}^*(t)$ is followed without correction in the presence of these disturbances, the resulting trajectory cannot be expected to be either optimal or to satisfy the terminal conditions. Again, if the control law $\mathbf{u}^*(\mathbf{x}, t)$ is known, this difficulty would not occur. Knowledge of the current state $\mathbf{x}$ and the time t would suffice to determine the optimal control without reference to whether the state $\mathbf{x}(t)$ resulted from a disturbance from the open-loop optimal trajectory $\mathbf{x}^*(t)$.

Both of these types of problems will be discussed in what follows. To introduce the subject it will be assumed that the terminal time t_f is specified and that there are no inequality constraints. For additional material the reader is referred to the Bibliography at the end of the chapter.

8-2 LINEAR SYSTEMS WITH A QUADRATIC PERFORMANCE INDEX—FORMULATION

As just discussed, many engineering problems require that the optimal solution be known over a wide range of initial conditions. For these problems a meaningful solution requires determination of the control law $\mathbf{u}^*(\mathbf{x}, t)$ over the range of states $\mathbf{x}$ likely to be encountered. To obtain such a solution we shall restrict our initial formulation to linear problems with a quadratic performance index. It will later be seen that the development used to obtain the solution for such problems provides a means of treating nonlinear systems with a general performance index.

Thus to start, the system behavior is assumed to be governed to a satisfactory degree of approximation by the set of linear system equations

$$\dot{\mathbf{x}} = \mathbf{A}(t)\mathbf{x} + \mathbf{B}(t)\mathbf{u} \tag{1}$$

where $\quad \mathbf{x} = n$ vector of state variables

$\mathbf{u} = m$ vector of control variables

$\mathbf{A}(t) = n \times n$ matrix of known functions of time

$\mathbf{B}(t) = n \times m$ matrix of known functions of time

The system is taken to be initially at a known state so that

$$t_i \text{ and } \mathbf{x}(t_i) \quad \text{specified} \tag{2}$$

From the initial state the system is to be transferred to the terminal state given by

$$t_f \quad \text{specified} \tag{3a}$$

$$\overline{\boldsymbol{\psi}}\mathbf{x}(t_f) = \boldsymbol{\psi}_0 \tag{3b}$$

where $\overline{\psi}$ $= p \times n$ matrix of known constants
$\psi_0 = p \times 1$ matrix of known constants

Finally, in going from the initial to the terminal state the optimal control vector $\mathbf{u}(t)$ will be determined so as to minimize the performance index

$$\text{IP} = \tfrac{1}{2}\mathbf{x}'(t_f)\overline{\mathbf{g}}\mathbf{x}(t_f) + \frac{1}{2}\int_{t_i}^{t_f} [\mathbf{x}'\mathbf{W}_1(t)\mathbf{x} + 2\mathbf{x}'\mathbf{W}_2(t)\mathbf{u} + \mathbf{u}'\mathbf{W}_3(t)\mathbf{u}]\,dt \qquad (4)$$

where
$\overline{\mathbf{g}} = n \times n$ symmetric matrix of known constants
$\mathbf{W}_1(t) = n \times n$ symmetric matrix of known functions of time
$\mathbf{W}_2(t) = n \times m$ matrix of known functions of time
$\mathbf{W}_3(t) = m \times m$ symmetric matrix of known functions of time;
$\mathbf{W}_3(t)$ is taken to be positive definite

Having formulated the problem as one with linear system equations, linear terminal constraints, and a quadratic performance index it will be shown that the control law $\mathbf{u}^*(\mathbf{x}, t)$ can be found.

We proceed by forming the necessary conditions to be satisfied by the minimizing solution. Introducing the multipliers $\lambda(t)$ to the system equations of (1), where λ is an $n \times 1$ matrix, the characteristic equations of Eq. 3-5(5) are found to be

$$\dot{\lambda} = -\mathbf{A}'\lambda + \mathbf{W}_1\mathbf{x} + \mathbf{W}_2\mathbf{u} \qquad (5a)$$

$$0 = \mathbf{B}'\lambda - \mathbf{W}_2'\mathbf{x} - \mathbf{W}_3\mathbf{u} \qquad (5b)$$

In addition, on introducing the constant multipliers $\mathbf{v}$ to the terminal constraints of Eq. (3b), where $\mathbf{v}$ is a $p \times 1$ matrix, the transversality condition of Eq. 3-5(12) is seen to require†

$$\lambda(t_f) = -[\overline{\mathbf{g}}\mathbf{x}(t_f) + \overline{\psi}'\mathbf{v}] \qquad (6)$$

Summarizing, determination of the optimal control requires that Eqs. (1) and (5) be solved subject to the initial conditions of Eq. (2) and the terminal conditions of Eqs. (3) and (6). To start, Eq. (5b) is used to solve for $\mathbf{u}$ in terms of $\mathbf{x}$ and λ, yielding

$$\mathbf{u} = \mathbf{W}_3^{-1}(\mathbf{B}'\lambda - \mathbf{W}_2'\mathbf{x}) \qquad (7)$$

That this can be done follows since $\mathbf{W}_3(t)$ has been assumed to be positive

---

† In finding Eqs. (5) and (6) from Eqs. 3-5(5, 12), the augmented integrand used was
$$F = \tfrac{1}{2}(\mathbf{x}'\mathbf{W}_1\mathbf{x} + 2\mathbf{x}'\mathbf{W}_2\mathbf{u} + \mathbf{u}'\mathbf{W}_3\mathbf{u}) + \lambda'(\dot{\mathbf{x}} - \mathbf{A}\mathbf{x} - \mathbf{B}\mathbf{u})$$
Equivalently if the variable $x_{n+1}(t)$ is introduced, satisfying the conditions
$$\dot{x}_{n+1} = \tfrac{1}{2}[\mathbf{x}'\mathbf{W}_1\mathbf{x} + 2\mathbf{x}'\mathbf{W}_2\mathbf{u} + \mathbf{u}'\mathbf{W}_3\mathbf{u}] \qquad x_{n+1}(0) = 0$$
then Eq. (4) becomes $\text{IP} = \tfrac{1}{2}\mathbf{x}'(t_f)\overline{\mathbf{g}}\mathbf{x}(t_f) + x_{n+1}(t_f)$ and the equations of 3-8(5, 6, 10) would yield the results of Eqs. (5) and (6).

definite; hence its inverse always exists. Equation (7) is next used to eliminate $\mathbf{u}$ from Eqs. (1) and (5a). The resulting equations are

$$\dot{\mathbf{x}} = (\mathbf{A} - \mathbf{BW}_3^{-1}\mathbf{W}_2')\mathbf{x} + \mathbf{BW}_3^{-1}\mathbf{B}'\boldsymbol{\lambda} \tag{8a}$$

$$\dot{\boldsymbol{\lambda}} = (\mathbf{W}_1 - \mathbf{W}_2\mathbf{W}_3^{-1}\mathbf{W}_2')\mathbf{x} - (\mathbf{A}' - \mathbf{W}_2\mathbf{W}_3^{-1}\mathbf{B}')\boldsymbol{\lambda} \tag{8b}$$

With t_i and t_f specified, integration of the $2n$ equations of (8) provides $2n$ constants of integration. These, along with the p constant multipliers $\boldsymbol{\nu}$ provide $2n + p$ constants to be determined by the $2n + p$ conditions consisting of the n conditions of Eq. (2), the p conditions of Eq. (3b) and the n conditions of Eq. (6).

8-3 LINEAR SYSTEMS WITH A QUADRATIC PERFORMANCE INDEX—SOLUTION

Recall now that it is our desire to find the control law vector $\mathbf{u}^*(\mathbf{x}, t)$ and not the control function vector $\mathbf{u}^*(t)$. From Eq. 8-2(7) it is seen that to find the control as a function of the state will require expressing the multiplier $\boldsymbol{\lambda}(t)$ in terms of the state $\mathbf{x}(t)$. To do this, the terminal relation of Eq. 8-2(6) can be used for guidance in determining the form the relation must take. Referring to Eq. 8-2(6) we assume a relation of the form

$$\boldsymbol{\lambda}(t) = -[\mathbf{P}(t)\mathbf{x}(t) + \mathbf{R}(t)\boldsymbol{\nu}] \tag{1}$$

where $\mathbf{P}(t) = n \times n$ matrix of functions of time
$\mathbf{R}(t) = n \times p$ matrix of functions of time

Then the terminal condition of Eq. 8-2(6) is immediately satisfied by requiring

$$\mathbf{P}(t_f) = \bar{\mathbf{g}} \tag{2a}$$

$$\mathbf{R}(t_f) = \bar{\boldsymbol{\psi}}' \tag{2b}$$

To find the equations governing the matrices $\mathbf{P}(t)$ and $\mathbf{R}(t)$, the relation of Eq. (1) is differentiated and used to replace $\dot{\boldsymbol{\lambda}}(t)$ in Eq. 8-2(8b). Then using Eq. 8-2(8a) to eliminate the $\dot{\mathbf{x}}$ term that results on differentiating Eq. (1), one obtains

$$0 = [\dot{\mathbf{P}} + \mathbf{P}(\mathbf{A} - \mathbf{BW}_3^{-1}\mathbf{W}_2') + (\mathbf{W}_1 - \mathbf{W}_2\mathbf{W}_3^{-1}\mathbf{W}_2')]\mathbf{x}$$
$$+ [\mathbf{P}(\mathbf{BW}_3^{-1}\mathbf{B}') + (\mathbf{A}' - \mathbf{W}_2\mathbf{W}_3^{-1}\mathbf{B}')]\boldsymbol{\lambda} + \dot{\mathbf{R}}\boldsymbol{\nu} \tag{3}$$

Finally, Eq. (1) is used to replace $\boldsymbol{\lambda}$ in terms of $\mathbf{x}$ and $\boldsymbol{\nu}$, yielding

$$0 = [\dot{\mathbf{P}} + \mathbf{P}(\mathbf{A} - \mathbf{BW}_3^{-1}\mathbf{W}_2') + (\mathbf{W}_1 - \mathbf{W}_2\mathbf{W}_3^{-1}\mathbf{W}_2')$$
$$- \mathbf{P}(\mathbf{BW}_3^{-1}\mathbf{B}')\mathbf{P} + (\mathbf{A}' - \mathbf{W}_2\mathbf{W}_3^{-1}\mathbf{B}')\mathbf{P}]\mathbf{x}$$
$$+ [\dot{\mathbf{R}} - \mathbf{P}(\mathbf{BW}_3^{-1}\mathbf{B}')\mathbf{R} + (\mathbf{A}' - \mathbf{W}_2\mathbf{W}_3^{-1}\mathbf{B}')\mathbf{R}]\boldsymbol{\nu} \tag{4}$$

For Eq. (4) to be satisfied it is sufficient that the coefficients of $\mathbf{x}$ and $\mathbf{v}$ vanish identically. Setting each to zero provides the governing equations for $\mathbf{P}(t)$ and $\mathbf{R}(t)$. These are

$$\dot{\mathbf{P}} = -\mathbf{P}\mathbf{A} - \mathbf{A}'\mathbf{P} - \mathbf{W}_1 + (\mathbf{P}\mathbf{B} + \mathbf{W}_2)\mathbf{W}_3^{-1}(\mathbf{W}_2' + \mathbf{B}'\mathbf{P}) \tag{5a}$$

$$\dot{\mathbf{R}} = (\mathbf{P}\mathbf{B}\mathbf{W}_3^{-1}\mathbf{B}' - \mathbf{A}' + \mathbf{W}_2\mathbf{W}_3^{-1}\mathbf{B}')\mathbf{R} \tag{5b}$$

Equation (5a) is a matrix Riccati equation for the determination of the components of the $n \times n$ matrix $\mathbf{P}(t)$. Once $\mathbf{P}(t)$ is known, the matrix $\mathbf{R}(t)$ is found by solution of the linear equation of (5b).

Note that if the transpose of Eq. (5a) is taken, the equation governing $\mathbf{P}'(t)$ is the same as that for $\mathbf{P}(t)$. Since $\mathbf{P}(t_f)$ is symmetric by Eq. (2), it follows that $\mathbf{P}(t)$ is symmetric. Hence only $[n(n + 1)]/2$ of the n^2 components of $\mathbf{P}(t)$ need to be determined.

The constant multipliers $\mathbf{v}$ remain to be determined so as to have the solution for $\mathbf{x}(t_f)$ satisfy the terminal constraints of Eq. 8-2(3b). For use in Eq. 8-2(3b) the state vector $\mathbf{x}(t_f)$ can be found from Eq. 8-2(8a). First Eq. (1) is used to eliminate $\boldsymbol{\lambda}(t)$ from Eq. 8-2(8a). In this manner the resulting equation for $\mathbf{x}(t)$ is seen to be of the form

$$\dot{\mathbf{x}} = \mathbf{C}_1(t)\mathbf{x} + \mathbf{C}_2(t)\mathbf{v} \tag{6}$$

Solving Eq. (6), one finds that the state $\mathbf{x}(t_f)$ is related to the state $\mathbf{x}(t)$ in the following way† :

$$\mathbf{x}(t_f) = \mathbf{D}_1(t)\mathbf{x}(t) + \mathbf{D}_2(t)\mathbf{v} \tag{7}$$

† To show this let us solve the inhomogeneous matrix equation of (14),

$$\frac{d\mathbf{x}}{dt} = \mathbf{C}_1(t)\mathbf{x} + \mathbf{C}_2(t)\mathbf{v} \tag{a}$$

Let $\mathbf{X}(t)$, an $n \times n$ matrix termed the fundamental matrix, be the solution of the corresponding homogeneous matrix equation

$$\frac{d\mathbf{X}}{dt} = \mathbf{C}_1(t)\mathbf{X} \qquad \mathbf{X}(t_f) = \mathbf{I} \tag{b}$$

where $\mathbf{I}$ is the identity matrix. To solve the original inhomogeneous equation of (a) we use the method of variation of parameters. Let

$$\mathbf{x}(t) = \mathbf{X}(t)\mathbf{v}(t) \tag{c}$$

where $\mathbf{v}(t)$ is an $n \times 1$ matrix of functions to be determined. Then

$$\frac{d\mathbf{x}}{dt} = \mathbf{X}\frac{d\mathbf{v}}{dt} + \frac{d\mathbf{X}}{dt}\mathbf{v} = \mathbf{C}_1(t)\mathbf{x} + \mathbf{C}_2(t)\mathbf{v}$$

from which one has

$$\mathbf{X}\frac{d\mathbf{v}}{dt} = \mathbf{C}_2(t)\mathbf{v}$$

Multiplying through by $\mathbf{X}^{-1}$ and integrating between t and t_f it follows that

$$\mathbf{v}(t_f) = \mathbf{v}(t) + \left[\int_t^{t_f} \mathbf{X}^{-1}(\tau)\mathbf{C}_2(\tau)\,d\tau\right]\mathbf{v} \tag{d}$$

Note that by Eqs. (b) and (c),

$$\mathbf{v}(t_f) = \mathbf{x}(t_f) \qquad \mathbf{v}(t) = \mathbf{X}^{-1}(t)\mathbf{x}(t) \tag{e}$$

Using Eq. (e) in Eq. (d) one obtains

$$\mathbf{x}(t_f) = \mathbf{X}^{-1}(t)\mathbf{x}(t) + \left[\int_t^{t_f} \mathbf{X}^{-1}(\tau)\mathbf{C}_2(\tau)\,d\tau\right]\mathbf{v}$$

which is of the form specified by Eq. (7).

Then multiplying Eq. (7) through by $\bar{\psi}$, the terminal condition of Eq. 8-2(3b) is seen to require

$$\psi_0 = \bar{\psi}\mathbf{D}_1(t)\mathbf{x}(t) + \bar{\psi}\mathbf{D}_2(t)\mathbf{v} \qquad (8)$$

Letting $\mathbf{S}(t) = \bar{\psi}\mathbf{D}_1(t)$ and $\mathbf{Q}(t) = \bar{\psi}\mathbf{D}_2(t)$, where $\mathbf{S}$ is a $p \times n$ matrix and $\mathbf{Q}$ a $p \times p$ matrix, the relation of Eq. (8) can be written more compactly as

$$\psi_0 = \mathbf{S}(t)\mathbf{x}(t) + \mathbf{Q}(t)\mathbf{v} \qquad (9)$$

Thus if we choose that at t_f the conditions

$$\mathbf{S}(t_f) = \bar{\psi} \qquad (10a)$$

$$\mathbf{Q}(t_f) = 0 \qquad (10b)$$

be satisfied, the terminal condition of Eq. 8-2(3b) will be met.

If $\mathbf{S}(t)$ and $\mathbf{Q}(t)$ can be determined for $t \leqslant t_f$ so as to satisfy the conditions of Eq. (10), Eq. (9) can be used to solve for the multipliers $\mathbf{v}$ in terms of the current state $\mathbf{x}(t)$ at any time. Let us assume for the moment that $\mathbf{Q}(t)$ and $\mathbf{S}(t)$ are known. Then solving Eq. (9) for $\mathbf{v}$ one finds†

$$\mathbf{v} = \mathbf{Q}^{-1}(t)\psi_0 - \mathbf{Q}^{-1}(t)\mathbf{S}(t)\mathbf{x}(t) \qquad (11)$$

With the relation of Eq. (11) the multipliers $\mathbf{v}$ can be eliminated from the expression of Eq. (1), giving $\lambda(t)$ directly in terms of $\mathbf{x}(t)$. One obtains

$$\lambda(t) = -\mathbf{R}(t)\mathbf{Q}^{-1}(t)\psi_0 - [\mathbf{P}(t) - \mathbf{R}(t)\mathbf{Q}^{-1}(t)\mathbf{S}(t)]\mathbf{x}(t) \qquad (12)$$

Finally the desired result, the control law vector, is found by eliminating $\lambda(t)$ from Eq. 8-2(7), yielding

$$\begin{aligned} \mathbf{u}^*(\mathbf{x}, t) = {} & -\mathbf{W}_3^{-1}(t)\mathbf{B}'(t)\mathbf{R}(t)\mathbf{Q}^{-1}(t)\psi_0 \\ & - \mathbf{W}_3^{-1}(t)\{\mathbf{W}_2'(t) + \mathbf{B}'(t)[\mathbf{P}(t) - \mathbf{R}(t)\mathbf{Q}^{-1}(t)\mathbf{S}(t)]\}\mathbf{x}(t) \end{aligned} \qquad (13)$$

Once $\mathbf{P}(t)$, $\mathbf{Q}(t)$, $\mathbf{R}(t)$, and $\mathbf{S}(t)$ are known, Eq. (13) provides the control law $\mathbf{u}^*(\mathbf{x}, t)$ to be applied given knowledge of the current state $\mathbf{x}(t)$.

The method of determining $\mathbf{P}(t)$ and $\mathbf{R}(t)$ has already been detailed: namely solution of Eq. (5) subject to the conditions of Eq. (2). The equations governing $\mathbf{S}(t)$ and $\mathbf{Q}(t)$ remain to be found. To determine these equations the relation of Eq. (9) in which $\mathbf{S}(t)$ and $\mathbf{Q}(t)$ are introduced is differentiated, yielding

$$0 = \dot{\mathbf{S}}(t)\mathbf{x}(t) + \mathbf{S}(t)\dot{\mathbf{x}}(t) + \dot{\mathbf{Q}}(t)\mathbf{v} \qquad (14)$$

Equation 8-2(8a), giving $\dot{\mathbf{x}}(t)$ in terms of $\mathbf{x}(t)$ and $\lambda(t)$, is first used to eliminate $\dot{\mathbf{x}}(t)$ from Eq. (14). Then $\lambda(t)$ is replaced by its value in terms of $\mathbf{x}(t)$

---

† Note that for Eq. (11) to be obtained requires that $\mathbf{Q}^{-1}(t)$ exist. More will be said later of the implications of this requirement.

and $\mathbf{v}$ through use of Eq. (1). After going through these two steps Eq. (14) becomes

$$0 = [\dot{\mathbf{S}} + \mathbf{S}(\mathbf{A} - \mathbf{BW}_3^{-1}\mathbf{W}_2') - \mathbf{SBW}_3^{-1}\mathbf{B}'\mathbf{P}]\mathbf{x} + [\dot{\mathbf{Q}} - \mathbf{SBW}_3^{-1}\mathbf{B}'\mathbf{R}]\mathbf{v} \qquad (15)$$

For Eq. (15) to be satisfied it is sufficient for the coefficients of $\mathbf{x}(t)$ and $\mathbf{v}$ to vanish identically, yielding the two equations

$$\dot{\mathbf{S}} = \mathbf{S}(\mathbf{BW}_3^{-1}\mathbf{B}'\mathbf{P} - \mathbf{A} + \mathbf{BW}_3^{-1}\mathbf{W}_2') \qquad (16a)$$

$$\dot{\mathbf{Q}} = \mathbf{SBW}_3^{-1}\mathbf{B}'\mathbf{R} \qquad (16b)$$

for determination of $\mathbf{S}(t)$ and $\mathbf{Q}(t)$.

A relation can now be shown between the matrices $\mathbf{S}(t)$ and $\mathbf{R}(t)$. The matrix $\mathbf{R}(t)$ is found as the solution of Eq. (5b) subject to the terminal condition of Eq. (2b), while $\mathbf{S}(t)$ must satisfy Eq. (16a) and the terminal condition of Eq. (10a). Comparing the equations and terminal conditions one finds the simple relation

$$\mathbf{S}(t) = \mathbf{R}'(t) \qquad (17)$$

Hence $\mathbf{S}(t)$ is known once $\mathbf{R}(t)$ is determined. In what follows $\mathbf{S}(t)$ will be replaced by $\mathbf{R}'(t)$ using Eq. (17).

To summarize, the optimal control law is given by

$$\mathbf{u}^*(\mathbf{x}, t) = -\mathbf{W}_3^{-1}(t)\mathbf{B}'(t)\mathbf{R}(t)\mathbf{Q}^{-1}(t)\boldsymbol{\psi}_0$$
$$- \mathbf{W}_3^{-1}(t)\{\mathbf{W}_2'(t) + \mathbf{B}'(t)[\mathbf{P}(t) - \mathbf{R}(t)\mathbf{Q}^{-1}(t)\mathbf{R}'(t)]\}\mathbf{x}(t) \qquad (18)$$

where the matrices $\mathbf{P}(t)$, $\mathbf{R}(t)$, and $\mathbf{Q}(t)$ are determined so as to satisfy the equations [see Eqs. (5) and (16)]

$$\dot{\mathbf{P}} = -\mathbf{PA} - \mathbf{A}'\mathbf{P} - \mathbf{W}_1 + (\mathbf{PB} + \mathbf{W}_2)\mathbf{W}_3^{-1}(\mathbf{W}_2' + \mathbf{B}'\mathbf{P}) \qquad (19a)$$

$$\dot{\mathbf{R}} = (\mathbf{PBW}_3^{-1}\mathbf{B}' - \mathbf{A}' + \mathbf{W}_2\mathbf{W}_3^{-1}\mathbf{B}')\mathbf{R} \qquad (19b)$$

$$\dot{\mathbf{Q}} = \mathbf{R}'\mathbf{BW}_3^{-1}\mathbf{B}'\mathbf{R} \qquad (19c)$$

subject to the terminal conditions [see Eqs. (2) and (10)]

$$\mathbf{P}(t_f) = \bar{\mathbf{g}} \qquad (20a)$$

$$\mathbf{R}(t_f) = \bar{\boldsymbol{\psi}}' \qquad (20b)$$

$$\mathbf{Q}(t_f) = 0 \qquad (20c)$$

From Eq. (19) it is seen that $\mathbf{P}(t)$ is determined as the solution of a Riccati equation. Once $\mathbf{P}(t)$ is known, $\mathbf{R}(t)$ follows from the solution of a linear equation, after which $\mathbf{Q}(t)$ is found by direct integration. The matrix $\mathbf{P}(t)$ has already been shown to be symmetric. Taking the transpose of Eq. (19c), it is seen that $\dot{\mathbf{Q}}' = \dot{\mathbf{Q}}$. Thus under the terminal condition of Eq. (20c) one finds that $\mathbf{Q}(t)$ is also symmetric.

8-4 THE INERTIA PLANT

To illustrate the theory just developed, let us treat the problem of the so-called inertia plant, that is, a plant governed by the system equations

$$\dot{x}_1 = x_2$$
$$\dot{x}_2 = u$$

(1)

Our task will be to take the system from any initial state to the origin in the fixed time t_f while minimizing the control energy expended. Thus

$$\text{IP} = \frac{1}{2} \int_0^{t_f} u^2 \, dt$$

(2)

and

$$\left.\begin{array}{c} x_1(0) \\ x_2(0) \end{array}\right\} \text{ arbitrary}$$

(3a)

$$\begin{array}{c} x_1(t_f) = 0 \\ x_2(t_f) = 0 \end{array} \quad t_f \text{ specified}$$

(3b)

To proceed, expressing the system equations of (1) in matrix form using the notation of Section 8-2, one has

$$\mathbf{A}(t) = \begin{bmatrix} 0 & 1 \\ 0 & 0 \end{bmatrix} \qquad \mathbf{B}(t) = \begin{bmatrix} 0 \\ 1 \end{bmatrix}$$

(4)

Doing the same for the performance index one finds

$$\mathbf{W}_1(t) \equiv 0 \qquad \mathbf{W}_2(t) \equiv 0 \qquad \mathbf{W}_3(t) = 1 \qquad \bar{\mathbf{g}} \equiv 0$$

(5)

while the terminal conditions of Eq. (3b) require

$$\bar{\boldsymbol{\psi}} = \begin{bmatrix} 1 & 0 \\ 0 & 1 \end{bmatrix} \qquad \boldsymbol{\psi}_0 = 0$$

(6)

The matrices $\mathbf{P}(t)$, $\mathbf{R}(t)$, and $\mathbf{Q}(t)$ must now be found from Eq. 8-3(19), subject to the terminal conditions of Eq. 8-3(20), so that the control law of Eq. 8-3(18) can be formed. As there are two state variables and two terminal constraints, for this problem $n = 2$ and $p = 2$. Then since $\mathbf{P}(t)$, $\mathbf{R}(t)$, and $\mathbf{Q}(t)$ are, respectively, $n \times n$, $n \times p$, and $p \times p$ matrices, each is a 2×2 matrix in this example.

Starting with determination of $\mathbf{P}(t)$, Eqs. 8-3(19a, 20a) require that the matrix equation

$$\dot{\mathbf{P}} = -\mathbf{P}\mathbf{A} - \mathbf{A}'\mathbf{P} + \mathbf{P}\mathbf{B}\mathbf{B}'\mathbf{P} \tag{7}$$

be solved subject to the condition

$$\mathbf{P}(t_f) = 0 \tag{8}$$

By inspection, the solution

$$\mathbf{P}(t) \equiv 0 \tag{9}$$

is seen to satisfy both the equation of (7) and the terminal condition of Eq. (8).

Continuing, using the result of Eq. (9), Eqs. 8-3(19b, 20b) require that $\mathbf{R}(t)$ satisfy the equation

$$\dot{\mathbf{R}} = -\mathbf{A}'\mathbf{R} \tag{10}$$

subject to the terminal condition

$$\mathbf{R}(t_f) = \begin{bmatrix} 1 & 0 \\ 0 & 1 \end{bmatrix} \tag{11}$$

In expanded form Eq. (10) becomes

$$\begin{bmatrix} \dot{R}_{11} & \dot{R}_{12} \\ \dot{R}_{21} & \dot{R}_{22} \end{bmatrix} = -\begin{bmatrix} 0 & 0 \\ 1 & 0 \end{bmatrix}\begin{bmatrix} R_{11} & R_{12} \\ R_{22} & R_{21} \end{bmatrix} \tag{12}$$

Thus by components one has

$$\begin{aligned} \dot{R}_{11} &= 0 \\ \dot{R}_{12} &= 0 \\ \dot{R}_{21} &= -R_{11} \\ \dot{R}_{22} &= -R_{12} \end{aligned} \tag{13}$$

The solution of Eq. (13) subject to the terminal condition of Eq. (11) yields

$$\mathbf{R}(t) = \begin{bmatrix} 1 & 0 \\ t_f - t & 1 \end{bmatrix} \tag{14}$$

$\mathbf{Q}(t)$ is determined as the solution of Eq. 8-3(19c) under the condition of Eq. 8-3(20c). Letting $\tau = t_f - t$, these require, respectively, that

$$\dot{\mathbf{Q}} = \begin{bmatrix} 1 & \tau \\ 0 & 1 \end{bmatrix}\begin{bmatrix} 0 \\ 1 \end{bmatrix}\begin{bmatrix} 0 & 1 \end{bmatrix}\begin{bmatrix} 1 & 0 \\ \tau & 1 \end{bmatrix} \tag{15a}$$

$$\mathbf{Q}(t_f) = 0 \tag{15b}$$

Carrying out the indicated matrix multiplications of Eq. (15) one obtains

$$\dot{\mathbf{Q}} = \begin{bmatrix} \dot{Q}_{11} & \dot{Q}_{12} \\ \dot{Q}_{21} & \dot{Q}_{22} \end{bmatrix} = \begin{bmatrix} \tau^2 & \tau \\ \tau & 1 \end{bmatrix} \tag{16}$$

Then on integrating Eq. (16) and meeting the condition of Eq. (15b), there results

$$\mathbf{Q}(t) = \begin{bmatrix} \dfrac{\tau^3}{3} & \dfrac{\tau^2}{2} \\[2ex] \dfrac{\tau^2}{2} & \tau \end{bmatrix} \tag{17}$$

For use in forming $\mathbf{u}^*(\mathbf{x}, t)$, $\mathbf{Q}^{-1}(t)$ is required. From Eq. (17) one finds†

$$\mathbf{Q}^{-1}(t) = -\frac{12}{\tau^4} \begin{bmatrix} \tau & -\dfrac{\tau^2}{2} \\[2ex] -\dfrac{\tau^2}{2} & \dfrac{\tau^3}{3} \end{bmatrix} \tag{18}$$

Hence $\mathbf{Q}^{-1}(t)$ is well behaved away from $t = t_f$.

With these results the control law for the problem, as given by Eq. 8-3(18), can be formed. For this example, since $\boldsymbol{\psi}_0 = 0$, $\mathbf{W}_2(t) \equiv 0$, $\mathbf{W}_3(t) = 1$, Eq. 8-3(18) reduces to

$$\mathbf{u}^*(\mathbf{x}, t) = -\mathbf{B}'(t)[\mathbf{P}(t) - \mathbf{R}(t)\mathbf{Q}^{-1}(t)\mathbf{R}'(t)]\mathbf{x}(t) \tag{19}$$

Then using $\mathbf{P}(t)$, $\mathbf{R}(t)$, and $\mathbf{Q}^{-1}(t)$ as given by Eqs. (9), (14), and (18) one has

$$\mathbf{u}^*(\mathbf{x}, t) = -\frac{12}{\tau^4}\begin{bmatrix} 0 & 1 \end{bmatrix}\begin{bmatrix} 1 & 0 \\ \tau & 1 \end{bmatrix}\begin{bmatrix} \tau & -\dfrac{\tau^2}{2} \\[2ex] -\dfrac{\tau^2}{2} & \dfrac{\tau^3}{3} \end{bmatrix}\begin{bmatrix} 1 & \tau \\ 0 & 1 \end{bmatrix}\begin{bmatrix} x_1 \\ x_2 \end{bmatrix} \tag{20}$$

yielding the optimal control law

$$\mathbf{u}^*(\mathbf{x}, t) = -\frac{6x_1}{(t_f - t)^2} - \frac{4x_2}{t_f - t} \tag{21}$$

Given knowledge of the time to go, $t_f - t$, and the state variables $x_1(t)$

---

† Given the matrix

$$\mathbf{A} = \begin{bmatrix} a_{11} & a_{12} \\ a_{21} & a_{22} \end{bmatrix}$$

where

$$\text{determinant } \mathbf{A} = |\mathbf{A}| = a_{11}a_{22} - a_{12}a_{21}$$

then

$$\mathbf{A}^{-1} = \frac{1}{|\mathbf{A}|}\begin{bmatrix} a_{22} & -a_{12} \\ -a_{21} & a_{11} \end{bmatrix}$$

and $x_2(t)$, Eq. (21) provides the optimal control to be applied to take the system to the terminal state $x_1(t_f) = x_2(t_f) = 0$ while minimizing the IP of Eq. (2).

8-5 IMPLEMENTATION OF THE SOLUTION

The solution obtained in Section 8-4, although useful in illustrating the theory, is not typical of the technique that would be used in practice. In general, it will not be possible to solve analytically for the matrices $\mathbf{P}(t)$, $\mathbf{R}(t)$, and $\mathbf{Q}(t)$. Rather, their determination will require numerical solution by computer of Eq. 8-3(19) subject to the terminal condition of Eq. 8-3(20).

In working numerically care must always be taken because the character of the solutions obtained cannot be easily studied. It has been shown (*1*), (*2*) that the grouping $\mathbf{P}(t) - \mathbf{R}(t)\mathbf{Q}^{-1}(t)\mathbf{R}'(t)$ which occurs in the optimal control law of Eq. 8-3(18) is of special significance. Defining the $n \times n$ symmetric† matrix

$$\mathbf{M}(t) \equiv \mathbf{P}(t) - \mathbf{R}(t)\mathbf{Q}^{-1}(t)\mathbf{R}'(t) \tag{1}$$

the finiteness of $\mathbf{M}(t)$ in the interval $t_i < t < t_f$ is found to be equivalent to satisfying the Jacobi condition for the problem. That is, if $\mathbf{M}(t)$ is finite in $t_i < t < t_f$ there are no conjugate points in the interval, which is a requirement for the optimal solution (see Section 2-11).

The condition on $\mathbf{M}(t)$ does not, however, preclude the possibility that $\mathbf{P}(t)$, $\mathbf{R}(t)$, and $\mathbf{Q}(t)$ become infinite at some time while $\mathbf{M}(t)$ remains finite there. That this can occur will be demonstrated by example in Section 8-6. Thus, since computationally it is impossible to deal with quantities which grow unbounded, it is to our advantage to determine $\mathbf{M}(t)$ directly, if possible, rather than by forming $\mathbf{M}(t)$ from its component parts. Further, if the $n \times n$ symmetric matrix $\mathbf{M}(t)$ can be found directly as the solution of the appropriate differential system, less numerical work will be required than if $\mathbf{P}(t)$ an $n \times n$ matrix, $\mathbf{R}(t)$ an $n \times p$ matrix and $\mathbf{Q}(t)$ a $p \times p$ matrix must be found from Eqs. 8-3(19, 20) to form $\mathbf{M}(t)$.

---

† That $\mathbf{M}(t)$ is symmetric follows from the fact that $\mathbf{P}(t)$ and $\mathbf{Q}(t)$ have been shown to be symmetric in Section 8-3; that is,

$$\mathbf{P}'(t) = \mathbf{P}(t) \qquad \mathbf{Q}'(t) = \mathbf{Q}(t) \tag{a}$$

and hence

$$\mathbf{Q}^{-1'}(t) = \mathbf{Q}^{-1}(t) \tag{b}$$

Thus forming $\mathbf{M}'(t)$ and using Eqs. (a) and (b), one has

$$\mathbf{M}'(t) = \mathbf{P}'(t) - \mathbf{R}(t)\mathbf{Q}^{-1'}(t)\mathbf{R}'(t)$$
$$= \mathbf{P}(t) - \mathbf{R}(t)\mathbf{Q}^{-1}(t)\mathbf{R}'(t) = \mathbf{M}(t)$$

which is to be shown.

To this end we determine next the governing differential equation for $\mathbf{M}(t)$. Differentiating Eq. (1) we obtain

$$\dot{\mathbf{M}} = \dot{\mathbf{P}} - \dot{\mathbf{R}}\mathbf{Q}^{-1}\mathbf{R}' - \mathbf{R}\dot{\mathbf{Q}}^{-1}\mathbf{R}' - \mathbf{R}\mathbf{Q}^{-1}\dot{\mathbf{R}}' \tag{2}$$

The relations of Eq. 8-3(19) are then used to eliminate $\dot{\mathbf{P}}$, $\dot{\mathbf{R}}$, and $\dot{\mathbf{Q}}^{-1}$ from Eq. (2), where†

$$\dot{\mathbf{Q}}^{-1} = -\mathbf{Q}^{-1}\mathbf{R}'\mathbf{B}\mathbf{W}_3^{-1}\mathbf{B}'\mathbf{R}\mathbf{Q}^{-1} \tag{3}$$

Combining terms the resulting equation can be written

$$\dot{\mathbf{M}} = -\mathbf{M}\mathbf{A} - \mathbf{A}'\mathbf{M} - \mathbf{W}_1 + (\mathbf{M}\mathbf{B} + \mathbf{W}_2)\mathbf{W}_3^{-1}(\mathbf{W}_2' + \mathbf{B}'\mathbf{M}) \tag{4}$$

Equation (4) is the governing equation for $\mathbf{M}(t)$. Note that it is the same as Eq. 8-3(19a) governing $\mathbf{P}(t)$.

To find $\mathbf{M}(t)$ from Eq. (4), the value of $\mathbf{M}(t_1)$, where t_1 can be chosen at our convenience, must be known. Since $\mathbf{Q}(t_f) = 0$ by Eq. 8-3(20c), it follows that $\mathbf{Q}^{-1}(t_f)$ will not exist. Hence from the definition of $\mathbf{M}(t)$ in Eq. (1), $\mathbf{M}(t_f)$ in general will not exist. Thus it is not possible in general to integrate Eq. (4) back from t_f to determine $\mathbf{M}(t)$ for $t < t_f$ since $\mathbf{M}(t_f)$ is not known.

However, it has been shown (*1*) that if the system equations of Eq. 8-2(1) are totally controllable,‡ $\mathbf{Q}^{-1}(t)$ will exist in the neighborhood of t_f; that is, $\mathbf{Q}^{-1}(t_f - \epsilon)$ can be found from $\mathbf{Q}(t_f - \epsilon)$. By inspection of Eqs. 8-3(19, 20), it is seen that $\mathbf{P}(t)$ and $\mathbf{R}(t)$ are regular in the neighborhood of t_f; hence $\mathbf{P}(t_f - \epsilon)$ and $\mathbf{R}(t_f - \epsilon)$ can also be found.

Our procedure is thus the following one. The matrices $\mathbf{P}(t_f - \epsilon)$, $\mathbf{R}(t_f - \epsilon)$, and $\mathbf{Q}(t_f - \epsilon)$ are found as the solution of Eq. 8-3(19) under the conditions of Eq. 8-3(20). From these quantities, using the definition of Eq. (1), $\mathbf{M}(t_f - \epsilon)$ is formed.

$$\mathbf{M}(t_f - \epsilon) = \mathbf{P}(t_f - \epsilon) - \mathbf{R}(t_f - \epsilon)\mathbf{Q}^{-1}(t_f - \epsilon)\mathbf{R}'(t_f - \epsilon) \tag{5}$$

---

† Forming

$$\mathbf{Q}\mathbf{Q}^{-1} = \mathbf{I}$$

and differentiating one obtains

$$\dot{\mathbf{Q}}\mathbf{Q}^{-1} + \mathbf{Q}\dot{\mathbf{Q}}^{-1} = 0$$

Solving for $\dot{\mathbf{Q}}^{-1}$ and eliminating $\dot{\mathbf{Q}}$ using Eq. 8-3(19c), one finds

$$\dot{\mathbf{Q}}^{-1} = -\mathbf{Q}^{-1}\mathbf{R}'\mathbf{B}\mathbf{W}_3^{-1}\mathbf{B}'\mathbf{R}\mathbf{Q}^{-1}$$

which is the relation given by Eq. (3).

‡ The system equations

$$\dot{\mathbf{x}} = \mathbf{A}(t)\mathbf{x} + \mathbf{B}(t)\mathbf{u}$$

are said to be totally controllable if it is possible for the control to transfer any finite initial state to any finite terminal state in any finite time interval. See E. Kreindler and P. E. Sarachik, "On the Concepts of Controllability and Observability of Linear Systems," *IEEE Trans. Automat. Control*, April, 1964, pp. 129–136.

For $t < t_f - \epsilon$, $\mathbf{M}(t)$ is then found as the solution of Eq. (4) subject to the condition of Eq. (5).

In terms of $\mathbf{M}(t)$ the optimal control law of Eq. 8-3(18) becomes

$$\mathbf{u}^*(\mathbf{x}, t) = -\mathbf{W}_3^{-1}(t)\mathbf{B}'(t)\mathbf{R}(t)\mathbf{Q}^{-1}(t)\boldsymbol{\psi}_0$$
$$- \mathbf{W}_3^{-1}(t)[\mathbf{W}_2'(t) + \mathbf{B}'(t)\mathbf{M}(t)]\mathbf{x}(t) \quad (6)$$

In the form of Eq. (6), determination of $\mathbf{u}^*(\mathbf{x}, t)$ still requires that the product $\mathbf{R}(t)\mathbf{Q}^{-1}(t)$ be known if the terminal conditions are inhomogeneous; that is, $\boldsymbol{\psi}_0 \neq 0$. As with $\mathbf{M}(t)$, it can be shown that $\mathbf{R}(t)\mathbf{Q}^{-1}(t)$ will be finite in $t_i < t < t_f$ if there are no conjugate points in the interval. However, individually $\mathbf{R}(t)$ and $\mathbf{Q}(t)$ need not exist outside of a region ϵ about t_f, as will be demonstrated by example in Section 8-6.

To overcome this computational difficulty we proceed, as we have just done for determination of $\mathbf{M}(t)$, by finding the product $\mathbf{R}(t)\mathbf{Q}^{-1}(t)$ directly. Define

$$\mathbf{N}(t) \equiv \mathbf{R}(t)\mathbf{Q}^{-1}(t) \quad (7)$$

Then differentiating Eq. (7) one finds

$$\dot{\mathbf{N}} = \dot{\mathbf{R}}\mathbf{Q}^{-1} + \mathbf{R}\dot{\mathbf{Q}}^{-1} \quad (8)$$

Eliminating $\dot{\mathbf{R}}(t)$ and $\dot{\mathbf{Q}}^{-1}(t)$ using Eq. 8-3(19b), Eq. (3) and the definition of $\mathbf{M}(t)$ as given by Eq. (1), the relation of Eq. (8) can be written

$$\dot{\mathbf{N}} = (\mathbf{MBW}_3^{-1}\mathbf{B}' - \mathbf{A}' + \mathbf{W}_2\mathbf{W}_3^{-1}\mathbf{B}')\mathbf{N} \quad (9)$$

Equation (9) is the governing equation for $\mathbf{N}(t)$. As discussed earlier in the section in connection with $\mathbf{M}(t)$, the quantities $\mathbf{R}(t)$ and $\mathbf{Q}(t)$ will be well behaved in an interval ϵ about t_f. Thus $\mathbf{R}(t_f - \epsilon)$ and $\mathbf{Q}^{-1}(t_f - \epsilon)$ can be found from solution of Eqs. 8-3(19, 20) and $\mathbf{N}(t_f - \epsilon)$ formed.

$$\mathbf{N}(t_f - \epsilon) = \mathbf{R}(t_f - \epsilon)\mathbf{Q}^{-1}(t_f - \epsilon) \quad (10)$$

For $t < t_f - \epsilon$, $\mathbf{N}(t)$ is found as the solution of Eq. (9) subject to the condition of Eq. (10). It is interesting to note that Eq. (9) for $\mathbf{N}(t)$ is the same as Eq. 8-3(19b) for $\mathbf{R}(t)$ with $\mathbf{M}(t)$ replacing $\mathbf{P}(t)$.

To summarize, in terms of $\mathbf{M}(t)$ and $\mathbf{N}(t)$, the optimal control law for the problem is given by

$$\mathbf{u}^*(\mathbf{x}, t) = -\mathbf{W}_3^{-1}(t)\mathbf{B}'(t)\mathbf{N}(t)\boldsymbol{\psi}_0 - \mathbf{W}_3^{-1}(t)[\mathbf{W}_2'(t) + \mathbf{B}'(t)\mathbf{M}(t)]\mathbf{x}(t) \quad (11)$$

In the interval $t_f - \epsilon < t < t_f$, $\mathbf{P}(t)$, $\mathbf{R}(t)$, and $\mathbf{Q}(t)$ are found as solutions of Eqs. 8-3(19) subject to the terminal condition of Eq. 8-3(20). The length of the ϵ interval is determined by the requirement that $\mathbf{Q}^{-1}(t_f - \epsilon)$ be sufficiently well behaved. For $t < t_f - \epsilon$ the control law of Eq. (11) is used with $\mathbf{M}(t)$

determined as the solution of the differential system of Eqs. (4, 5). Once $\mathbf{M}(t)$ is known, solution of the differential system of Eqs. (9, 10) provides $\mathbf{N}(t)$.

Finally, finiteness of $\mathbf{M}(t)$ and $\mathbf{N}(t)$ is equivalent to satisfying the Jacobi condition for the problem. Should any component of $\mathbf{M}(t)$ and $\mathbf{N}(t)$ become infinite at a time t_1 in the interval $t_i < t < t_f$, there is a conjugate point at this time and no minimizing solution exists to the problem as formulated.

8-6 A TUTORIAL EXAMPLE

Consider a system defined by the scalar equation

$$\dot{x}(t) = x(t) + u(t) \tag{1}$$

Our problem is that of transferring the system from any initial state

$$t_i \text{ and } x(t_i) \qquad \text{arbitrary} \tag{2}$$

to the terminal state

$$t_f \text{ and } x(t_f) = x_f \qquad \text{specified} \tag{3}$$

so as to minimize the performance index

$$\text{IP} = \frac{1}{2} \int_{t_i}^{t_f} (u^2 - 2x^2)\, dt \tag{4}$$

In the general notation of Section 8-2 we have

$$\begin{aligned} n &= 1 &&\text{a single state variable exists} \\ m &= 1 &&\text{a single control variable exists} \\ p &= 1 &&\text{a single terminal constraint exists} \end{aligned}$$

Thus for this problem $\mathbf{P}(t)$, $\mathbf{R}(t)$, $\mathbf{Q}(t)$, which are $n \times n$, $n \times p$, and $p \times p$ matrices, respectively, reduce to single functions. Further,

$$\begin{aligned} \mathbf{A}(t) &= 1 & \mathbf{B}(t) &= 1 \\ \overline{\boldsymbol{\psi}} &= 1 & \boldsymbol{\psi}_0 &= x_f \\ \bar{\mathbf{g}} &= 0 & \mathbf{W}_2(t) &= 0 \\ \mathbf{W}_1(t) &= -2 & \mathbf{W}_3(t) &= 1 \end{aligned} \tag{5}$$

The solution of Eq. 8-3(19) subject to Eq. 8-3(20) for $\mathbf{P}(t)$, $\mathbf{R}(t)$, and $\mathbf{Q}(t)$ then requires that the equations

$$\begin{aligned} \dot{P} &= 2 - 2P + P^2 \\ \dot{R} &= (P - 1)R \\ \dot{Q} &= R^2 \end{aligned} \tag{6}$$

be solved subject to the conditions

$$P(t_f) = 0$$
$$R(t_f) = 1 \tag{7}$$
$$Q(t_f) = 0$$

On satisfying the equations and conditions of Eqs. (6) and (7) one finds analytically that

$$P(t) = \frac{2 \tan(t - t_f)}{1 + \tan(t - t_f)} = \frac{2 \sin(t - t_f)}{\sin(t - t_f) + \cos(t - t_f)} \tag{8a}$$

$$R(t) = \frac{1}{\sin(t - t_f) + \cos(t - t_f)} \tag{8b}$$

$$Q(t) = \frac{\sin(t - t_f)}{\sin(t - t_f) + \cos(t - t_f)} \tag{8c}$$

However, note that as t approaches $t_f - \pi/4$ from above, $P(t)$, $R(t)$, and $Q(t)$ become unbounded. That is,

$$\lim_{\epsilon \to 0} P\left(t_f - \frac{\pi}{4} + \epsilon\right) = -\infty$$

$$\lim_{\epsilon \to 0} R\left(t_f - \frac{\pi}{4} + \epsilon\right) = \infty$$

$$\lim_{\epsilon \to 0} Q\left(t_f - \frac{\pi}{4} + \epsilon\right) = -\infty$$

Thus, if the solution for $P(t)$, $R(t)$, and $Q(t)$ is being found by numerical integration of Eq. (6) back from the terminal condition of Eq. (7), it would not be possible to proceed beyond $t = t_f - \pi/4$. Nonetheless, the problem does have a solution for $t_f - t_i > \pi/4$. The difficulty encountered is precisely that discussed in Section 8-5. For if we form the functions $M(t)$ and $N(t)$ of Eqs. 8-5(1, 7) from the solution of Eq. (8), letting $\tau = t_f - t$, one finds

$$M(t) = P - RQ^{-1}R' = \frac{2 \sin \tau}{\sin \tau - \cos \tau} - \frac{1}{\sin \tau(\sin \tau - \cos \tau)}$$
$$= 1 + \cot(t_f - t) \tag{9}$$

$$N(t) = RQ^{-1} = -\frac{1}{\sin(t_f - t)}$$

from which it is seen that both $M(t)$ and $N(t)$ are well behaved at $t = t_f - \pi/4$. In fact, $M(t)$ and $N(t)$ are well behaved in the interval $t_f - t < \pi$. At $t = t_f - \pi$, both $M(t)$ and $N(t)$ become infinite, indicating that a conjugate point exists there. Based on this it can be concluded that a solution to the problem only exists provided $t_f - t_i < \pi$.

Imagine that it had been necessary to solve the differential system of Eqs. (6) and (7) numerically. The computations would proceed back from t_f a time ϵ until $Q^{-1}(t_f - \epsilon)$ is properly behaved. At $t_f - \epsilon$, the quantities $M(t_f - \epsilon)$ and $N(t_f - \epsilon)$ would be formed from $P(t_f - \epsilon)$, $R(t_f - \epsilon)$, and $Q(t_f - \epsilon)$, yielding by Eq. (9) the values

$$M(t_f - \epsilon) = 1 + \cot \epsilon$$

$$N(t_f - \epsilon) = -\frac{1}{\sin \epsilon} \tag{10}$$

Then for $t < t_f - \epsilon$, by Eqs. 8-5(4, 9), $M(t)$ and $N(t)$ would be found as the solution of the equations

$$\dot{M} = 2 - 2M + M^2$$

$$\dot{N} = (M - 1)N \tag{11}$$

subject to the conditions of Eq. (10). In this manner the result of Eq. (9) would be obtained computationally, avoiding the singularities which occur in $P(t)$, $R(t)$, and $Q(t)$ individually.

With $M(t)$ and $N(t)$ determined, for $t_f - t_i < \pi$ the optimal control law for the problem is given by

$$u^*(x, t) = \frac{x_f}{\sin(t_f - t)} - [1 + \cot(t_f - t)]x(t) \tag{12}$$

8-7 THE NONLINEAR PROBLEM—A GUIDANCE SOLUTION

To start, let us formulate the general problem for which the optimal control law is desired. The problem is that of minimizing the performance index

$$\text{IP} = g[\mathbf{x}(t_f)] + \int_{t_i}^{t_f} \bar{f}[\tau, \mathbf{x}(\tau), \mathbf{u}(\tau)]\, d\tau \tag{1}$$

subject to the differential constraints

$$\dot{\mathbf{x}}(t) = \mathbf{f}[t, \mathbf{x}(t), \mathbf{u}(t)] \tag{2}$$

while transferring the system from the initial state

$$t_i \text{ and } \mathbf{x}(t_i) \qquad \text{arbitrary} \tag{3}$$

to the terminal manifold

$$t_f \text{ specified}$$

$$\boldsymbol{\psi}[\mathbf{x}(t_f)] = 0 \tag{4}$$

where $\quad\mathbf{x}(t) = n \times 1$ matrix, the state vector

$\mathbf{u}(t) = m \times 1$ matrix, the control vector

$g[\mathbf{x}(t_f)] = $ a known function of $\mathbf{x}(t_f)$

$\bar{f}[t, \mathbf{x}(t), \mathbf{u}(t)] = $ a known function of t, $\mathbf{x}(t)$, $\mathbf{u}(t)$

$\mathbf{f}[t, \mathbf{x}(t), \mathbf{u}(t)] = n \times 1$ matrix of known functions of t, $\mathbf{x}(t)$, $\mathbf{u}(t)$

$\mathbf{\psi}[\mathbf{x}(t_f)] = p \times 1$ matrix of known functions of $\mathbf{x}(t_f)$

Having formulated the problem for which the control law $\mathbf{u}^*(\mathbf{x}, t)$ is desired, it should be noted that, independent of desire, it is usually not possible to find an explicit representation of $\mathbf{u}^*(\mathbf{x}, t)$ for the nonlinear problem as stated. Although special problems exist for which solutions have been found, they are the exception, not the rule. Instead of attempting to recount these special cases we shall turn our attention to a different but related problem which can be treated in a general manner.

Let us assume that rather than requiring the solution over a wide range of initial conditions as in Eq. (3), the initial states that need be considered are limited and may be taken as specified. The condition

$$t_i \text{ and } \mathbf{x}(t_i) \qquad \text{specified} \tag{5}$$

now replaces that of Eq. (3).

For the problem as defined by Eqs. (1), (2), (4), and (5) it is a relatively straightforward procedure to find the optimal open-loop solution. Numerical integration by computer of the governing equations is available and search routines, as we have seen in Sections 4-2 and 4-3, usually provide convergence of the solution to the required conditions of the two-point boundary-value problem resulting from the optimization. The optimal open-loop solution thus obtained will be termed the nominal solution and denoted by $\mathbf{u}^*(t)$ and $\mathbf{x}^*(t)$.

In the absence of errors the control $\mathbf{u}^*(t)$ would be called for and generated. With this control, excluding system disturbances, the trajectory $\mathbf{x}^*(t)$ would result. However, in any realistic engineering system it is never possible to totally eliminate either errors or disturbances. Thus it is unlikely that the system will be on the optimal trajectory $\mathbf{x}^*(t)$ at any time. Rather it can be expected that the system will be at some neighboring state $\mathbf{x}(t_1)$ to the optimal state $\mathbf{x}^*(t_1)$ at time t_1, where

$$\mathbf{x}(t_1) = \mathbf{x}^*(t_1) + \delta\mathbf{x}(t_1)$$

Our problem is now one of taking the system from the neighboring state $\mathbf{x}(t_1)$ to the desired terminal manifold so as to minimize the IP of Eq. (1) over the time interval $t_f - t_1$ remaining before t_f. To achieve this, the nominal control $\mathbf{u}^*(t)$ over $t_f - t_1$ will have to be modified by an amount dependent on the magnitude of the state perturbation. In what follows we shall seek to find an explicit relation between the optimal control change at any instant and the state perturbation $\delta\mathbf{x}(t)$ about the nominal state at that time. In essence we are looking for the optimal control perturbation law; that is, $\delta\mathbf{u}^*(\delta\mathbf{x}, t)$. The problem of determining $\delta\mathbf{u}^*(\delta\mathbf{x}, t)$ is termed the *guidance problem*.

The procedure used is to insert the perturbed quantities

$$\mathbf{x}(t) = \mathbf{x}^*(t) + \delta\mathbf{x}(t)$$
$$\mathbf{u}(t) = \mathbf{u}^*(t) + \delta\mathbf{u}(t) \tag{6}$$

into the IP of Eq. (1) and then seek the control perturbation $\delta\mathbf{u}(t)$ which minimizes the dominant terms in the expansion of the resulting expression about the value obtained using the nominal solution $\mathbf{u}^*(t)$, $\mathbf{x}^*(t)$.

The interval $t_i < t < t_1$ is assumed to be past. Hence the integral over that interval in the IP of Eq. (1) is an integral over known history, and may be replaced by the constant C_1. The differential constraints of Eq. (2) and the terminal constraints of Eq. (4) are next adjoined to the performance index using the multipliers obtained from the nominal solution for $\mathbf{u}^*(t)$, $\mathbf{x}^*(t)$. These are the $n \times 1$ matrix of multipliers $\boldsymbol{\lambda}^*(t)$ to the differential constraints and the $p \times 1$ matrix of multipliers $\boldsymbol{\nu}^*$ to the terminal constraints. The augmented performance index is then given by

$$\mathrm{IP} = C_1 + g[\mathbf{x}(t_f)] + \boldsymbol{\nu}^{*\prime}\boldsymbol{\psi}[\mathbf{x}(t_f)]$$
$$+ \int_{t_1}^{t_f} \{\bar{f}[\tau, \mathbf{x}(\tau), \mathbf{u}(\tau)] + \boldsymbol{\lambda}^{*\prime}(\tau)\{\dot{\mathbf{x}}(\tau) - \mathbf{f}[\tau, \mathbf{x}(\tau), \mathbf{u}(\tau)]\}\} \, d\tau \tag{7}$$

For convenience in writing the equations which follow, the Hamiltonian function [see Eq. 7-5(17)]

$$H(t, \mathbf{x}, \mathbf{u}, \boldsymbol{\lambda}^*) = \boldsymbol{\lambda}^{*\prime}(t)\mathbf{f}[t, \mathbf{x}(t), \mathbf{u}(t)] - \bar{f}[t, \mathbf{x}(t), \mathbf{u}(t)] \tag{8}$$

as well as the function

$$G[\boldsymbol{\nu}^*, \mathbf{x}(t_f)] = g[\mathbf{x}(t_f)] + \boldsymbol{\nu}^{*\prime}\boldsymbol{\psi}[\mathbf{x}(t_f)] \tag{9}$$

are introduced into Eq. (7). The result is

$$\mathrm{IP} = C_1 + G[\boldsymbol{\nu}^*, \mathbf{x}(t_f)] - \int_{t_i}^{t_f} \{H[\tau, \mathbf{x}(\tau), \mathbf{u}(\tau), \boldsymbol{\lambda}^*(\tau)] - \boldsymbol{\lambda}^{*\prime}(\tau)\dot{\mathbf{x}}(\tau)\} \, d\tau \tag{10}$$

To put into effect the previously indicated procedure we consider the value of the IP of Eq. (10) evaluated using the perturbed state and control vectors of Eq. (6). Thus we consider

$$\mathrm{IP} = C_1 + G[\boldsymbol{\nu}^*, \mathbf{x}^*(t_f) + \delta\mathbf{x}(t_f)]$$
$$- \int_{t_1}^{t_f} [H(\tau, \mathbf{x}^* + \delta\mathbf{x}, \mathbf{u}^* + \delta\mathbf{u}, \boldsymbol{\lambda}^*) - \boldsymbol{\lambda}^{*\prime}(\dot{\mathbf{x}}^* + \delta\dot{\mathbf{x}})] \, d\tau \tag{11}$$

which after integration by parts of the $\delta\dot{\mathbf{x}}$ term in the integral becomes

$$\text{IP} = C_1 + G[\mathbf{v}^*, \mathbf{x}^*(t_f) + \delta\mathbf{x}(t_f)] + \boldsymbol{\lambda}^{*\prime}(t_f)\,\delta\mathbf{x}(t_f) - \boldsymbol{\lambda}^{*\prime}(t_1)\,\delta\mathbf{x}(t_1)$$

$$- \int_{t_1}^{t_f} [H(\tau, \mathbf{x}^* + \delta\mathbf{x}, \mathbf{u}^* + \delta\mathbf{u}, \boldsymbol{\lambda}^*) - \boldsymbol{\lambda}^{*\prime}\dot{\mathbf{x}}^* + \dot{\boldsymbol{\lambda}}'\,\delta\mathbf{x}]\,d\tau \quad (12)$$

The IP of Eq. (12) is expanded about the nominal value it attains for $\delta\mathbf{x}(t) = \delta\mathbf{u}(t) \equiv 0$, where the notation

$$G^* = G[\mathbf{v}^*, \mathbf{x}^*(t_f)] \qquad \mathbf{G}_x^* = \frac{\partial G}{\partial\mathbf{x}}[\mathbf{v}^*, \mathbf{x}^*(t_f)] \qquad \mathbf{G}_{xx}^* = \frac{\partial^2 G}{\partial\mathbf{x}^2}[\mathbf{v}^*, \mathbf{x}^*(t_f)]$$

$$H^* = H(t, \mathbf{x}^*, \mathbf{u}^*, \boldsymbol{\lambda}^*) \qquad \mathbf{H}_x^* = \frac{\partial H}{\partial\mathbf{x}}(t, \mathbf{x}^*, \mathbf{u}^*, \boldsymbol{\lambda}^*)$$

$$\mathbf{H}_u^* = \frac{\partial H}{\partial\mathbf{u}}(t, \mathbf{x}^*, \mathbf{u}^*, \boldsymbol{\lambda}^*) \qquad \mathbf{H}_{xx}^* = \frac{\partial^2 H}{\partial\mathbf{x}^2}(t, \mathbf{x}^*, \mathbf{u}^*, \boldsymbol{\lambda}^*) \quad \text{etc.}$$

will be used. One finds through second order in the perturbations that

$$\text{IP} = C_1 - \boldsymbol{\lambda}^{*\prime}(t_1)\,\delta\mathbf{x}(t_1) + G^* - \int_{t_1}^{t_f}[H^* - \boldsymbol{\lambda}^{*\prime}\dot{\mathbf{x}}^*]\,d\tau$$

$$+ [\mathbf{G}_x^* + \boldsymbol{\lambda}^{*\prime}(t_f)]\,\delta\mathbf{x}(t_f) - \int_{t_1}^{t_f}[(\mathbf{H}_x^* + \dot{\boldsymbol{\lambda}}^{*\prime})\,\delta\mathbf{x} + \mathbf{H}_u^*\,\delta\mathbf{u}]\,d\tau \quad (13)$$

$$+ \frac{1}{2}\delta\mathbf{x}_f'\mathbf{G}_{xx}^*\,\delta\mathbf{x}_f - \frac{1}{2}\int_{t_1}^{t_f}[\delta\mathbf{x}'\mathbf{H}_{xx}^*\,\delta\mathbf{x} + 2\,\delta\mathbf{x}'\mathbf{H}_{xu}^*\,\delta\mathbf{u} + \delta\mathbf{u}'\mathbf{H}_{uu}^*\,\delta\mathbf{u}]\,d\tau$$

Now, by the transversality condition of 3-8(10d) for the nominal solution, the coefficient of $\delta\mathbf{x}(t_f)$ in Eq. (13) vanishes:

$$\boldsymbol{\lambda}^{*\prime}(t_f) = -\mathbf{G}_x^* \quad (14)$$

Similarly, the Euler equations of Eq. 7-5(4) for the nominal solution are seen to require the coefficients of $\delta\mathbf{x}$ and $\delta\mathbf{u}$ in the second integral of Eq. (13) to vanish:

$$\dot{\boldsymbol{\lambda}}^{*\prime} = -\mathbf{H}_x^*$$

$$\mathbf{H}_u^* = 0 \quad (15)$$

All the remaining terms in Eq. (13), except the second-order ones, are fixed quantities. Thus to minimize the IP of Eq. (13), $\delta\mathbf{u}(t)$ must be chosen so as to minimize the second variation of the IP, which we will denote by IP_2:

$$\text{IP}_2 = \frac{1}{2}\delta\mathbf{x}_f'\mathbf{G}_{xx}^*\,\delta\mathbf{x}_f$$

$$- \frac{1}{2}\int_{t_1}^{t_f}[\delta\mathbf{x}'\mathbf{H}_{xx}^*\,\delta\mathbf{x} + 2\,\delta\mathbf{x}'\mathbf{H}_{xu}^*\,\delta\mathbf{u} + \delta\mathbf{u}'\mathbf{H}_{uu}^*\,\delta\mathbf{u}]\,d\tau \quad (16)$$

To proceed, the equation governing the propagation of the state perturbations must be known. By taking the variation of the system equations of Eq. (2) about the nominal solution, the desired relation is obtained,

$$\delta\dot{\mathbf{x}} = \mathbf{f}_x^* \, \delta\mathbf{x} + \mathbf{f}_u^* \, \delta\mathbf{u} \tag{17}$$

where

$$\mathbf{f}_x^* = \frac{\partial\mathbf{f}}{\partial\mathbf{x}}\,(t, \mathbf{x}^*, \mathbf{u}^*) \qquad \mathbf{f}_u^* = \frac{\partial\mathbf{f}}{\partial\mathbf{u}}\,(t, \mathbf{x}^*, \mathbf{u}^*)$$

Finally, the minimization by $\delta\mathbf{u}$ must be such that an admissible terminal perturbation $\delta\mathbf{x}(t_f)$ is achieved. Taking the variation of the terminal constraint of Eq. (4) about the nominal solution one finds that this requires

$$\boldsymbol{\psi}_x^* \, \delta\mathbf{x}(t_f) = 0 \tag{18}$$

where

$$\boldsymbol{\psi}_x^* = \frac{\partial\boldsymbol{\psi}}{\partial\mathbf{x}}\,[\mathbf{x}^*(t_f)]$$

In Eqs. (16), (17), and (18) all the starred quantities are either functions of time alone or constants since they have been evaluated along the known nominal solution.

Minimization of the quadratic performance index of Eq. (16) by proper choice of $\delta\mathbf{u}(t)$, subject to the linear system equations of Eq. (17) and the linear terminal constraints of Eq. (18), is required. But this is just the problem for which the solution was obtained and discussed in Sections 8-2 through 8-5 with $\delta\mathbf{x}$ and $\delta\mathbf{u}$ replacing $\mathbf{x}$ and $\mathbf{u}$ there. Repeating the results of Eqs. 8-3(19) and 8-3(20) as applied to this problem, one finds that the matrices $\mathbf{P}(t)$, $\mathbf{R}(t)$, and $\mathbf{Q}(t)$ must satisfy the equations†

$$\dot{\mathbf{P}} = -\mathbf{P}\mathbf{f}_x^* - \mathbf{f}_x^{*\prime}\mathbf{P} + \mathbf{H}_{xx}^* - (\mathbf{P}\mathbf{f}_u^* - \mathbf{H}_{xu}^*)\mathbf{H}_{uu}^{*-1}(-\mathbf{H}_{ux}^* + \mathbf{f}_u^{*\prime}\mathbf{P})$$

$$\dot{\mathbf{R}} = (-\mathbf{P}\mathbf{f}_u^*\mathbf{H}_{uu}^{*-1}\mathbf{f}_u^{*\prime} - \mathbf{f}_x^* + \mathbf{H}_{xu}^*\mathbf{H}_{uu}^{*-1}\mathbf{f}_u^{*\prime})\mathbf{R} \tag{19}$$

$$\dot{\mathbf{Q}} = -\mathbf{R}^\prime\mathbf{f}_u^*\mathbf{H}_{uu}^{*-1}\mathbf{f}_u^{*\prime}\mathbf{R}$$

and the terminal conditions

$$\mathbf{P}(t_f) = \mathbf{G}_{xx}^* = \mathbf{g}_{xx}^* + \boldsymbol{\nu}^{*\prime}\boldsymbol{\psi}_{xx}^*$$

$$\mathbf{R}(t_f) = \boldsymbol{\psi}_x^{*\prime} \tag{20}$$

$$\mathbf{Q}(t_f) = 0$$

---

† Recall that in Section 8-2 $\mathbf{W}_3(t)$ was required to be a positive definite matrix. For this problem $\mathbf{W}_3(t) = -\mathbf{H}_{uu}^*(t)$. Now the optimal nominal control $\mathbf{u}^*(t)$ is determined from the maximum principle by the requirement that it maximize $H(t, x^*, u, \lambda^*)$; from which it follows that along the optimal $\mathbf{H}_u^* = 0$ and $\delta\mathbf{u}^\prime\mathbf{H}_{uu}^*\,\delta\mathbf{u} < 0$ (strong form of Legendre–Clebsch condition). The latter condition shows that $\mathbf{H}_{uu}^*(t)$ is negative definite; hence $-\mathbf{H}_{uu}^*(t)$ is positive definite, as required.

The desired optimal control perturbation law is then given from Eq. 8-5(11) as

$$\delta\mathbf{u}^*(\delta\mathbf{x}, t) = \mathbf{H}_{uu}^{*-1}[-\mathbf{H}_{uu}^* + \mathbf{f}_u^{*\prime}\mathbf{M}]\,\delta\mathbf{x}(t) \tag{21}$$

where

$$\mathbf{M}(t) = \mathbf{P}(t) - \mathbf{R}(t)\mathbf{Q}^{-1}(t)\mathbf{R}'(t) \tag{22}$$

The discussion concerning $\mathbf{M}(t)$ in Section 8-5 applies here also. If $\mathbf{M}(t)$ becomes unbounded in $t_i < t < t_f$, a conjugate point exists in the original problem and the nominal solution is not a minimizing solution. Further, the advantage noted in Section 8-5 of working with $\mathbf{M}(t)$ directly after some interval about t_f applies to the present problem. Thus after $\mathbf{M}(t_f - \epsilon)$ has been determined from Eq. (22) by the solution of Eq. (19) subject to the conditions of Eq. (20), $\mathbf{M}(t)$ for $t < t_f - \epsilon$ should be found as the solution of the equation

$$\dot{\mathbf{M}} = -\mathbf{M}\mathbf{f}_x^* - \mathbf{f}_x^{*\prime}\mathbf{M} + \mathbf{H}_{xx}^* - (\mathbf{M}\mathbf{f}_u^* - \mathbf{H}_{xu}^*)\mathbf{H}_{uu}^{*-1}(-\mathbf{H}_{ux}^* + \mathbf{f}_u^{*\prime}\mathbf{M}) \tag{23}$$

subject to the condition found at $t_f - \epsilon$.

For realistic problems determination of $\mathbf{P}(t)$, $\mathbf{R}(t)$, $\mathbf{Q}(t)$, and $\mathbf{M}(t)$ from Eqs. (19) and (23) in general will require numerical solution of the equations. Once $\mathbf{M}(t)$ has been found computationally, it and the elements of $\mathbf{H}_{uu}^{*-1}(t)$, $\mathbf{H}_{ux}^*(t)$, and $\mathbf{f}_u^*(t)$ would be stored to compute the optimal control perturbation $\delta\mathbf{u}^*(\delta\mathbf{x}, t)$ from Eq. (21) as required by the process.

8-8 A TRANSFER PROBLEM

We treat here the two-dimensional nonlinear problem of transferring between initial and final states a vehicle governed by the system equations

$$\begin{aligned} \dot{x} &= V\cos\theta \\ \dot{y} &= V\sin\theta \end{aligned} \tag{1}$$

The vehicle's velocity V is assumed to be constant. Control of the vehicle depends on the velocity orientation angle θ, which is the control variable for the problem. The nonlinearities in the problem arise from the $\sin\theta$ and $\cos\theta$ terms in Eq. (1).

Given the initial state

$$t_i = 0 \qquad x(0) = 0 \qquad y(0) = 0 \tag{2}$$

our task is to transfer between this initial state and the required terminal manifold†

$$\psi[x(t_f)] = x(t_f) - x_f = 0 \qquad t_f \text{ and } x_f \text{ specified} \tag{3}$$

† For the problem to have a solution, this necessitates that $|x_f| \leq Vt_f$.

in such a manner as to maximize $y(t_f)$. Thus

$$\text{IP} = -y(t_f) \tag{4}$$

The nominal solution is found first. This requires that the Euler equations

$$\dot{\lambda}_1 = 0$$
$$\dot{\lambda}_2 = 0 \tag{5}$$
$$-\lambda_1 \sin \theta + \lambda_2 \cos \theta = 0$$

be solved subject to the transversality condition

$$\lambda_2(t_f) = 1 \tag{6}$$

The variables $\lambda_1(t)$ and $\lambda_2(t)$ are the multipliers to the $\dot{x}$ and $\dot{y}$ equations of (1), respectively.

From Eqs. (5) and (6) it is seen that

$$\lambda_1(t) = \bar{\lambda}_1, \text{ a const.} \tag{7a}$$

$$\lambda_2(t) = 1 \tag{7b}$$

$$\tan \theta = \frac{1}{\bar{\lambda}_1} \tag{7c}$$

where Eq. (7c) is equivalent to the two relations†

$$\sin \theta = \frac{1}{\sqrt{1 + \bar{\lambda}_1^2}} \qquad \cos \theta = \frac{\bar{\lambda}_1}{\sqrt{1 + \bar{\lambda}_1^2}} \tag{8}$$

Let

$$\rho = \sqrt{1 + \bar{\lambda}_1^2} \tag{9}$$

Then inserting the extremal control of Eq. (8) into the system equations of Eq. (1), one obtains the equations

$$\dot{x} = \frac{V\bar{\lambda}_1}{\rho}$$

$$\dot{y} = \frac{V}{\rho} \tag{10}$$

which are to be solved subject to the initial and terminal conditions of Eqs. (2) and (3).

---

† Actually all Eq. (7c) requires is that

$$\sin \theta = \frac{\pm 1}{\sqrt{1 + \bar{\lambda}_1^2}} \qquad \cos \theta = \frac{\pm \bar{\lambda}_1}{\sqrt{1 + \bar{\lambda}_1^2}}$$

The fact that θ must maximize the Hamiltonian for the problem, $H = \lambda_1 V \cos \theta + \lambda_2 V \sin \theta$, eliminates the negative sign.

Integrating Eq. (10) using Eq. (2) one has

$$x(t) = \frac{V\bar{\lambda}_1 t}{\rho}$$

$$y(t) = \frac{Vt}{\rho}$$
(11)

Then to satisfy the terminal condition of Eq. (3) one finds

$$\bar{\lambda}_1 = \frac{x_f}{[V^2 t_f^2 - x_f^2]^{1/2}}$$
(12)

Using Eq. (12), the nominal control of Eq. (8) becomes

$$\sin \theta^* = \left[1 - \left(\frac{x_f}{Vt_f}\right)^2\right]^{1/2} \qquad \cos \theta^* = \frac{x_f}{Vt_f}$$
(13)

while the nominal trajectory is

$$x^*(t) = x_f \frac{t}{t_f}$$

$$y^*(t) = V\left[1 - \left(\frac{x_f}{Vt_f}\right)^2\right]^{1/2} t$$
(14)

The guidance problem can now be formulated and solved. As applied to this problem, Eqs. 8-7(8, 9) yield

$$H = \lambda_1 V \cos \theta + \lambda_2 V \sin \theta$$

$$G = -y(t_f) + v[x(t_f) - x_f]$$
(15)

From Eq. (15) one finds that

$$\mathbf{H}^*_{xu} = \mathbf{H}^*_{xx} = \mathbf{H}^*_{x} = \mathbf{G}^*_{xx} \equiv 0$$
(16)

and

$$\mathbf{H}^*_{uu} = -\frac{V}{\sin \theta^*}$$
(17)

With Eqs. (16), (17) and Eq. 8-7(16), the second variation of our IP over the interval t_1 to t_f, where t_1 is arbitrary, is seen to be

$$\mathrm{IP}_2 = \frac{1}{2} \frac{V}{\sin \theta^*} \int_{t_1}^{t_f} \delta\theta^2(\tau) \, d\tau$$
(18)

Taking the variation of our system equations of (1) about the nominal solution, the perturbation propagation equations are found to be

$$\delta\dot{x}_1 = -V \sin \theta^* \, \delta\theta$$

$$\delta\dot{x}_2 = V \cos \theta^* \, \delta\theta$$
(19)

In the notation of Eq. 8-7(17), one has from Eq. (19) that

$$\mathbf{f}_x^* \equiv 0 \qquad \mathbf{f}_u^* = \begin{bmatrix} -V \sin \theta^* \\ V \cos \theta^* \end{bmatrix} \tag{20}$$

Taking the variation of our terminal constraint of Eq. (3) about the nominal, one finds

$$\delta x(t_f) = 0 \tag{21}$$

so that, in the notation of Eq. 8-7(18),

$$\boldsymbol{\psi}_x^* = [1 \quad 0] \tag{22}$$

To find the feedback perturbation control law the differential system of Eqs. 8-7(19, 20) must be solved in the neighborhood of t_f for $\mathbf{P}(t)$, $\mathbf{R}(t)$, and $\mathbf{Q}(t)$. For this problem, since $n = 2$ and $p = 1$, $\mathbf{P}(t)$ is a 2×2 matrix, $\mathbf{R}(t)$ a 2×1 matrix, and $\mathbf{Q}(t)$ a scalar quantity. Using Eqs. (16), (17), (20), and (22) the differential system to be solved becomes

$$\dot{\mathbf{P}} = \mathbf{P}\mathbf{f}_u^* \frac{\sin \theta^*}{V} \mathbf{f}_u^{*\prime}\mathbf{P} \tag{23a}$$

$$\dot{\mathbf{R}} = \left(\mathbf{P}\mathbf{f}_u^* \frac{\sin \theta^*}{V} \mathbf{f}_u^{*\prime} \right) \mathbf{R} \tag{23b}$$

$$\dot{Q} = \mathbf{R}'\mathbf{f}_u^* \frac{\sin \theta^*}{V} \mathbf{f}_u^{*\prime}\mathbf{R} \tag{23c}$$

where

$$\mathbf{P}(t_f) = 0 \tag{24a}$$

$$\mathbf{R}(t_f) = \begin{bmatrix} 1 \\ 0 \end{bmatrix} \tag{24b}$$

$$Q(t_f) = 0 \tag{24c}$$

The solution of Eq. (23a) subject to the condition of Eq. (24a) is seen to be

$$\mathbf{P}(t) = \begin{bmatrix} P_{11} & P_{12} \\ P_{21} & P_{22} \end{bmatrix} \equiv 0 \tag{25}$$

It then follows from Eqs. (23b, 24b) that $\mathbf{R}(t)$ must satisfy the system

$$\dot{\mathbf{R}} = \begin{bmatrix} \dot{R}_{11} \\ \dot{R}_{21} \end{bmatrix} = \begin{bmatrix} 0 \\ 0 \end{bmatrix} \qquad \mathbf{R}(t_f) = \begin{bmatrix} 1 \\ 0 \end{bmatrix}$$

Hence

$$\mathbf{R}(t) = \begin{bmatrix} R_{11}(t) \\ R_{21}(t) \end{bmatrix} = \begin{bmatrix} 1 \\ 0 \end{bmatrix} \tag{26}$$

With $\mathbf{R}(t)$ known, Eq. (23c) for $Q(t)$ becomes

$$\dot{Q} = [1 \quad 0] \begin{bmatrix} -V \sin \theta^* \\ V \cos \theta^* \end{bmatrix} \frac{\sin \theta^*}{V} [-V \sin \theta^* \quad V \cos \theta^*] \begin{bmatrix} 1 \\ 0 \end{bmatrix}$$

or

$$\dot{Q} = V \sin^3 \theta^* \tag{27}$$

The solution of Eq. (27) subject to the terminal condition of Eq. (24c) yields

$$Q(t) = -V(t_f - t) \sin^3 \theta^* \tag{28}$$

Then from Eqs. (25), (26), and (28) the matrix $\mathbf{M}(t)$ of Eq. 8-7(22) can be formed:

$$\mathbf{M}(t) = \begin{bmatrix} 1 \\ 0 \end{bmatrix} \frac{1}{V(t_f - t) \sin^3 \theta^*} [1 \quad 0]$$

or

$$\mathbf{M}(t) = \begin{bmatrix} M_{11} & M_{12} \\ M_{21} & M_{22} \end{bmatrix} = \frac{1}{V(t_f - t) \sin^3 \theta^*} \begin{bmatrix} 1 & 0 \\ 0 & 0 \end{bmatrix} \tag{29}$$

Since $\mathbf{M}(t)$ is finite for all $t < t_f$, there are no conjugate points in the interval $t_i < t < t_f$.

Finally, the optimal control perturbation law of Eq. 8-7(21) can now be formed for the problem, yielding

$$\delta\theta^* = \frac{-\sin \theta^*}{V} [-V \sin \theta^* \quad V \cos \theta^*] \frac{1}{V(t_f - t) \sin^3 \theta^*} \begin{bmatrix} 1 & 0 \\ 0 & 0 \end{bmatrix} \begin{bmatrix} \delta x \\ \delta y \end{bmatrix}$$

or

$$\delta\theta^*(\delta x, t) = \frac{\delta x(t)}{V(t_f - t) \sin \theta^*} \tag{30}$$

which is the desired result.

REFERENCES

1. W. E. Schmitendorf and S. J. Citron, "On the Applicability of the Riccati Transformation Technique for Solution of Optimal Control Problems," *AA and ES Rept.* 67–10, Purdue Univ., Lafayette, Ind., 1967.
2. S. E. Dreyfus, "Control Problems with Linear Dynamics, Quadratic Criterion, and Linear Terminal Constraints," *IEEE Trans. Automatic Control*, Vol. AC–12, No. 3, June, 1967, pp. 303–304.

BIBLIOGRAPHY

Breakwell, J. V., J. L. Speyer, and A. E. Bryson, "Optimization and Control of Nonlinear Systems Using the Second Variation," *SIAM J. Control*, Ser. A, Vol. 1, No. 2, 1963, pp. 193–223.

Bryson, A. E., and W. F. Denham, "Multivariable Terminal Control for Minimum Square Deviation from a Nominal Path," *Proceedings of the Institute for Aerospace Science Symposium on Vehicle Systems Optimization*, Garden City, N.Y., Nov. 1961, pp. 91–97.

Kelley, H. J., "Guidance Theory and Extremal Fields," *IRE Trans. Auto. Control*, October 1962, pp. 75–82.

Lee, I., "Optimal Trajectory, Guidance and Conjugate Points," *Information Control J.*, Vol. 8, 1965, pp. 589–606.

McReynolds, S. R., and A. E. Bryson, "A Successive Sweep Method for Solving Optimal Programming Problems," Preprint Volume, *Joint Automatic Control Conference*, 1965, pp. 551–555.

Index

Index

A

Accessory differential equation, 45
Accessory minimum problem, 45
Adjoint differential form, 188
Adjoint equations, 89, 116, 145, 190, 196, 198
  on control boundary, 117
  on state boundary, 145
  relation to dynamic programming, 231
  use of in examples, 103, 106, 124, 149, 161, 204
  See also Characteristic equations; Euler equations
Adjoint variables, 189
  relation to dynamic programming, 231
Allocation problem, 210
Augmented integrand (*see* Integrand)

B

Bang-bang control, 133, 150
Bolza, problem of, 62, 77

C

Calculus of variations, 4, 33
  relation to control formulation, 88
  relation to dynamic programming, 231
Characteristic equations, 37, 58, 67, 72, 89, 120, 141
  use of in examples, 38, 54, 59, 74, 79, 94, 132
  See also Adjoint equations; Euler equations
Chemical reactor, 106
  bounded control, 123
Comparison functions, 33, 51, 64, 81
Conjugate gradient, 209
Conjugate point, 50, 245, 247–249, 255
Conjugate point condition, 47, 245, 248
Conjugate point definition, 47
Constraint equations, 30
Control, classical, 11
  closed-loop, 2, 233
  conventional, 3, 11
  feedback, 2
  open-loop, 1

Control (*continued*)
  optimal, 3
Control boundary, 114, 117
Control condition, 117, 145
Control energy, 242
Control equation, 89, 95, 103, 107, 117
  relation to dynamic programming, 231
Control formulation, 88, 175
Control function, 12, 99, 133, 235, 238
Control law, 12, 100, 235, 238, 240
Control perturbation, 13
  See also Optimal control perturbation
  law
Control variables, 5, 88, 175
Control-variable inequality constraints,
  112, 132, 153, 157, 234
Controllability, 14
  totally controllable, 102, 246
Controller, 3
Corners, 40, 55
Corner conditions, 42, 52, 53, 57, 58, 72,
  119, 142, 145
  See also Weierstrass–Erdmann corner
  conditions; Juncture conditions

D

Delta notation, 35
Dependent differential, 31
Differential constraints, 55, 74
Direct methods, 167, 169
Discontinuous control, 146
Disturbances, 2, 13, 100, 236
Dynamic programming, 5, 210, 213,
  216, 235
  as an *N-stage* decision process, 225
  computational method, 223, 233
  discrete form of, 224
  embedding, 212, 216, 223
  optimal control sequence for, 226
  polynomial approximation for, 233
  principle of optimality, 213, 217
  recursion relation for, 212, 225
  relation to calculus of variations, 229,
  231

E

Eigenfunctions, 104
Embedding (*see* **Dynamic** program-
  ming)
End conditions, 63, 74, 79
Equations of motion, 88, 93
  See also System equations
Errors, 13, 100, 236
Euler equations, 35, 52, 229, 253, 256
  See also Adjoint equations; Charac-
  teristic equations
Extremals, 37, 80
Extremal solutions, 41
Extremal trajectory, 100

F

First integral, 40, 53, 57, 59, 72, 90, 134
First variation, 28, 35, 36, 51, 56, 62,
  65, 115, 140
Frequency-domain, 3
Frequency response, 3
Fundamental matrix, 239

G

Guidance problem, 101, 252, 257

H

H function, 92, 122, 162, 231, 232
  See also Hamiltonian
Hamiltonian, 232, 252, 256
Hamilton–Jacobi equation, 233

I

Independent, differentials, 31, 68
  variations, 67
Index of performance, 4, 11, 13, 62, 78,
  89
  augmented, 31, 56, 63, 140
  quadratic, 237
Indirect methods, 167

Inertia plant, 131, 242
Influence coefficients, 110
Integrand, augmented, 56, 58, 59, 63, 71, 74, 79, 89, 140, 229
Intermediate conditions, 139, 142, 161
  See also Juncture conditions

J

Jacobi condition (*see* Conjugate point condition)
Jacobi differential equation, 45
Juncture conditions, 119, 124, 142, 145, 161

L

Lagrange, problem of, 62, 77
Lagrange multipliers, 31–33, 56, 69, 71, 89, 172, 183–185, 200, 204, 237
Legendre–Clebsch condition, 80, 85, 86, 88
Lumped-parameter systems, 5
Lunar ascent problem, 92, 202

M

Matrix notation, 62
Maximum principle, 5, 92, 95, 107, 133, 149, 232
  See also Weierstrass condition
Mayer, problem of, 62, 77, 89, 193
Minima, of definite integrals, 33
  of functions, 17, 168
  of functionals, 33, 193
Minimum, absolute, 18, 21, 27
  local, 18, 26, 27
Minimum time problem, 93
Minimum-weight beam, 154
Multipliers (*see* Lagrange multipliers)

N

Nominal solution, 191
  optimal open-loop solution, 251

Nominal trajectory, 236
Numerical integration, 109
Numerical solution, 245, 255

O

Observability, 14
Optimal control function, 12
Optimal control law, 12, 220, 233, 234, 244, 245, 247, 250
Optimal control perturbation law, 252, 255, 259
Optimal feedback control, 235
Optimal policy, 213
Optimal return function, 213, 216, 222

P

Penalty-function, 208
Perturbation equations, 182, 195, 204, 254, 257
Phase variables, 11
Plant, 3
Pontryagin's maximum principle, 92
  See also Maximum principle; Weierstrass condition

R

Regulator problem, 226, 235
Riccati equation, 221, 239, 241

S

Second order perturbations, 253
Second variation, 39, 43, 46, 253, 257
Singular control, 150
Specific optimal control, 13
State variables, 5, 8, 88, 175
State-variable boundary, 143, 160
  initial conditions, 144
State-variable formulation, 5
State-variable inequality constraint, 143, 153, 157
  qth-order, 144
State perturbation, 251

266   INDEX

Steepest descent, 4, 179, 201
   constraints, 170
   control metric, 179, 184
   control step size, 179, 184, 199
   "safe" step size, 171
   stopping condition, 194, 195, 202, 203
Stochastic optimal control, 5
Suboptimal control, 13
Switching boundary, 135
Switching function, 133, 150
System constraint, 59
System design, 11
System equations, 5, 88, 114, 194
   linear, 236
System output, 15
System parameters, 4

T

Terminal conditions, 140
Terminal constraints, 199
Terminal perturbation, 254
Time constant, 2
Transfer problem, 255
Transversality condition, 68, 71–73, 90, 142, 237, 253
   relation to dynamic programming, 229, 231

use of in examples, 73, 74, 79, 96, 103, 107, 132, 149, 161, 256
Two-point boundary-value problem, 109, 128, 166, 167, 190

U

Undetermined multipliers (*see* Lagrange multipliers)

V

Variation, strong, 34, 81, 85
   weak, 34, 80, 85
Vertical sounding rocket, 8

W

Weierstrass condition, 80, 85, 86, 88, 91, 120, 121, 232
   See also H function; Maximum principle
Weierstrass E function, 85
Weierstrass–Erdmann corner conditions, 41, 90
   See also Corner conditions